PRACTICAL
BUSINESS
MATH

AN APPLICATIONS
APPROACH

D1435749

BRIEF EDITION

PRACTICAL BUSINESS MATH

AN APPLICATIONS APPROACH

EIGHTH EDITION

MICHAEL D. TUTTLE
NORTHWOOD UNIVERSITY

Prentice Hall

Upper Saddle River, New Jersey
Columbus, Ohio

Property of Library
Cape Fear Comm. College
Wilmington, N. C.

To Cody, Danielle,
Emilie, and Sigurd

Library of Congress Cataloging-in-Publication Data
Tuttle, Michael D.
    Practical business math : an applications approach / Michael D. Tuttle.—8th ed.
      p. cm.
    Includes index.
    ISBN 0-13-025667-6 (pbk.)
     1. Business mathematics—Problems, exercises, etc.   I. Title.

    HF5694.T88 2001
    650′.01′513—dc21

                                            99-089103

Vice President and Publisher: Dave Garza
Senior Editor: Elizabeth Sugg
Associate Editor: Michelle Churma
Production Editor: Louise N. Sette
Production Supervision: Clarinda Publication Services
Design Coordinator: Robin G. Chukes
Cover Designer: Rod Harris
Cover Art: © Kathy Hanley
Production Manager: Brian Fox
Marketing Manager: Shannon Simonsen

This book was set in Optima by The Clarinda Company. It was printed and bound by Banta Book Group. The cover was printed by Phoenix Color Corp.

Earlier editions copyright © 1994, 1990, 1987, 1983, 1982, 1978 by Wm. C. Brown Communications, Inc.

Copyright © 2001, 1998 by Prentice-Hall, Inc., Upper Saddle River, New Jersey 07458. All rights reserved. Printed in the United States of America. This publication is protected by Copyright and permission should be obtained from the publisher prior to any prohibited reproduction, storage in a retrieval system, or transmission in any form or by any means, electronic, mechanical, photocopying, recording or likewise. For information regarding permission(s), write to: Rights and Permissions Department.

10 9 8 7 6 5 4 3 2 1
ISBN: 0-13-025667-6 (Standard Edition)
         0-13-025660-9 (Brief Edition)
         0-13-026486-5 (Annotated Instructor's Edition)

Contents

TO THE INSTRUCTOR

The eighth edition of *Practical Business Math: An Applications Approach* continues the successful format of previous editions with an added emphasis on the integrated presentation that makes teaching and learning easier. This emphasis is sure to improve student understanding of business math topics. The explanation, example, exercise, assignment, and mastery test format has been continued. This format is enhanced with a complete cross-reference system of chapter objectives, student mastery test answers, essential calculator keystroke solutions, and topic page location. Clearly identified chapter objectives, practical small business examples, and understandable explanations make *Practical Business Math* an excellent learning tool.

We live in a global economy. New and developing businesses are providing challenging opportunities for those prepared to compete in the world of business. *Practical Business Math* prepares the student for the dramatic changes that are taking place and those that are yet unannounced. The practical nature of the presentation, the realistic business scenarios, the problem-solving techniques, and the measurement of success prepare the student for additional courses in business and the world of work.

About This Book

Practical Business Math uses the fundamental tools of basic arithmetic to solve everyday business problems. You may select *Practical Business Math,* Brief Edition (chapters 1–17), or you may select *Practical Business Math,* Standard Edition (chapters 1–23).

Each chapter in *Practical Business Math* consistently follows this format:

Chapter Objectives list the specific goals so students know what is expected. The page reference is provided for easy location of the instructional material.

Sport-A-Merica is a case study that illustrates real-world use of math skills in a business setting. Some specific problems are introduced here and solved in the chapter examples.

Explanation-Example-Exercise format provides clear explanation immediately followed by an illustrative example. Exercises are practice problems designed to reinforce the topics just learned. This material is organized around the chapter objectives. This easy-to-follow format facilitates the learning process.

Assignments check the student's understanding of the concepts covered in the chapter and are divided into skill problems, business application problems, and a challenge problem. All sets of problems are arranged in order of increasing difficulty. Answers to the odd-numbered problems are given in the back of the book and essential keystroke solutions are given in the instructor's edition.

Mastery Test is designed for take-home or in-class examination. The answers to all mastery test problems are in the back of the book.

Chapter 23 provides a review for the student. Fifty problems are keyed to the chapter/page where the concept is introduced.

Writing Across the Curriculum is integrated into assignment and mastery test problems in all chapters. Students are expected to provide answers to some problems in sentence form.

Suggested Classroom Time and Perceived Difficulty bar graphs are included in all chapter openings in the Instructor's Edition to aid in planning. This will assist you in determining the amount of time needed for each of the chapters in the course.

Margin Notations for the instructor and the student are found in all chapters. The notations for the instructor (in the Annotated Instructor's Edition only) will provide additional information that may be useful to introduce topics to the class. The notations to the student (in both the student and annotated instructor's editions) are in the form of alerts, warnings, and reminders to watch for possible errors in solving problems and to clarify misconceptions.

Essential keystrokes are included in the Annotated Instructor's Edition (shown within parentheses in italics). These keystrokes will provide the important values to solve exercises and problems and possibly aid in the instruction process.

Answers to all exercises and problems are shown in the Annotated Instructor's Edition in boldfaced color and italics. Answers to all exercises and odd-numbered Skill and Business Application problems are provided in the back of the student edition. Answers along with essential keystroke solutions and the chapter objective for all end-of-chapter Mastery Tests in the student edition are also shown in the back of the book. The answer to the first exercise is shown along with the essential keystrokes in the student edition.

The Annotated Instructor's Edition of the text places all of the information that you will need for the classroom in a single source. There is no need to carry several items to class and refer from one to the other to lecture, to provide salient comments, and to give detailed solutions to exercises and problems. No more need to fumble around. It is all right here, in this one item!

New to This Edition

All material in each chapter has been updated to include the most recent business practices such as international business scenarios, where applicable, and government regulations, including the minimum wage law changes, payroll tax deductions, and income taxes regulations.

In addition, the chapter on Insurance has been enhanced on the topic of life insurance, including discussion of universal life and variable life. The basis of life insurance rates is explained.

The chapter on Payroll includes an expanded coverage on the topic of F.U.T.A. and state unemployment taxes for the employer.

The chapter on Business Measurements addresses the Euro as a measure of exchange and measure of economic activity.

This edition introduces the term *amount* in the chapter newly titled "Percent." The use of this term will clarify the confusion that surrounds the use of words such as percentage, piece, part, or rate. The term *amount* has been class-tested and has proven to be successful.

The chapter on Installment Loans has been moved ahead of the chapter on Consumer Credit for improved understanding.

Algebriac solutions have been placed in the margin in the chapters on Percent and Markup to aid those who prefer an alternative approach.

A Note about the Sport-A-Merica Case

Though Sport-A-Merica, Inc. is a fictional business, it is based on an actual store operated in the Midwest. The case is designed to provide the student with a scenario of the day-to-day problems faced by small-business owners. The case, introduced chapter by chapter, discusses business problems that pertain specifically to the major emphasis in each chapter. One or more business problems are introduced in each chapter's case discussion. Later, as the chapter material is presented, the problems are solved as part of the worked examples. This will draw the student into the reality of the case and the application of math to business. The chapter can also be presented by itself, without the case.

Supplements for the Instructor

The Annotated Instructor's Edition of *Practical Business Math* provides a complete supplemental package designed to reduce preparatory work. The teaching aids in this package are extensive and will enhance classroom activity and learning.

The Annotated Instructor's Edition provides:

- Answers to all problems at the location of the exercise or problem.
- Essential keystrokes for all exercises, problems, and Mastery Tests at their location to assist in the instruction process.

- Suggested Classroom Time bar graph in each chapter to assist in determining where to spend the most amount of classroom time in the course.
- Perceived Student Difficulty bar graph in each chapter to indicate the areas of the material that have historically been more difficult for students.
- Margin Notations that contain additional information that can be used to enhance the classroom instruction and separately boxed helpful warnings, alerts, and reminders for students.

The Instructor's Resource Section provides:

- A Chapter Interdependencies table that demonstrates the interrelationship of each chapter to the other chapters in the book. This table is very useful for planning purposes, especially for new or adjunct faculty.
- A set of four suggested course outlines that emphasize different areas of study. These outlines will assist in planning which chapters to use in a course outline or syllabus.
- A set of transparency/copy masters that includes the various tables that students would use on tests. These perforated pages can be removed from the book for reproduction purposes. They are ready to copy!
- Mastery Tests A, B, and C in a tear-out ready-to-use form.
- A Final Exam consisting of fifty questions that cover the entire text in a tear-out ready-to-use form.

PH Custom Test for Windows

- A computerized Test Item File (PH Custom Test for Windows) contains 2,200 problems (100 per chapter covering all of the chapter objectives). The problem types include:
  Objective (multiple choice)
  Skill
  Business Application
  Essay

Test Item File

The Test Item File, a printed version of the PH Custom Test for Windows, is also available. This visual tool will assist you in selecting questions you may wish to include on a test. Ask your representative for details.

Companion Website

Invite your students to investigate the Companion Website at www.prenhall.com/tuttle for additional preparation using another vehicle for learning.

Acknowledgments

The creation of a new edition requires the input from many sources. The eighth edition of *Practical Business Math* reflects suggestions and comments from students, adopters, and reviewers.

Additional thanks go to the many instructors who suggested improvements and volunteered valuable suggestions and the many students, especially my own, who provided suggestions to further improve this book.

The four individuals who reviewed this edition deserve a special note of appreciation for their diligent and focused efforts:

Mary Young Bowers, College of Eastern Utah
Dr. James Farris, Golden West College (California)
Arthur Migala
Tom Wiener, Iowa Central Community College

Because I am ever striving to find even the smallest point for clarification, detailed solutions, and accuracy, the final accuracy review by James J. McGinnis, PhD, was extremely valuable. Adopters will appreciate, as I do, his thoroughness.

The editorial staff at Prentice Hall provided the guidance and attention to detail that all authors need. Each member, at every stage of development, was helpful as well as supportive and very organized. Thank you, team!

The revision workload was eased by the valued and extensive contributions of Kathy Tuttle, on her third edition, and, for the fifth time, the proofing skills of Nona Carroll. They have worked to make this the best-ever edition of *Practical Business Math*.

The Sport-A-Merica case involved a lot of work and cooperation from many people. I appreciate the help of the following people in making the story and photos come alive:

Joseph McColly, owner of the Bicyclery
John Molite, manager of Babies•Я•Us
Mary Smith, owner of Wojtusik-Smith & Associates
Donald C. Jacobs, CFP, Prudential
Anthony Marino, Jr., Internal Revenue Service
Decebal Dumitrescu, photo model
Maria Martinez, photo model
Beth Ann Miller, photo model

I know that you will enjoy using this book in your classroom.

TO THE STUDENT

This book is all about taking very basic math and applying that math to everyday business problems. This is realistic, practical material for future business managers.

If you are one of many people who never felt strong in math, don't feel threatened by this book. This book has been used successfully by business students for more than twenty years. It has a proven track record for making the subjects understandable. The easy-to-follow explanations, worked examples, and exercises work together to help you learn the material without overdoing it! You are first introduced to a bite-size concept, shown how it works, and then you have a chance to put the concept to use. Once you understand that concept, you can move on to the next concept. At the end of the chapter is an assignment section of mixed problems. You can find out how you are doing on the problems at any point by comparing your answers to the answers in the back of the book. There you will also find the essential calculator keystrokes to the Student Mastery Test as well as the objective for the problem. This will allow you to refer back to the part of the chapter that will give you the review you need before you take your test from the instructor. There is even a review chapter to get you ready for the final exam!

As you ready yourself for this course, take the few minutes necessary to learn how this book is designed to make you successful in Business Math.

1. Each chapter opens with a discussion about a business called Sport-A-Merica. In this discussion, some background is provided and a problem is introduced. These are worth reading to give you a feeling of the practical nature of the chapter material.

2. The objectives for the chapter and the page where they appear are shown on the first page of the chapter. This will give you a good cross-reference to use when reviewing the chapter material.

3. Look for the notations in the margin in the chapter. These questions, helpful hints, and warnings will help you solve problems, avoid potential errors, and clarify possible misconceptions. You will also notice business forms located within the chapter. These are business forms of actual companies. They will give you the exposure to the forms that you will use on the job.

4. The chapter is set up to introduce a topic, show an example, and give you an opportunity to practice on some exercises. This piece of the chapter is completed right there in one place.

5. You can assure yourself that you understand the material by checking your answers in the back of the book. You can then go on to the next piece of the chapter.

6. At the end of the chapter you will find the Assignment Section. The problems are arranged in three levels. The Skill problems are similar to those that you completed in the exercises. The Business Application problems are presented in a sentence form. The last problem in the chapter is a Challenge Problem. This problem will be the most difficult problem in the chapter and will challenge you! Answers to the odd-numbered problems are found in the back of the book. Check over your work to see how you are doing.

7. This is a text/workbook. Its design provides adequate room for you to work the exercises and problems in the book. There is no need to carry an extra notebook.

8.   At the end of the chapter is a mastery test. This test is a combination of all of the problems in the chapter. It is a good review of the chapter in preparation for the test that your instructor will give you.

This design is used throughout the book. It is a proven approach to learning business math.
    Your approach to using this book will make a difference in your academic success. Though your instructor may have some specific instructions for you to follow in the course, you should also consider the following:

1.   Read the chapter prior to the class session that will discuss the material. You will have an idea of the topic, a feel for the material, and probably some questions to ask. You may even decide to try a few of the exercises prior to the class session.

2.   When your instructor presents the chapter material, you will have a second presentation of the topic. This may clear up the questions that you had when you read the chapter by yourself. If questions remain, now is the time to ask.

3.   Do the Assignment Section of the chapter thoroughly. The practice will build confidence and identify any additional questions that you may need to review from the chapter material.

4.   Go back to the first page of the chapter. Look at each of the objectives listed on the page. Are you able to do each of the objectives? Do you feel confident that you can pass a test that covers all of the objectives?

5.   If you are sure that you understand the chapter material, take the Mastery Test at the end of the chapter without looking back in the chapter for help. You can then check your answers in the back of the book to confirm your understanding.

6.   How much time should you spend doing all of this? What is normal? A good place to start is to spend two hours outside of class for every hour that you spend in class. If you do that for yourself, you will probably do the best job academically that you can do.

7.   Don't assume that because the arithmetic in this course is very basic that the material is equally easy. This course deals with everyday business activities. Many people who have taken courses similar to this one are doing those activities. They are successful because they learned this material. You can be successful also.

Best wishes to you in Business Math!

WORKING WITH WHOLE NUMBERS

1

OBJECTIVES

After mastering the material in this chapter, you will be able to:

1. Read and write whole numbers. p. 2
2. Round off whole numbers. p. 4
3. Perform the four fundamental arithmetic operations (addition, subtraction, multiplication, and division) of whole numbers with more accuracy and speed. p. 5
4. Work word problems using the four fundamental arithmetic operations. p. 6
5. Understand and use the following terms:
   Place Values
   Addend
   Sum
   Difference
   Minuend
   Subtrahend
   Borrowing
   Multiplicand
   Multiplier
   Product
   Dividend
   Divisor
   Quotient
   Remainder

SPORT-A-MERICA

 Kerry Finn and Hal Balmer are forming Sport-A-Merica, Inc., a retail sports center, specializing in equipment for sports enthusiasts of all ages. Kerry and Hal will form their business as a corporation and will begin in a building located on the main street of the downtown area. The store is near city parking, other retail stores, and city office buildings.

Kerry was first inspired to participate in athletics by her brother Rob. She followed his footsteps in track and cross-country in high school and went on to college on an athletic scholarship in cross-country. Kerry surpassed Rob's performances and competed in many long-distance races, even after college. She was a top finisher in the New York and Boston marathons, in other regional races, and became a hometown heroine.

Kerry first met Hal at a 10-kilometer race. Hal had been active in sports in school and was now a representative of a sportswear firm. Kerry bought a piece of clothing from him. While writing up the sale, they discovered they had attended the same college, although at different times. Hal earned a degree in business and had been in marketing and retail for several years, with most of his experience in sports clothing and athletic shoes.

Following further conversations, Kerry and Hal decided to start their own business in retail sporting goods. Kerry had $50,000 in prize money from some of the races she won. Hal had saved $38,000 from his commissions. They bought a small building for $26,000, purchased merchandise for $41,000, and started their business. Kerry acts as president and makes all the critical decisions—with Hal's advice, of course. She is responsible for sales and public relations. Hal deals with the day-to-day operations of the business. He hires the staff, prices the merchandise, and makes most of the buying decisions. And there are so many decisions to make! With Kerry, Hal, and one sales clerk, Sport-A-Merica is open for business! Kerry and Hal know it will take a great deal of determination to make the business successful.

Kerry and Hal both recognize the importance of good records and the analysis of those records—selling expenses, travel expenses, sales for the month, cost of sales, taxes due, depreciation expenses, checkbook balancing—the list goes on. They realize they need to be on top of the records, as do any employees they may hire. Hal must complete his expense report for the week of October 10, for example. He has the receipts for his meals, travel, and lodging. He expects his reimbursement to be nearly $600. He is also concerned about the check he needs to write to Cindi White for $64. Is there enough money in the checking account? Correct calculations are essential to profits, customer satisfaction, and the success of Sport-A-Merica, Inc.

This chapter will develop your skill in the four fundamental arithmetic operations. These fundamentals are the building blocks for a better understanding of material to be presented later in the text.

Reading and Writing Whole Numbers

place value

Business communications often include numerical information that must be read. To be able to read and write numbers, you must be able to use place values. Each digit in a number, because of its position, has a place value. Once mastered, these **place values** will aid you in reading numbers. The following illustration gives a number with all of its place values labeled.

Objective 1

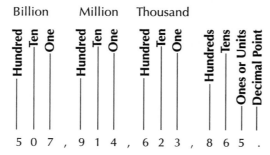

Notice that the number has a comma between each *group* of three digits to the left of the decimal place. The commas break the number into thousands, millions, etc., so that it can be easily read.

The digit in the ten billions place is 0.

The digit in the hundred thousands place is 6.

The digit in the tens place is 6.

The digit in the hundred millions place is 9.

The digit in the ones or units place is 5.

Exercise A

In the number 562,704,831

1. The digit in the hundreds place is ____8____ .

2. The digit in the millions place is _____ .

3. The digit in the tens place is _____ .

4. The digit in the ten thousands place is _____ .

5. The digit in the units place is _____ .

Exercise B

In the number 814,036,725

1. The digit in the __thousands__ place is 6.

2. The digit in the _____ place is 0.

3. The digit in the _____ place is 2.

4. The digit in the _____ place is 7.

5. The digit in the _____ place is 3.

(Check your progress by turning to the answers at the back of the book.)

With your knowledge of place values, you can read and write numbers using the word description. In the preceding illustration, each group of three digits to the left of the decimal point has a name (i.e., thousands, millions, billions, etc.). To read a number, first read the digit in the leftmost group followed by that group name and continue with this process until all of the digits are read.

The ability to read numbers will enable you to communicate with others in business more successfully. It will also improve your image as a competent student, employee, or manager.

Example 1

The number 46,002 is read: 46 thousand, 2 *or* forty-six thousand, two.
Notice that there are no "ands" used in reading the number.

Example 2

The number 693,924,806 is read: Six hundred ninety-three million, nine hundred twenty-four thousand, eight hundred six.

Read the following numbers to yourself.

123,066,231	960,012
7,839,850	3,140
7,016	90,008

When reading numbers that are in dollars, add the word "dollars" after the last digit is read.

Example 3

The amount $33,108 is read: 33 thousand, 108 dollars *or* thirty-three thousand, one hundred eight dollars.

Read the following dollar amounts to yourself.

$3,509	$7,750,000
$3,602	$19,436,502,500
$987,503	$51,437

○ *How many times will you receive or provide monetary values through voice mail or over the phone? This is an important skill!*

Now that you are able to read whole numbers it will be easier for you to write numbers with words. This knowledge will be useful when writing checks. The check form has a place for the amount to be written with words as well as with numbers.

When writing the number, a *comma* is placed between each written portion corresponding to each group of numbers (i.e., thousands, millions, etc.). Values twenty-one through ninety-nine are *hyphenated*.

Example 4

Courtesy of Market Approach, Inc., Evanston IL.

Exercise C

Write in the dollar amounts in the following checks using the value shown on each check. Follow the procedure shown in example 4, check number 1268.

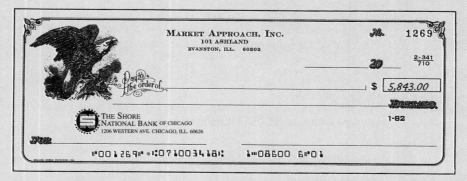

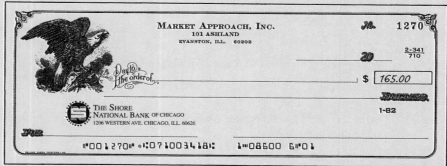

Courtesy of Market Approach, Inc., Evanston IL.

Rounding Off
Objective 2

Accuracy in arithmetic operations is essential and must be a goal of every business student. There are occasions, however, when the answer may need to be rounded off in order for it to be a more useful figure. The rounding off process does not eliminate the need for accurate arithmetic but relies on accuracy in order that the answer may be correctly rounded off. Directions for rounding off usually ask you to round off to the nearest ten, hundred, etc. (See example 5.)

Example 5

"Round off your answer to the nearest ten thousand."

Rounding Rules

If the instructions are to "round off to the nearest ten thousand," you must be concerned with the digit in the ten thousands place and the digit immediately to the right of it, that is, the thousands place. In the number 627,584, the digit 2 is in the ten thousands place. The 2 will be changed to a 3 if the digit immediately to the right of it is a 5 or larger. In this case the digit is a 7; therefore, the answer is 630,000. Notice that all the digits to the right of the rounded number are changed to zeros. When the digit immediately to the right is a 4 or smaller, the digit in question remains the same and the digits to the right are changed to zeros.

Example 6

┌─ **Digit in question (to be rounded)**

┌─ **Decision-making digit (5 or larger ?)**

627,584 = 630,000

If the same number were rounded off to the nearest hundred thousand, then the answer would be 600,000.

┌─ **Digit in question (to be rounded)**

┌─ **Decision-making digit (5 or larger ?)**

627,584 = 600,000

Rounding rules also apply to values on the other side of the decimal point and will be discussed in chapter 3, Working with Decimal Numbers.

Exercise D

Round off the following numbers to the place value indicated.

1. 97,806 to the nearest thousand is _____98,000_____.

2. 2,637 to the nearest ten is _____.

3. 278,901 to the nearest hundred is _____.

Exercise E

1. $1,927,853 rounded to the nearest hundred thousand dollars is _____$1,900,000_____.

2. $627,557 rounded to the nearest ten thousand dollars is _____.

3. $323,306,902 rounded to the nearest ten million dollars is _____.

Addition
Objective 3   *sum*

When adding numbers that have more than one column, you must start with the column farthest to the right, then proceed to the left. The answer is the **sum.**

Example 7

Add the 7 and 4; the sum is 11. Write the 1 under the right column and carry the 10 to the left column. Then add 1 plus 5 plus 9 for a total of 15 tens, giving a **sum,** or answer, of 151.

		1
57	Addend	57
+94	Addend	+94
	Sum	151

(57 + 94 =)

Exercise F

Complete the following problems.

A.

1. 46 +78 124	2. 36 +17	3. 38 +74	4. 163 +56	5. 19 +71	6. 65 +41	7. 82 +17

B.

1. 568 +204	2. 2,509 +984	3. 8,480 +6,735	4. 735 +1,294	5. 9,450 +1,205

C.

1. $5,432 +2,684	2. $86,756 +5,375	3. $9,874 +6,543	4. $62,747 +14,154	5. $76,344 +12,796

D.

1. $3,798 6,500 47,950 +56,387	2. $1,699 184 4,050 +5,632	3. $450 7,699 995 +75,600	4. $14,750 6,390 2,795 +4,000	5. $9,950 1,795 4,350 +7,885

Horizontal Addition

Occasionally records are kept in such a way that the information can be totaled horizontally as well as vertically. This approach allows for an analysis of the totals for several uses.

Objective 4

Exercise G

 Hal has submitted information on the travel expense report for the week of October 10. He has not totaled the amounts for each day, nor for each category. Complete the travel expense report from the information on the form, doing horizontal and vertical addition. How much does Sport-A-Merica owe Hal for his expenses?

SPORT-A-MERICA, INC.
Travel Expense Report

Date: Oct., 10, 1997
For: Hal Balmer

	Monday	Tuesday	Wednesday	Thursday	Friday	Saturday	Sunday	Total
Breakfast		$ 4	$ 5	$ 4				
Lunch		3	3					
Dinner	$ 18	15	16					
Travel	186	27	4	136				
Lodging	38	38	38					
Tips	3	4	4					
Misc.				9				
Total								

Horizontal addition is a good way to check your work.

Subtraction

difference
minuend
subtrahend

Even though addition is undoubtedly the most frequently used arithmetic process, in business we must also be concerned with our skill in other fundamental processes. Subtraction is the process of finding the **difference** between two numbers. The larger of the two numbers is termed the **minuend,** and the smaller number is termed the **subtrahend.** The subtrahend is subtracted from the minuend to find the difference.

Example 8

```
  9    Minuend
 −5    Subtrahend
  4    Difference
```

(9 − 5 =)

borrowing

The subtraction of larger numbers requires an additional operation, namely, **borrowing.** Borrowing is required when the subtrahend is larger than the minuend in the place value considered. When this occurs, it is necessary to borrow a unit from the place value to the left, adding the unit or ten to the minuend, and subtracting a unit from the value from which it is borrowed. Subtraction then can proceed as in the previous example. An example will help clarify the terms.

Example 9

```
                          2 1        2 1
   34    Minuend           3̷4         3̷4
  −18    Subtrahend      −18        −18
         Difference                  16
```

(34 − 18 =)

Checking Addition Answers by Subtraction

In order to be sure of your addition accuracy, it is best to check your answers before proceeding with another problem. The best way to check your answer is by subtraction. After you have found the sum or total of two addends, subtract one addend from the sum or total. Your answer should be the remaining addend.

```
    642    Addend
  +927    Addend
  1,569   Sum
  −927    Addend
    642    Addend
```

Checking Subtraction Answers by Addition

The accuracy of your subtraction can be improved by checking your answers by the use of addition. Although this method is not foolproof, it will help to detect most of the simple errors. This method requires only one step. After you have found the difference between the minuend and the subtrahend, add the difference to the subtrahend. The result should be the minuend.

```
     ⎧   854    Minuend
  + ⎨  −687    Subtrahend
     ⎩  +167    Difference
         854    Minuend (check)
```

Exercise H

Sport-A-Merica recently fired its accounts payable clerk because she did not keep records in the check register. You are asked, as the new employee, to bring up-to-date some of the records that are not in order. The former employee recorded the amounts of the checks but failed to keep a running balance. Hal, the office manager, explains that the amount of the deposit must be added to the previous balance and that the amount of each check is to be subtracted from the previous balance. The check stubs through check number 1273 are in order. Bring the checkbook up-to-date by following the example shown on check stub number 1268. Is there enough money in the account to pay Cindi White?

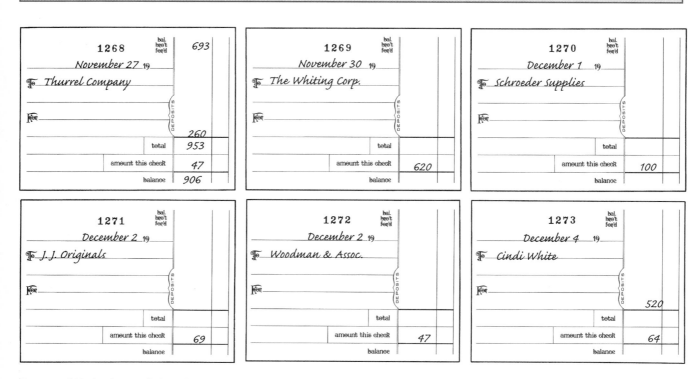

Courtesy of Market Approach, Inc., Evanston IL.

Exercise I

Find the difference in each of the problems below.

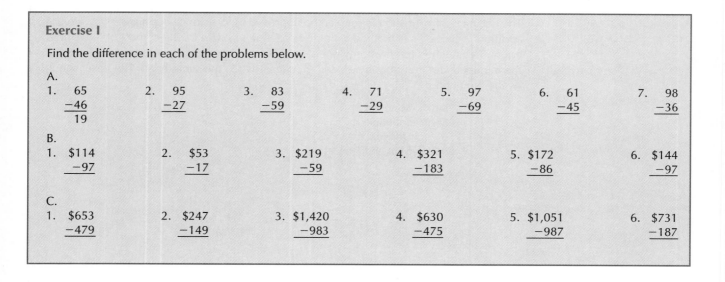

A.
1. 65
   −46
   19
2. 95
   −27
3. 83
   −59
4. 71
   −29
5. 97
   −69
6. 61
   −45
7. 98
   −36

B.
1. $114
   −97
2. $53
   −17
3. $219
   −59
4. $321
   −183
5. $172
   −86
6. $144
   −97

C.
1. $653
   −479
2. $247
   −149
3. $1,420
   −983
4. $630
   −475
5. $1,051
   −987
6. $731
   −187

Multiplication

Multiplication is used in business to price stock, compute payroll, and determine the interest due, taxes due, depreciation, and markup amount, along with a variety of other applications. It is really a shortcut approach to addition. Instead of adding a series of values that are all alike, we can multiply the value by the number of repetitions.

multiplicand The two numbers used in multiplying are called the **multiplicand** (the number on the top)
multiplier and the **multiplier** (the number on the bottom). The result of this process is called the
product **product.**

Example 10

$$
\begin{array}{r}
427 \\
\times 53 \\
\hline
1{,}281 \\
21\ 35 \\
\hline
22{,}631
\end{array}
$$

427   Multiplicand
×53   Multiplier
1,281   Partial product (3 × 427)
21 35   Partial product (5 × 427)
22,631   Product (53 × 427)

(427 × 53 =)

Multiplying by 10, 100, 1,000, etc.

Frequently the multiplier or the multiplicand is a 10, 100, or 1,000. The problem can be
set up and the product found by the normal multiplication process, but it is much easier
and faster to use the following technique.

Example 11

645   Multiplicand
× 100   Multiplier
64,500   Product (645 × 100)

○ *You **can** look good at math!—
Note: Count the number of
zeros in the multiplier. The
same number of additional
zeros is found in the product.*

(645 × 100 =)

Notice in the example that the product has the same number of zeros in it as the
multiplier has.

Example 12

462 × 1,000 = 462,000

(462 × 1,000 =)

Example 13

$98 per share × 100 shares = $9,800 total price

(98 × 100 =)

Exercise J

Complete the following problems using the methods just described.

A.

1.	2.	3.	4.	5.
673	892	546	891	894
×100	×100	×10	×1,000	×10
67,300				

B.

1.	2.	3.	4.	5.
819	961	570	469	162
×10	×100	×10	×100	×1,000

C.

1.	2.	3.	4.
$2,871	$4,712	$6,534	$1,927
×47	×122	×35	×125

D.

1.	2.	3.	4.
$8,965	$83,935	$29,171	$4,782
×17	×3	×24	×29

Exercise K

George's Hardware Store took inventory on June 30 and found the following merchandise on hand in the paint department. Find the total cost of merchandise on hand in the paint department.

Item Description	Quantity	Cost	Total Cost
Paint			
Pint cans	47	$ 2	$94
Quart cans	93	3	
Gallon cans	89	6	
Stain			
Pint cans	23	2	
Quart cans	36	3	
Brushes			
½-Inch	27	1	
1-Inch	51	2	
2-Inch	40	3	
3-Inch	36	4	
4-Inch	32	6	
Thinner			
Quart cans	41	2	
Gallon cans	26	4	
Rollers			
9-Inch	42	1	
11-Inch	50	2	
Drop Cloths			
9′ × 9′ Plastic	35	2	
12′ × 12′ Plastic	20	3	
12′ × 12′ Cloth	4	16	
12′ × 16′ Cloth	7	18	
Can Holders	25	2	
Cheesecloth	20	1	
Handles	12	1	
Brush Cleaners	16	3	
Wallpapering Tools	12	3	
Paste	17	4	

Total value of merchandise on hand in paint department _____

Division

dividend
divisor
quotient

Division is the fourth and last of the fundamental operations in arithmetic. Like multiplication is to addition, division is a time-saving approach for subtraction. The number to be divided is called the **dividend,** the number that is used to divide by is called the **divisor,** and the answer or result is the **quotient.**

Example 14

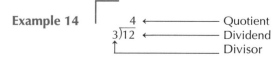

$$(12 \div 3 =)$$

The symbol $\overline{)}$ signifies division. $3\overline{)12}$ is read as "3 divided into 12." The division sign ($\div$) can also be used in the same problem and would be written $12 \div 3$ or read "12 divided by 3." Notice the key words "divided into" and "divided by." This will become even more important in chapter 3, Working with Decimal Numbers.

Long Division

The problem $5,301 \div 57$ requires that 57 be first divided into the group of digits farthest to the left of the decimal point. Fifty-seven cannot be divided into 5 or 53, but 57 will divide into 530. Fifty-seven will divide into 530 nine times. The nine is written directly over the last digit to the right in the group (530).

Example 15

```
        9
57)5,301
   5 13
     17
```

(5,301 ÷ 57 =)

remainder   Seventeen is left over, or is a **remainder** from dividing 57 into 530. The 1 is brought down as the next digit in the dividend. Now 57 can be divided into 171. We find that 57 divides into 171 three times.

```
       93
57)5,301
   5 13
     171
     171
```

In this example the remainder is zero. If there had been a remainder of, say, 14, the answer would have been stated as the quotient "93 with a remainder of 14."

Example 16

```
      3,076
18)55,385
   54
    1 38
    1 26
      125
      108
       17
```

Answer: 3,076 remainder 17

(55,385 ÷ 18 =)

Dividing by 10, 100, 1,000, etc.

○ *Now you can look even* ***better*** *at math! Note: Count the number of zeros in the divisor. The same number of zeros is removed from the dividend.*

When the divisor is a 10, 100, or 1,000, the technique of multiplication by 10, 100, or 1,000 may be used. The example of $9,853 \times 100$ was used to show that the zeros are added to the multiplicand in order to find the product 985,300. If the problem was changed to read $985,300 \div 100$, the answer could be found by removing the two zeros from the divisor 100 and from the dividend and writing the quotient 9,853.

Example 17 $64,820,400 \div 10 = 6,482,040$

(64,820,400 ÷ 10 =)

Example 18 $98,800 \div 100 = 988$

(98,800 ÷ 100 =)

Example 19 $1,600,000 \div 1,000 = 1,600$

(1,600,000 ÷ 1,000 =)

Exercise L

Complete the following problems.

$$
\begin{array}{r} 157 \\ 627\overline{)98,439} \end{array}
$$

1. $627\overline{)98,439}$ 2. $49\overline{)65,366}$ 3. $801\overline{)94,518}$ 4. $183\overline{)638,519}$

5. $867\overline{)365,007}$ 6. $37\overline{)75,254}$ 7. $389\overline{)87,500}$ 8. $34\overline{)657}$

Exercise M

Complete the following problems by the shortcut method. A check can always be made by dividing in the normal manner.

1. $793,000 \div 100 = $ __7,930__   2. $47,000 \div 100 = $ _____

3. $90,600 \div 100 = $ _____   4. $891,000 \div 1,000 = $ _____

5. $600,500 \div 10 = $ _____

Checking Answers

Division and multiplication problems can be checked easily by using the opposite process to find the answer.

To check a divisor of 7, a dividend of 58, a quotient of 8, and a remainder of 2, multiply the divisor by the quotient and add the remainder to the product.

$$
\begin{array}{r} 8 \\ 7\overline{)58} \\ 56 \\ \hline 2 \end{array}
\qquad
\begin{array}{r} \text{Check} \quad 7 \\ \times 8 \\ \hline 56 \\ +2 \\ \hline 58 \end{array}
$$

Example 20

$$
\begin{array}{r} 6,054 \text{ remainder } 7 \\ 8\overline{)48,439} \end{array}
\qquad
\begin{array}{r} \text{Check} \quad 6,054 \\ \times 8 \\ \hline 48,432 \\ +7 \\ \hline 48,439 \end{array}
$$

$(48,439 \div 8 =)$

Multiplication can be checked by dividing the product by the multiplier or the multiplicand. The quotient will be the multiplier or multiplicand, depending on which was used as a divisor.

$$
\begin{array}{r} 602 \\ \times 9 \\ \hline 5,418 \end{array}
\qquad
\begin{array}{r} 602 \\ 9\overline{)5,418} \\ 5\ 4 \\ \hline 18 \\ 18 \end{array}
\quad or \quad
\begin{array}{r} 9 \\ 602\overline{)5,418} \\ 5,418 \end{array}
$$

Use of this check method will help you find errors. Accuracy is important!

ASSIGNMENT Chapter 1

Name _____ Date _____ Score _____

Skill Problems

In the number 865,903,427

1. The digit in the hundreds place is _____.

2. The digit in the tens place is _____.

3. The digit in the units place is _____.

4. The digit in the millions place is _____.

In the number 6,423,570,198

5. The digit in the _____ is 6.

6. The digit in the _____ is 1.

7. The digit in the _____ is 5.

8. The digit in the _____ is 4.

Fill in the dollar values written in words for the following check amounts.

9. $148.00 _____ DOLLARS.

10. $3,579.00 _____ DOLLARS.

11. $250,000.00 _____ DOLLARS.

Round off the following numbers to the place value indicated.

12. 56,278 to the nearest ten thousand is _____.

13. 5,449 to the nearest ten is _____.

14. 654,729 to the nearest hundred thousand is _____.

Select the answer from the choices below.

A. Place Values   B. Addend   C. Sum   D. Difference   E. Minuend
F. Subtrahend   G. Borrowing   H. Multiplicand   I. Multiplier   J. Product
K. Dividend   L. Divisor   M. Quotient   N. Remainder

15. _____ The result of subtraction.

16. _____ Numbers being added.

17. _____ Using a larger place value to provide enough to subtract from.

18. _____ The amount resulting from the addition of two or more numbers.

19. _____ The bottom number of two numbers that are to be multiplied.

Complete the travel expense report doing horizontal and vertical addition.

	Monday	Tuesday	Wednesday	Thursday	Total
Breakfast	$5	$6	$4	—	**20.** $
Lunch	—	4	6	—	**21.**
Dinner	19	14	11	$25	**22.**
Travel	257	35	26	281	**23.**
Lodging	65	65	65	—	**24.**
Tips	4	5	4	—	**25.**
Misc.	2	—	3	—	**26.**
Totals	**27.** $	**28.** $	**29.** $	**30.** $	**31.** $

Complete the following problems. Check your answers.

32.   532
    +89

33.   9,650
    + 425

34. $87
    −59

35.   1,028
    −584

36.   753
    ×1,000

37. $206
    × 27

38. 6,603
      560
    1,450
      935
    +451

39. 847 ÷ 62

40. $634
        ×42

41. $658
        ×42

42. $82,861
     −17,973

43. 42)984

44. $568
       3,904
       +257

45. $82,564
     −12,593

46. $3,549,560 ÷ 25

47. Round off your answer to the nearest ten.
      492 ÷ 6
48. Round off your answer to the nearest hundred.
       5,680
         427
      11,473
      +1,865

49. Round off your answer to the nearest hundred dollars.
          $875
          ×37

50. Write the answer to this problem in words.
          $860
         ×562

51.   239
          ×16

52. Find the difference between sixty-two thousand, eight hundred fifty-seven and twenty-nine thousand, six hundred eight.

53. 472,800 ÷ 10

54. $6,547
      −898

55. $568,950
     −425,982

56. 76)34,656

57. $693
         ×47

58. 8)192

59. 64)1,985

60. 1,000)1,900,000,000

Business Application Problems
Complete the following problems using the appropriate fundamental operation.

1. The shipping department sent out six packages on Friday. The weights of each of the six packages were 15, 16, 43, 5, 12, and 25 pounds. What was the total weight of the six packages sent?

2. The Gift Shop received 40 boxes. Each box contained 144 cards. How many cards did they receive?

3. Amy sold a customer $420 of dishes for use in a restaurant. Each dish was priced at $6. How many dishes were sold?

4. Patrick sold 62,431 items the first week of the new year. He also indicated that 1,834 units were returned. How many items were kept by customers?

5. Bill estimates the expenses for the month to be as follows for his company: $218 for electricity, $2,200 for rent, $8,950 for labor, $180 for telephone, and $350 for equipment repair. What is the total of the estimated expenses for the month?

6. A total of 3,592 customers entered The Pet Shop in the month of June. Sales for June totaled $50,288. How many dollars of sales were made for each customer that entered the shop in the month?

7. A delivery van used 32 gallons of fuel on Thursday. The van is able to travel 17 miles on each gallon. How many miles did the van travel?

8. One of every six units of product produced by Raco, Inc. is destined to be shipped out of the country. Each unit is sold for $23. Raco, Inc. produced 8,412 units in one week. What is the dollar value of the product shipped out of the country in the week?

9. There are 27 employees at the Viril Spa. The average wage is $620 a week. What is the total of the wages paid for the week?

10. The voice mail message from the eastern regional sales representative stated that sales were "fifty-eight thousand, seven hundred sixty dollars" for the month of October. Not considered in the message were $985 of merchandise returns by customers. Using words, indicate below the amount of actual sales, after the returns, by "voice-mail" message to the sales representative.

11. The income from a design studio must be split evenly among seven partners. The income was $115,486 for the year. What was the share for Patsy, a partner?

12. A lawsuit has been decided in favor of five complainants in the amount of $225,000. Before they can receive their share, the lawyer is entitled to half. What is the amount each complainant will receive after the lawyer's cut?

13. The ABC Party Store had $1,723 in the cash register at the end of a day's business. When they opened for business that morning, there was $437 in the register. Mr. Lee said that $86 of the money was sales tax they had collected for the state. How much did they sell during the day?

14. Kerry contributed $50,000 in cash and Hal contributed $38,000 in cash to start Sport-A-Merica. The company bought a building for $26,000 and purchased $41,000 in merchandise, paying cash for both. How much cash does the company have left?

15. At the beginning of the month, Mr. Harold's business car had 15,975 miles on it. At the end of the month, the odometer read 17,193 miles. He estimated that 147 miles were for personal use. How many miles did he travel on business during the month?

16. An employee produced 682 parts in an eleven-hour day. How many parts were produced each hour?

17. Janie decides to purchase her books for the fall semester. At the College Book Store she finds the prices as follows: math, $65; English, $50; history, $78; and accounting, $90. She decides that the prices are too high and goes across the street to the Campus Book Store. She finds the prices there as follows: math, $60; English, $50; history, $81; and accounting, $95. At which store should she buy all her books to get the lowest total price?

18. The production schedule for the Seldon Manufacturing Corporation calls for 1,645 units to be produced on each shift. The company has a three-shift operation. How many units will be produced in five days of production?

19. You are given the task of determining the cost of a banquet for ninety-seven community leaders. The hall will cost $290. The meal will consist of a steak at an estimated cost of $10; a baked potato at $3; vegetable, relish dish, beverage, and roll at $6. Determine the total cost of the banquet rounded off to the nearest hundred dollars.

20. Alex sold three building lots for $16,000 each. He also sold a house for $62,500. What was the total value of the real estate he sold?

21. The Municipal Transportation Department bought seven new minibuses at a cost of $27,300 each. They also bought eleven cars for $12,200 each. What was the total expenditure?

22. The Retail Trade Association has decided to create a package to be used by managers for the training of their employees. The cost will be shared equally by each of the 42 members of the association. Each package will consist of 7 CDs at $14 each; a set of 5 booklets at $10 each set; stationery supplies at $1; a hardbound container at $4; and a set of practice formats at $42 each. The association has requested that twenty training packages be assembled for each association member. What will be the total cost for twenty training packages?

23. A salesperson must travel a route of 611 miles in order to cover the sales territory. There are twenty-eight stops to make on the route; each stop takes about one hour. Traveling time for the complete route is about eleven hours. If the salesperson wishes to work only eight hours a day, how many days will it take to complete the route?

24. The Stuben Oil Company read the meter on the gasoline storage tank. On Monday, the meter read 4,736,902 gallons; on Wednesday, it read 4,019,456. On Thursday morning a delivery of 550,000 gallons was added to the tank. On Saturday at closing time, the meter read 3,457,995. How many gallons of gas were sold during the week?

25. The town of Norden had a population of 47,867 people on January 1 of last year. During the year 427 babies were born; 381 residents died; 188 people, composing 62 families, moved in; and 73 people moved out. What was the population of Norden at the end of the year?

26. The inventory of Spectrum, Inc., totaled 842 parts of item J926. The cost of the inventory was $13,472. What was the cost per unit?

27. A resort found that 473 clients had been there last year. This year, they lost 239 customers from last year but recruited 357 new customers. How many customers did they have this year?

28. Write a complete sentence in answer to this question: Each of the 49 restaurants in the town of Witon served 167 meals per hour for the five hours that preceded the homecoming game. How many meals were served in Witon in the five-hour period?

29. Write a complete sentence in answer to this question: New lounge chairs for a pool area will cost $127 each. The Recreation Department has decided on a purchase of twenty-three chairs. What will be the cost of the purchase?

30. Write a complete sentence in answer to this question: The Food Mart has decided to add 42 items to their existing inventory of 1,219 items. At the same time, they will discontinue 81 items. How many items will they carry in inventory?

★ **Challenge Problem**

At the beginning of the month, the Gene Robins Company had 422 units of a product, "Talson," on hand. The company purchased another 960 pieces during the month at a cost of $1.40 per unit. Forty pieces of the new shipment were found to be defective and were returned to the seller. During the month, the company sold 865 units after having 14 returned by customers because of poor packing. If the company made a profit of $1.23 on each unit sold, (a) how much will be made on the units already sold and (b) how much would be made if all of the acceptable units remaining were sold?

Name _____ Date _____ Score _____

1. Find the product of the following numbers: 64 and 850.

2. $ 1,457
   5,800
   3,575
   +   41

3. $ 42,803
   + 787,276

4. 18)75,859

5. The profit of the Edrunton Company must be evenly distributed between the six department managers. The profit for the quarter was $64,638. What will each department manager receive as his or her share of the profits?

6. 549 × 187 =

7. 768 ÷ 40 =

8. The Sales Department of the Wolton Corporation reported the following figures for May: marine equipment—$6,185,005; automotive equipment—$54,980; service repair parts—$7,681; and miscellaneous—$1,822. What were the total sales for the month?
   $6,185,005
      54,980
       7,681
   +   1,822

9. Jay Moore produced 891 parts on his machine for the week. One hundred forty-seven of those were determined to be oversize and, therefore, not part of the acceptable lot. Another 63 failed a strength test set up by the Quality Department. How many of the parts were acceptable (to the nearest ten)?
   891
   −147
   − 63

10. The answering service tape has recorded a message from the Western Sales Region Manager, A. F. Litona. His statement indicates that sales of the "Aferton" product were "eighty-five thousand, six hundred five dollars," the "Virmonon" product "two hundred thousand, thirty-eight dollars," and the "Olsand" product "forty-one thousand, eighty dollars." What were the total sales for the period?
    $ 85,605
     200,038
      41,080

11. Find the total difference between the following: 651 units at $387 each and 120 units at $456 each.

12. Write the following value using words: $43,679.

13. Subtract and round off your answer to the nearest ten thousand dollars.
    $459,073
    − 91,752

14. Mrs. Jesop knows that the salespeople in her group will be successful on one out of every two sales calls that they make. If each call takes two hours and each successful call develops a $98 order, what will be the volume of sales from a sales force of twelve people, each working forty hours a week?

15. Round off 685,224 to the nearest hundred.

(If you missed problems 2, 3, and 8, you need to review addition; problems 9, 11, or 13, you need to review subtraction; problems 1, 6, or 14, you need to review multiplication; problems 4, 5, or 7, you need to review division; problems 10 or 12, you need to review reading and writing whole numbers; and problems 13 or 15, you need to review rounding off.) Record your score on the Record of Performance on Mastery Tests in the back of the book.

WORKING WITH FRACTIONS

2

OBJECTIVES

After mastering the material in this chapter, you will be able to:

1. Identify fraction types. p. 22
2. Change mixed numbers to improper fractions, and improper fractions to mixed numbers. p. 23
3. Reduce fractions to their lowest terms. p. 24
4. Perform the basic fundamental processes (addition, subtraction, multiplication, and division) with accuracy in problems dealing with fractions. p. 25
5. Develop a ratio from data provided. p. 34
6. Understand and use the following terms:
   Denominator
   Numerator
   Common Fraction
   Improper Fraction
   Mixed Numbers
   Prime Numbers
   Common Denominator
   Lowest Common Denominator (LCD)
   Ratio

SPORT-A-MERICA

The selection of the store Kerry and Hal agreed to purchase took into consideration the many types of equipment Sport-A-Merica will handle. Hal knew there would need to be space for seasonal merchandise. Other space would have to be set aside for equipment, end-of-season sale merchandise, an office area, a receiving area, and an employee lounge.

When Kerry saw the first building Hal located, she asked, "How much space will we have for baseball equipment?" Kerry knew that there were a lot of baseball leagues in the immediate area. By knowing that one-fifth of the space will be set aside for baseball equipment, Hal was able to give her an answer. As a result, they turned down the first building because the floor space would not allow enough room for baseball equipment once the other space needed was also taken into consideration. Knowing that three-fourths of the baseball business would involve the baseball leagues, this building would not have been a good choice.

Kerry and Hal looked further. Hal found the right place—one that offered 1,000 square feet, the proper space to allow for each area needed in the store. Hal is happy that he had considered the floor space and had applied fractions to the planning. Now he must determine the number of square feet to be used for baseball leagues. He will also want to know the ratio of office floor space to sales floor. Kerry is sure to ask.

Description and Types
Objective 1

A fraction can be described as a part of a whole. This description fits many business situations. Mastery of fractions and their applications is therefore required.

Example 1

The fraction $\frac{3}{4}$ states that the whole is divided into four equal parts. Three of the four parts make up the portion that the fraction is describing. The fraction can be presented graphically in a variety of forms, each describing the same part of a whole, $\frac{3}{4}$.

$$3/4 = \frac{3}{4} = \quad = \quad = \quad =$$

denominator
numerator

The bottom portion of the fraction that describes the number of equal parts is called the **denominator.** The top portion that specifies the number of equal parts that the fraction is describing is called the **numerator.**

A fraction can be read in two different ways. The fraction $\frac{7}{8}$ can be read as "seven eighths" or as "seven divided by eight," since the line between the numerator and the denominator signifies division. Either a diagonal line / or a horizontal line — can be used to separate the numerator and the denominator.

Example 2

$\frac{72}{146}$ can be read as "seventy-two one hundred forty-sixths" or as "seventy-two divided by one hundred forty-six."

A third way to read a fraction is to describe how it looks (i.e., "seventy-two over one hundred forty-six"). This method is also acceptable.

Read the following fractions to yourself.

$$\frac{9}{16} \qquad \frac{41}{90} \qquad \frac{5}{8} \qquad \frac{4}{7} \qquad \frac{16}{21} \qquad \frac{193}{216}$$

$$\frac{17}{20} \qquad \frac{1}{4} \qquad \frac{10}{17} \qquad \frac{1}{3} \qquad \frac{3}{4} \qquad \frac{7}{16}$$

$$\frac{17}{651} \qquad \frac{5}{16} \qquad \frac{5}{55} \qquad \frac{3}{8} \qquad \frac{21}{109} \qquad \frac{7}{21}$$

common fraction
improper fraction

There are three types of fractions. The type that we have been working with is called a **common fraction.** This type has a denominator (bottom number) larger than the numerator (top number). An **improper fraction** has a numerator larger than or equal to the denominator, as shown in example 3.

Example 3

$$\frac{19}{6} \qquad \frac{40}{10} \qquad \frac{93}{16} \qquad \frac{3}{3} \qquad \frac{19}{9} \qquad \frac{8}{4} \qquad \frac{6}{5}$$

mixed numbers

Mixed numbers contain a whole number as well as a common fraction, as shown in example 4.

Example 4

$$4\frac{1}{3} \qquad\qquad 7\frac{1}{2} \qquad\qquad 14\frac{63}{91} \qquad\qquad 5\frac{1}{7} \qquad\qquad 1\frac{9}{16}$$

In most instances, fractions will be presented as common fractions or as mixed numbers. Improper fractions are found in arithmetic operations and as temporary answers. They are not acceptable as final answers unless specifically asked for in a problem.

Exercise A

Identify the following by type of fraction: common, mixed, or improper.

1. $\frac{9}{13}$ _____common_____   2. $\frac{46}{17}$ _____   3. $\frac{1,107}{1,000}$ _____

4. $\frac{41}{35}$ _____   5. $16\frac{2}{9}$ _____   6. $\frac{4}{7}$ _____

7. $\frac{11}{9}$ _____   8. $\frac{76}{81}$ _____   9. $\frac{1}{4}$ _____

10. $\frac{4}{5}$ _____   11. $4\frac{1}{10}$ _____   12. $16\frac{2}{3}$ _____

Interchanging Fraction Types
Objective 2

It is sometimes necessary to interchange fraction types in order to perform the fundamental processes or to find the correct answer. All answers in the improper fraction form should be changed to a mixed number or a whole number.

Example 5

$$\begin{array}{r} 2 \\ 4\overline{)9} \\ 8 \\ \hline 1 \end{array}$$

The improper fraction $\frac{9}{4}$ can be changed to a mixed number by dividing nine by four, just as the fraction is read. The result is that there are two whole number integers (two 4s) in $\frac{9}{4}$ and one-fourth left over. The whole number is written. The remainder is then written over the divisor. $2\frac{1}{4}$

$(9 \div 4 =)$

Example 6

$$\frac{197}{9} = 9\overline{)197} = 21\frac{8}{9}$$
$$\begin{array}{r} 18 \\ \hline 17 \\ 9 \\ \hline 8 \end{array}$$

$(197 \div 9 =)$

Exercise B

Change the following improper fractions to whole or mixed numbers.

1. $\frac{13}{8} =$ $1\frac{5}{8}$ 2. $\frac{67}{12} =$ _____   3. $\frac{14}{5} =$ _____   4. $\frac{47}{14} =$ _____

5. $\frac{19}{6} =$ _____   6. $\frac{5}{2} =$ _____   7. $\frac{19}{9} =$ _____   8. $\frac{93}{10} =$ _____

9. $\frac{43}{10} =$ _____   10. $\frac{9}{8} =$ _____   11. $\frac{16}{4} =$ _____   12. $\frac{57}{8} =$ _____

If an improper fraction can be changed to a mixed number, then we should be able to do the reverse process (i.e., change a mixed number to an improper fraction).

Example 7 | The mixed number $4\frac{3}{5}$ can be changed to an improper fraction by multiplying the whole number (4) by the number of equal parts in each unit (the denominator 5) and adding the numerator to the product.

$$4\frac{3}{5} = \frac{(5 \times 4) + 3}{5} = \frac{23}{5}$$

(4 × 5 = + 3 =)

Example 8 | $7\frac{4}{10} = \frac{(7 \times 10) + 4}{10} = \frac{74}{10}$

(7 × 10 = + 4 =)

Exercise C

Change the following mixed numbers to improper fractions.

1. $6\frac{3}{4} =$ $\frac{27}{4}$

2. $9\frac{5}{6} =$

3. $7\frac{4}{5} =$

4. $1\frac{7}{8} =$

5. $110\frac{1}{5} =$

6. $3\frac{1}{2} =$

7. $18\frac{3}{8} =$

8. $6\frac{2}{3} =$

9. $4\frac{3}{4} =$

10. $1\frac{1}{2} =$

11. $5\frac{7}{16} =$

12. $10\frac{1}{2} =$

Reducing Fractions
Objective 3

Answers to problems in fractions are correct only if they are reduced to lowest terms. The part of a whole that is described as $\frac{3}{4}$ can also be described as $\frac{6}{8}, \frac{12}{16}, \frac{24}{32}$, etc. In each of these examples, the numerator has the same relationship to the denominator as it does in the fraction $\frac{3}{4}$. The fraction $\frac{3}{4}$ can be changed to $\frac{6}{8}$ by doubling both the numerator and the denominator. The fraction $\frac{6}{8}$ can be changed to $\frac{12}{16}$ by following the same process again.

In each case, the numerator and the denominator are being multiplied by an equal amount (2). Any fraction can be changed in this manner. The fraction $\frac{3}{7}$ can be changed to $\frac{15}{35}$ by multiplying both the numerator and denominator by 5.

You should also be able to do the opposite operation (i.e., divide both the numerator and denominator by the same number). When both the numerator and denominator are divided by 5, $\frac{15}{35}$ equals $\frac{3}{7}$. When the numerator and denominator no longer can be divided by a whole number other than 1, the fraction has been reduced to lowest terms.

How do you know when the fraction has been reduced to lowest terms? The best

prime numbers   approach is to use **prime numbers** as divisors. When the prime numbers cannot be evenly divided into the numerator and denominator, the fraction is fully reduced. A prime number is divisible only by itself and by one (1). The smaller prime numbers are 2, 3, 5, 7, 11, and 13.

Trial-and-Error Approach

A trial-and-error approach can be used to reduce fractions. This method states that the prime numbers are used to find a number that will divide into both the numerator and denominator. It is helpful to use a system that will make this approach easy to follow. Start with the smallest prime number and divide it into both the numerator and the denominator. The first prime number that divides evenly is a common divisor. If you are unsuccessful with the first prime number, go to the next highest prime number until you find one that works. Repeat this process until no prime number will divide evenly. The fraction will then be reduced to its lowest terms.

Example 9 | The fraction $\frac{12}{45}$ must be reduced.

Try prime number 2. 12 ÷ 2 = 6   45 ÷ 2 = 22 r 1

Two does not divide evenly and therefore is not a common divisor. Try prime number 3. 12 ÷ 3 = 4   45 ÷ 3 = 15

Three is a prime number that evenly divides into 12 and 45, and therefore, it is a common divisor. The fraction is reduced to $\frac{4}{15}$.

(12 ÷ 3 =   45 ÷ 3 =)

Exercise D

Reduce the following fractions to their lowest terms.

1. $\frac{5}{10} = \underline{\quad \frac{1}{2} \quad}$ 2. $\frac{9}{12} = \underline{\qquad}$ 3. $\frac{14}{20} \underline{\qquad}$ 4. $\frac{40}{100} = \underline{\qquad}$

5. $\frac{18}{36} = \underline{\qquad}$ 6. $\frac{14}{16} = \underline{\qquad}$ 7. $\frac{22}{50} = \underline{\qquad}$ 8. $\frac{3}{6} = \underline{\qquad}$

9. $\frac{6}{8} = \underline{\qquad}$ 10. $\frac{9}{15} = \underline{\qquad}$ 11. $\frac{16}{22} = \underline{\qquad}$ 12. $\frac{6}{16} = \underline{\qquad}$

Exercise E

Change the following fractions to fractions of equal value using the denominator indicated.

1. $\frac{3}{4} = \frac{15}{20}$ 2. $\frac{3}{5} = \frac{}{55}$ 3. $\frac{9}{11} = \frac{}{110}$ 4. $\frac{7}{20} = \frac{}{80}$

5. $\frac{3}{7} = \frac{}{35}$ 6. $\frac{9}{10} = \frac{}{30}$ 7. $\frac{4}{5} = \frac{}{20}$ 8. $\frac{1}{2} = \frac{}{10}$

9. $\frac{41}{100} = \frac{}{1,000}$ 10. $\frac{6}{7} = \frac{}{14}$ 11. $\frac{5}{12} = \frac{}{36}$ 12. $\frac{1}{4} = \frac{}{20}$

Addition and Subtraction
Objective 4

Fractions, like whole numbers, can be added in order to find the sum, or total.

Example 10 | $\dfrac{5}{13} + \dfrac{3}{13} = \dfrac{8}{13}$

(5 + 3 =)

The denominator must be the same in each fraction in order to add them. Only the numerators are added.

Common Denominator

In the previous example the addends had **common denominators** (both denominators were 13s in the example given). When the denominators of a fraction to be added are not the same, they must be made alike before the addition can begin.

Product Approach

A common denominator can always be found by multiplying the denominators. The product, or common denominator, becomes the denominator of both fractions. The next step is to change the numerator to give an equivalent fraction.

Example 11 $\dfrac{3}{4} + \dfrac{1}{5} = (4 \times 5 = 20,$ the common denominator)

To find the numerator, divide the old denominator into the common denominator and multiply the quotient by the numerator for each fraction.

$$\left[\dfrac{3 \times 5}{4 \times 5}\right] + \left[\dfrac{1 \times 4}{5 \times 4}\right] = \dfrac{15}{20} + \dfrac{4}{20} = \dfrac{19}{20}$$

Note: Remember to reduce the answer to lowest terms.

$(20 \div 4 =$ $20 \div 5 =)$

lowest common denominator (LCD)   Another way to determine a common denominator is to find the **lowest common denominator (LCD).** This will reduce the need for dealing with unnecessarily large fractions when performing additional functions. To use this approach, select the largest denominator and test multiples of it against the other denominators until the other denominators divide equally into the multiple of the first denominator.

Example 12   Find the lowest common denominator for the following fractions.

$$\dfrac{5}{6} + \dfrac{3}{8} =$$

The largest denominator is 8. Six will not divide equally into 8; therefore 8 is added to the original 8 to make 16. Six will not divide equally into 16; therefore 8 is added to 16 to make 24. Six will evenly divide into 24; therefore, 24 is the lowest common denominator.

$$\dfrac{20}{24} + \dfrac{9}{24} = \dfrac{29}{24} = 1\dfrac{5}{24}$$

$(24 \div 6 =$ $24 \div 8 =)$

The same approach is used when more than two fractions are to be added.

Example 13 $\dfrac{2}{3} + \dfrac{1}{2} + \dfrac{3}{5}$ 30 is the common denominator.

$$\left[\dfrac{2 \times 10}{3 \times 10}\right] + \left[\dfrac{1 \times 15}{2 \times 15}\right] + \left[\dfrac{3 \times 6}{5 \times 6}\right] =$$

$$\dfrac{20}{30} + \dfrac{15}{30} + \dfrac{18}{30} = \dfrac{53}{30}$$

$\dfrac{53}{30}$ reduces to $1\dfrac{23}{30}$

that is, $30\overline{)53}$
$\quad\quad\quad \dfrac{1}{}$
$\quad\quad\quad \dfrac{30}{23}$ or $\dfrac{23}{30}$

$(30 \div 3 =$ $30 \div 2 =$ $30 \div 5 =)$

Exercise F

Find the sum in the following problems.

1. $\frac{1}{6} + \frac{2}{3} + \frac{1}{4} =$

$\frac{2}{12} + \frac{8}{12} + \frac{3}{12} =$

$\frac{13}{12} = 1\frac{1}{12}$

2. $\frac{7}{8} + \frac{1}{2} + \frac{3}{4} =$

3. $\frac{2}{7} + \frac{1}{5} + \frac{2}{3}$

4. $\frac{7}{8} + \frac{3}{4} + \frac{3}{4} =$

5. $\frac{1}{10} + \frac{1}{2} + \frac{3}{7} =$

6. $\frac{9}{10} + \frac{3}{8} =$

7. $\frac{7}{20} + \frac{1}{3} + \frac{1}{2} =$

8. $\frac{2}{5} + \frac{5}{16} + \frac{3}{4}$

9 $\frac{7}{16} + \frac{3}{8} + \frac{1}{4} =$

10. $\frac{3}{7} + \frac{5}{6} + \frac{2}{3} =$

The subtraction of common fractions is similar to addition, because both require common denominators. Subtraction requires finding the difference of the numerators.

Example 14 $\frac{3}{4} - \frac{1}{4} = \frac{2}{4}$, which reduces to $\frac{1}{2}$

$(4 \div 2 =)$

Example 15 $\frac{1}{3} - \frac{2}{9} =$

$\left[\frac{1 \times 3}{3 \times 3}\right] - \frac{2}{9} =$

$\frac{3}{9} - \frac{2}{9} = \frac{1}{9}$

$(1 \times 3 =)$

Example 16 $\frac{3}{4} - \frac{1}{5} =$

$\left[\frac{3 \times 5}{4 \times 5}\right] - \left[\frac{1 \times 4}{5 \times 4}\right] =$

$\frac{15}{20} - \frac{4}{20} = \frac{11}{20}$

$(3 \times 5 = \quad 1 \times 4 =)$

Exercise G

Find the differences in the following problems.

1. $\dfrac{3}{4} - \dfrac{1}{8} =$

$\dfrac{6}{8} - \dfrac{1}{8} = \dfrac{5}{8}$

2. $\dfrac{5}{8} - \dfrac{1}{4} =$

3. $\dfrac{9}{10} - \dfrac{2}{3} =$

4. $\dfrac{1}{4} - \dfrac{1}{5} =$

5. $\dfrac{17}{20} - \dfrac{2}{9} =$

6. $\dfrac{7}{8} - \dfrac{5}{12} =$

7. $\dfrac{4}{9} - \dfrac{1}{4} =$

8. $\dfrac{2}{3} - \dfrac{1}{4} =$

9. $\dfrac{7}{10} - \dfrac{1}{16} =$

10. $\dfrac{1}{2} - \dfrac{3}{16} =$

Mixed Numbers

Mixed numbers, like common fractions, can be added to find the sum. This process consists of three steps.

Example 17

In the mixed numbers $8\frac{3}{8} + 2\frac{1}{8}$, (1) the whole numbers are added; (2) then the fractions are added; and, as a last step, (3) the mixed-number answer is reduced.

$$8\frac{3}{8}$$
$$+2\frac{1}{8}$$
$$10\frac{4}{8} = 10\frac{1}{2} \text{ reduced}$$

(8 + 2 =   3 + 1 =)

This three-step process is lengthened when the fractions of the mixed numbers have different denominators.

Example 18

$$4\frac{2}{3} + 6\frac{5}{8} =$$

A common denominator must be found, as it is when adding common fractions.

$$4\frac{2}{3} \quad = \quad 4\frac{16}{24}$$
$$+ 6\frac{5}{8} \quad = \quad +6\frac{15}{24}$$
$$10\frac{31}{24} = 11\frac{7}{24} \text{ reduced}$$

(4 + 6 =   16 + 15 =)

Exercise H

Find the sum in the following problems. Remember to reduce your answers to lowest terms.

1. $6\frac{3}{4} + 8\frac{1}{2} =$

$6\frac{3}{4} + 8\frac{2}{4} = 14\frac{5}{4}$

$= 15\frac{1}{4}$

2. $9\frac{2}{3} + 4\frac{5}{8} =$

3. $3\frac{5}{16} + 1\frac{3}{4} =$

4. $8\frac{1}{3} + 2\frac{5}{8} =$

5. $7\frac{1}{2} + 4\frac{3}{10} =$

6. $5\frac{3}{8} + 3\frac{5}{9} =$

7. $17\frac{3}{50} + 7\frac{3}{4} =$

8. $1\frac{7}{8} + 6\frac{5}{6} =$

9. $2\frac{3}{4} + 8\frac{1}{2} + 4\frac{2}{3} =$

10. $9\frac{1}{6} + 5\frac{3}{8} + 3\frac{1}{4} =$

The subtraction of mixed numbers uses a similar process.

Example 19

$$9\frac{5}{6} = \quad 9\frac{5}{6}$$
$$-1\frac{1}{3} = -1\frac{2}{6}$$
$$\overline{\qquad \quad 8\frac{3}{6}} = 8\frac{1}{2} \text{ reduced}$$

$(9 - 1 = \quad 5 - 2 =)$

Example 20

$$19\frac{1}{10}$$

In this case, a 1, or ten tenths, must be borrowed from the 19 in order to complete the subtraction.

$$-16\frac{7}{10}$$

$$18\frac{11}{10} \quad \left(18 + \frac{10}{10} + \frac{1}{10}\right)$$
$$-16\frac{7}{10}$$
$$\overline{\quad 2\frac{4}{10}} \quad = 2\frac{2}{5} \text{ reduced}$$

$(18 - 16 = \quad 11 - 7 =)$

Exercise I

Complete the following subtraction problems. Reduce your answers to lowest terms.

1. $14\frac{2}{3} - 7\frac{1}{3} = 7\frac{1}{3}$

2. $3\frac{3}{8} - 1\frac{1}{8} =$

3. $6\frac{3}{8} - 1\frac{1}{10} =$

4. $59\frac{1}{3} - 6\frac{3}{8} =$

5. $5\frac{7}{10} - 3\frac{1}{3} =$

6. $6\frac{1}{8} - 1\frac{1}{10} =$

7. $7\frac{3}{4} - 1\frac{1}{6} =$

8. $173\frac{1}{2} - \frac{1}{3} =$

9. $9\frac{3}{4} - 6\frac{2}{3} =$

10. $8\frac{1}{9} - 1\frac{1}{2} =$

Multiplication and Division

The multiplication of common fractions is simpler than addition or subtraction. There is no need for a common denominator.

Common Fractions

The denominators are multiplied, and the numerators are multiplied. The product should be reduced as in the previous fraction problems.

Example 21
$$\frac{5}{6} \times \frac{4}{7} = \frac{20}{42} = \frac{10}{21} \text{ reduced}$$

(5 × 4 =   6 × 7 =)

Example 22

Hal has decided to set aside one-fifth of the Sport-A-Merica store floor space for baseball equipment. Three-fourths of that space will be used for baseball leagues. How much of the 1,000-square-foot store will be used for baseball leagues?

$$\frac{1}{5} \times \frac{1,000}{1} \times \frac{3}{4} = \frac{3,000}{20} = 150 \text{ square feet}$$

(1 × 3 × 1,000 =   5 × 4 × 1 =)

○ *A whole number can be changed into a fraction by placing a denominator of one under the whole number. After all, the whole number is only an expression of so many units or ones.*

A shortcut or work-saving step can be used to reduce the amount of multiplication. By reducing the fractions before the multiplication begins, the factors are made smaller and the product is also much smaller.

Example 23
$$\frac{9}{\overset{}{\underset{5}{\cancel{10}}}} \times \frac{\overset{1}{\cancel{2}}}{5} = \frac{9}{25} \qquad \text{(Both 10 and 2 are divisible by 2.)}$$

(9 × 1 =   5 × 5 =)

Reducing can take place any time you can divide any numerator and any denominator by the same number.

Example 24 $\dfrac{3}{4} \times \dfrac{4}{9} = \dfrac{\overset{1}{\cancel{3}}}{4} \times \dfrac{4}{\underset{3}{\cancel{9}}} = \dfrac{1}{\cancel{4}} \times \dfrac{\overset{1}{\cancel{4}}}{3} = \dfrac{1}{3}$

(1 × 1 =   1 × 3 =)

Example 25 $\dfrac{6}{11} \times \dfrac{1}{42} = \dfrac{\overset{1}{\cancel{6}}}{11} \times \dfrac{1}{\underset{7}{\cancel{42}}} = \dfrac{1}{77}$

(1 × 1 =   11 × 7 =)

A series of fractions may be multiplied just as in the two previous examples. Reducing before multiplication is still advisable.

Example 26 $\dfrac{3}{4} \times \dfrac{7}{8} \times \dfrac{1}{9} = \dfrac{\overset{1}{\cancel{3}}}{4} \times \dfrac{7}{8} \times \dfrac{1}{\underset{3}{\cancel{9}}} = \dfrac{7}{96}$

(1 × 7 × 1 =   4 × 8 × 3 =)

Example 27 $\dfrac{9}{10} \times \dfrac{3}{4} \times \dfrac{8}{9} = \dfrac{\overset{1}{\cancel{9}}}{\underset{5}{\cancel{10}}} \times \dfrac{3}{\underset{1}{\cancel{4}}} \times \dfrac{\overset{\overset{1}{\cancel{2}}}{\cancel{8}}}{\underset{1}{\cancel{9}}} = \dfrac{3}{5}$

○ *Fractions will be used to compute the solutions of interest problems and depreciation problems.*

(1 × 3 × 1 =   5 × 1 × 1 =)

Exercise J

Find the product in the following problems. Reduce your answers.

1. $\dfrac{3}{16} \times \dfrac{8}{9} =$

2. $\dfrac{7}{16} \times 2\dfrac{1}{5} =$

3. $\dfrac{7}{8} \times \dfrac{5}{7} =$

$\dfrac{\overset{1}{\cancel{3}}}{\underset{2}{\cancel{16}}} \times \dfrac{\overset{1}{\cancel{8}}}{\underset{3}{\cancel{9}}} = \dfrac{1}{6}$

4. $\dfrac{5}{7} \times \dfrac{1}{6} =$

5. $\dfrac{2}{3} \times \dfrac{3}{4} =$

6. $\dfrac{5}{9} \times \dfrac{7}{10} =$

(Continued on next page)

Exercise J Continued

7. $\dfrac{7}{16} \times 8 \times \dfrac{13}{14} =$ 8. $4\dfrac{3}{4} \times 5 \times \dfrac{5}{6} =$ 9. $\dfrac{5}{8} \times \dfrac{5}{9} \times \dfrac{6}{7} =$

10. $\dfrac{13}{21} \times \dfrac{3}{10} \times \dfrac{5}{16} =$

Division

○ *No common denominator is necessary to multiply **or** to divide fractions.*

○ *To divide, turn the divisor upside down and then multiply!*

Division of common fractions requires a few additional instructions. The first step is to review the way in which a division problem is read. In chapter 1 on whole numbers, you found that a problem such as $18 \div 6$ is read "eighteen divided by six" or "six divided into eighteen." Eighteen is the dividend and six is the divisor. In the problem $\frac{3}{5} \div \frac{1}{2}$, $\frac{3}{5}$ is the dividend and $\frac{1}{2}$ is the divisor. You must be able to identify the divisor in order to be able to do the additional steps required in the division of fractions.

In the example $\frac{3}{5} \div \frac{1}{2}$, the divisor $\frac{1}{2}$ must be inverted (i.e., tipped upside down) and the ($\div$) sign changed to a ($\times$) sign. You then continue as you did in the multiplication problems.

Example 28

$$\frac{3}{5} \div \frac{1}{2} =$$
$$\frac{3}{5} \times \frac{2}{1} = \frac{6}{5} = 1\frac{1}{5}$$

$(3 \times 2 = \qquad 5 \times 1 =)$

Example 29

$$\frac{9}{10} \div \frac{2}{5} =$$
$$\frac{9}{\overset{}{\underset{2}{10}}} \times \frac{\overset{1}{\cancel{5}}}{2} = \frac{9}{4} = 2\frac{1}{4}$$

$(9 \times 1 = \qquad 2 \times 2 =)$

Exercise K

Find the quotient in the following problems. Reduce your answers.

1. $\dfrac{6}{7} \div \dfrac{1}{2} =$ 2. $\dfrac{9}{10} \div \dfrac{3}{5} =$ 3. $\dfrac{3}{4} \div \dfrac{7}{16} =$

$\dfrac{6}{7} \times \dfrac{2}{1} = \dfrac{12}{7} = 1\dfrac{5}{7}$

4. $\dfrac{5}{8} \div \dfrac{1}{4} =$ 5. $\dfrac{3}{8} \div \dfrac{2}{9} =$ 6. $\dfrac{6}{7} \div \dfrac{2}{3}$

Exercise K Continued

7. $\dfrac{7}{10} \div \dfrac{1}{5} =$ 　　　　　　 8. $\dfrac{3}{4} \div \dfrac{1}{8} =$ 　　　　　　 9. $\dfrac{5}{7} \div \dfrac{1}{6} =$

10. $\dfrac{5}{6} \div \dfrac{2}{3} =$

Mixed Numbers

The multiplication of mixed numbers requires an additional step (i.e., changing the mixed numbers to improper fractions).

Example 30

$$1\dfrac{1}{6} \times 3\dfrac{1}{2} =$$

$$\dfrac{7}{6} \times \dfrac{7}{2} = \dfrac{49}{12} = 4\dfrac{1}{12}$$

(7 × 7 =   6 × 2 =)

Example 31

$$2\dfrac{9}{10} \times 5\dfrac{1}{6} =$$

$$\dfrac{29}{10} \times \dfrac{31}{6} = \dfrac{899}{60} = 14\dfrac{59}{60}$$

(29 × 31 =   10 × 6 =)

The division of mixed numbers requires the same step (i.e., changing the mixed numbers to improper fractions).

Example 32

$$4\dfrac{1}{2} \div 3\dfrac{3}{8} =$$

$$\dfrac{9}{2} \div \dfrac{27}{8} =$$

$$\overset{1}{\underset{1}{\cancel{\dfrac{9}{2}}}} \times \overset{4}{\underset{3}{\cancel{\dfrac{8}{27}}}} = \dfrac{4}{3} = 1\dfrac{1}{3}$$

(1 × 4 =   1 × 3 =)

Exercise L

Find the product or quotient in the following problems. Reduce your answers to lowest terms.

1. $6\dfrac{1}{2} \times 3\dfrac{3}{16} \times 5\dfrac{1}{8} =$ 　　　　 2. $8\dfrac{1}{4} \div 1\dfrac{3}{4} =$ 　　　　 3. $6\dfrac{3}{4} \div 1\dfrac{5}{8} =$

$\dfrac{13}{2} \times \dfrac{51}{16} \times \dfrac{41}{8} =$

$\dfrac{27,183}{256} = 106\dfrac{47}{256}$

(Continued on next page)

Exercise L Continued

4. $4\frac{2}{3} \div 1\frac{1}{6} =$

5. $5\frac{1}{6} \div 2\frac{4}{5} =$

6. $10\frac{1}{4} \div 3\frac{4}{5} =$

7. $5\frac{6}{7} \div \frac{5}{9} =$

8. $5\frac{4}{9} \div 3\frac{3}{5} =$

Increases and Decreases

○ *Watch the wording on these examples!*

Business and government leaders frequently make comparisons of sales, revenues, employees, etc. The way in which the comparison is made is very important. The wording is the key to understanding the change, or difference.

Example 33

The sales of Sport-A-Merica were $400,000 in 1999. In 2000 the sales were $500,000, a $100,000 difference. This could be expressed as "an increase of $\frac{1}{4}$" or "$1\frac{1}{4}$ of last year's sales." Both mean that sales increased by $100,000. Be careful to watch the wording in problems that mention increase or decrease.

Example 34

If the sales of Sport-A-Merica were to go from $500,000 in 2000 to $400,000 in 2001, the information could be communicated by stating that sales decreased "by $\frac{1}{5}$" or "to $\frac{4}{5}$ of last year's sales." Again, in either method of stating the information, the sales decreased by $100,000 compared to the earlier period's sales of $500,000.

Ratios
Objective 5

A **ratio** is the comparison of one value to another. This may be expressed as a fraction such as $\frac{5}{1}$. In this case, 5 is being compared to 1. The number on the bottom is the value that is being compared to; it is the basis of comparison. The ratio may also be written as 5:1. The colon replaces the — sign. The ratio is still read as "5 to 1."

This ability to compare one value to another is useful in many business situations. A comparison can be made of the number of male to female employees in a firm, the amount of profits to sales over a period, and the number of hours of machine time to the number of hours of labor to perform a task.

Example 35

Sport-A-Merica had the following initial allocation of space in their first store layout.

Soft goods display = 340 square feet
Equipment display = 560 square feet
Office space = 100 square feet

What is the ratio of office space to the sales display area?

Total display area = 900 square feet
Total office area = 100 square feet

The ratio is 100:900, or 1:9 (reduced)

(100 ÷ 100 =   900 ÷ 100 =)

Example 36   There are eighty employees in the Accounting Department. Ten of the employees are in managerial positions. What is the ratio of nonmanagerial employees to managerial employees?

80 − 10 = 70

Ratio of 7 to 1

(70 ÷ 10 =   10 ÷ 10 =)

Exercise M

Use the following data to develop the ratios in this exercise.
Number of employees in firm = 400
Number of employees 40 years old or older = 60
Number of males = 240
Number of employees with 10 or more years of experience = 320
Number of employees that are in clerical positions = 80
Number of nonclerical employees in the firm = 320
Number of employees 50 years old or older = 20

1. Ratio of males to females = ____3:2____. (240:160)

2. Ratio of those between 40 and 50 years old to the total = _____.

3. Ratio of nonclerical to clerical employees = _____.

4. Ratio of all employees to male employees = _____.

5. Ratio of all employees to female employees = _____.

6. Ratio of employees who have fewer than 10 years to those who have 10 or more years of experience = _____.

ASSIGNMENT Chapter 2

Name _____ Date _____ Score _____

Skill Problems

Identify the following type of fraction: common, mixed, or improper.

1. 8/5 _____

2. 1/4 _____

3. 2 1/2 _____

4. 101/52 _____

Change the improper fraction to a common or mixed fraction.

5. 34/4 _____

6. 12/5 _____

7. 50/4 _____

8. 146/144 _____

Change the mixed numbers to improper fractions.

9. 5 1/3 _____

10. 16 2/9 _____

11. 6 6/7 _____

12. 15 3/4 _____

Reduce the fractions to the lowest terms.

13. 18/24 _____

14. 25/50 _____

15. 40/100 _____

16. 30/90 _____

17. 5/4 _____

18. 84/8 _____

19. 24/10 _____

20. 220/25 _____

Change the following fractions to fractions of equal value using the denominator indicated.

21. 3/4 = ____/16

22. 3/7 ____/21

23. 6/7 ____/21

24. 2/3 ____/33

Match the letter with the definition.

A. Denominator
B. Numerator
C. Common Fraction
D. Improper Fraction
E. Mixed Numbers
F. Prime Numbers
G. Common Denominator
H. Ratio
I. Lowest Common Denominator (LCD)

25. _____ The bottom portion of the fraction that describes the number of equal parts.

26. _____ Fraction with a denominator larger than the numerator.

27. _____ The comparison of one value to another.

Complete the following time card. (Hint: Use number of minutes over 60 for fraction.)

Day	Time In	Time Out	Time In	Time Out	Hours for Day
Monday	8:00 am	11:30 am	12:00 pm	5:00 pm	8 1/2 hours
Tuesday	8:30 am	11:30 am	12:30 pm	5:00 pm	**28.** _____
Wednesday	8:00 am	12:00 pm	12:30 pm	5:45 pm	**29.** _____
Thursday	8:15 am	12:00 pm	1:00 pm	5:00 pm	**30.** _____
Friday	8:00 am	1:00 pm	2:00 pm	7:15 pm	**31.** _____
Total Week	*********	*********	*********	*********	**32.** _____

Answer the following questions using the time card data and the following information.

The employee is supposed to punch in work at 8:00 am, receive an hour for lunch, and punch out at 5:00.

33. What is the ratio of days worked in excess of 8 hours to less than 8 hours? _____

34. What is the ratio of full lunch hours to partial lunch hours? _____

35. What is the ratio of timely departures to late departures? _____

Complete the following problems. Reduce your answer to lowest terms.

36. 1/6 + 2/9 = _____

37. 8/11 − 7/13 = _____

38. 5 3/4 + 8 2/3 = _____

39. 6/7 × 4/5 = _____

40. 9/5 ÷ 3/10 = _____

41. $\dfrac{3}{4} + \dfrac{2}{3} =$

42. $\dfrac{9}{10} \times \dfrac{7}{8} =$

43. $\dfrac{5}{8} + \dfrac{5}{6} =$

44. $\dfrac{3}{4} + \dfrac{7}{8} + \dfrac{1}{3} =$

45. $\dfrac{7}{10} - \dfrac{2}{3} =$

46. $1\dfrac{1}{4} \times \dfrac{7}{8} =$

47. $2\dfrac{5}{8} - \dfrac{2}{5} =$

48. $\dfrac{7}{8} + \dfrac{1}{4} + 1\dfrac{3}{5} =$

49. $5\dfrac{3}{10} - 1\dfrac{5}{8} =$

50. $6\dfrac{1}{3} + 5 + 3\dfrac{3}{8} =$

51. $2\dfrac{1}{4} - \dfrac{7}{8} =$

52. $\dfrac{4}{5} + \dfrac{3}{8} + \dfrac{1}{2} + \dfrac{3}{5} =$

53. $\dfrac{3}{4} \times \dfrac{7}{8} =$

54. $3\dfrac{1}{6} \div 2\dfrac{3}{5} =$

55. $2\dfrac{1}{6} - \dfrac{4}{5} =$

56. $\dfrac{4}{5} \times \dfrac{3}{8} \times \dfrac{1}{2} =$

57. $\dfrac{5}{8} \times \dfrac{4}{5} =$

58. $3\dfrac{2}{5} \times \dfrac{3}{4} =$

59. $\dfrac{3}{4} \div \dfrac{5}{8} =$

60. $\dfrac{3}{8} \times \dfrac{5}{6} \times \dfrac{3}{4} =$

61. $\dfrac{1}{5} \times \dfrac{3}{4} \times \dfrac{8}{9} \times \dfrac{4}{5} =$

62. $4\dfrac{1}{2} + 3\dfrac{2}{5} + \dfrac{7}{8} =$

63. $2\dfrac{5}{6} - \dfrac{3}{4} =$

64. $6 + \dfrac{5}{8} + \dfrac{2}{3} =$

Business Application Problems

1. The fleet of trucks averaged 12 1/4 miles a gallon. The fleet used 316 gallons of fuel for the week. How many miles could be driven by the fleet?

2. The value of an item in stock increased by 1/5 over one year. The value of the item at the beginning of the year was $42,800. What is the amount of the increase in value?

3. Don's Upholstery used 4 3/4 yards of material to cover a chair. Another 1 2/3 yards were used to cover a foot stool. How much material was used in total?

4. There are 44 men in the sales division in addition to 64 women. What is the ratio of men to women?

5. Of the 560 runners entered in a race, 3/7 are women. How many women are entered?

6. Sonny will need to purchase 38 packages of potato salad to be used at a luncheon for a group of seniors. Each package of potato salad will weigh 2/3 of a pound. How much potato salad will he need to purchase for the event?

7. A wedding gown will require 7 yards of material at $36 a yard plus 4 1/3 yards of lace at $81 a yard. What is the cost of the material for the gown?

8. Returned products totaled 52 units of the 3,796 units that were initially sold last year. What is the ratio of returns to initial sales for the year?

9. If there were approximately two million "millionaires" out of approximately 272,000,000 people in the United States, what is the ratio of millionaires to the general population?

10. Allison earns an overtime rate of 1 1/2 times her regular rate of pay. Her regular rate of pay is $9.50. What is her overtime rate?

11. A "1/4 off" sale has been offered by a storage company for the month of July. If the regular price is $28 for the 4' × 8' space, what is the sale price?

12. Pat Hough increased her sales by 1/10 over last month. Sales last month were $14,630. How much was the increase?

13. Lillian McTole found she should devote 1/4 of her available time to studying her business courses. She has three business courses that she feels deserve equal attention. Lillian has twenty-four hours of free time over the weekend. How much time should she spend on each business course?

14. Amy Coulter worked the following hours during last week: Monday, 6 3/10 hours; Tuesday, 8 3/10 hours; Wednesday, 9 hours; Thursday, 8 1/10 hours; and Friday, 9 1/10 hours. How many total hours did she work?

15. A desk and accessories sold for $300 five years ago. A replacement would cost $2,100 today. What is the ratio of today's price to the price of five years ago?

16. A man's coat was priced at $188. The proprietor of Jule and Justin's Men's Store decided to have a 1/4-off sale of all the merchandise in the store for the month. What would be the sale price on the coat?

17. Big Time Movers increased their revenue by 2 1/2 times over the revenue of last year. Last year's revenue was $560,000. What is the increase in the revenue for this year in dollars?

18. Jim worked 2 3/10 hours in the morning and later returned to work for 5 1/2 hours. How many total hours did he work for the day?

19. There are 488 employees at Filmer Corporation. Of the total, 61 are more than 50 years old. What is the ratio of all employees to those who are over 50?

20. An item increased in value to five times its price in one year. The current price is $465. To the nearest ten dollars, what was the price one year ago?

21. Bert earns $12 per hour. When she works overtime, she is paid at 1 1/2 times her regular rate. What is her overtime rate?

22. The delivery truck for Teller's Tuxedo & Tailoring ran the following route Tuesday morning: to the cleaners, 4/10 mile; to the post office, 7 1/10 miles; to the trucking company, 5 3/10 miles; to the store, 6 9/10 miles; to the gas station, 1/10 mile; to the coffee shop, 2 miles; back to the shop, 1 1/10 miles. What was the total mileage for the delivery truck that morning?

23. The electronics department uses 1/5 of the space of a store that pays $2,500 rent per month. What portion of the rent should be charged to the electronics department?

24. A piece of yard goods must be cut into lengths of 2 1/4 feet each. The bolt of material is 500 feet long. How many whole pieces can be cut from the bolt?

25. One-fifth of the college's enrollment is made up of juniors and seniors. How many are freshmen and sophomores if the total enrollment is 38,110?

26. According to recent U.S. Census data, a city's population has grown to 1 2/7 of its previous size. At the previous census, the population was 187,684. What is the present size?

27. The hours worked by employees in the shipping department were as follows:

A. Rimshaw	39 1/10
P. Borgstun	36 3/10
W. Billings	38 3/10
F. Stalworth	37 7/10
B. Feller	39 9/10
H. Tolunmar	36 4/10

Total the hours for the department.

28. The cost of manufacturing a flavoring used in the food industry is made up of 1/4 labor cost and 3/4 material cost. The flavoring in its finished form sells for $180 per ton. What is the dollar amount of the labor factor in the production of 81 tons of the substance?

29. Write a complete sentence in answer to this question. Of the 650 units of a housing development, four out of five were pre-sold prior to the start of construction. How many remain to be sold?

30. Write a complete sentence in answer to this question. The turnover rate of employees at C.D.C. is 9 out of 10 over a period of one year. If the year began with 450 employees, how many will still be employed at the end of the year?

★ **Challenge Problem**

The sales of Bilkton Marketing Corporation are three times the sales of a year ago. The increase in sales, one-half of which is due to Internet advertising, was welcome. The Board of Directors indicated that 1/4 of the increase will be paid out in dividends. The dividend would amount to $2.20 a share for each of the 1,630,000 shares. What was the amount of sales last year?

Name _____ Date _____ Score _____

1. Label the fraction parts below.

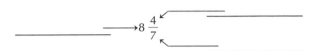

$$\longrightarrow 8\frac{4}{7}$$

2. Identify the following types of numbers.

_____ 1. $6\frac{1}{3}$ A. Common Fraction

_____ 2. $\frac{3}{4}$ B. Mixed Number

                     C. Improper Fraction

_____ 3. $\frac{5}{4}$

_____ 4. $\frac{7}{8}$

_____ 5. $4\frac{2}{3}$

_____ 6. $\frac{15}{10}$

3. Add $\frac{3}{4} + \frac{4}{5}$.

4. Subtract $\frac{2}{3}$ from $1\frac{1}{2}$.

5. Multiply $15 \times 3\frac{1}{3}$.

6. Divide $\frac{7}{8}$ by $\frac{1}{5}$.

7. Reduce $\frac{16}{3}$.

8. Sales for Campbell & Associates have increased by $\frac{1}{2}$ over last quarter's sales of $98,480. What are the sales for this quarter?

9. Nine hundred eighty-six citizens of a small city are more than sixty-two years old. The city's population is 8,874. What is the ratio of all of the citizens to those over sixty-two?

10. James Roberts worked $8\frac{1}{2}$ hours on Wednesday. During the day he spent $3\frac{1}{10}$ hours on the phone following up on orders that he had placed earlier in the month. How many hours did he spend off of the phone?

11. One-fifth of the inventory of the Cason Company is more than one month old. The total inventory value is $57,960. What is the value of the inventory that is dated more than one month old?

12. Ted and Kay each worked $6\frac{1}{2}$ hours on Thursday. They were paid $8 per hour. What was their combined total earnings?

13. There are 800 employees in the accounting department of a software company. One-half of these people will be retrained due to new technology. Of the remainder, $\frac{1}{4}$ are women, and $\frac{1}{10}$ of the men are to retire within the next year. How many men will there be in the accounting department a year from now?

(If you missed problems 1 or 2, you need to review Description and Types; problems 3, 4, or 10, you need to review Addition and Subtraction; problems 5, 6, 8, 11, 12, or 13, you need to review Multiplication and Division; problem 7, you need to review Reducing Fractions; and problem 9, you need to review Ratios.) Record your score on the Record of Performance on Mastery Tests in the back of the book.

1.01	1.28	35.1	151.7	0.4
4.58	2.24	248.5	1,116.9	1.6
1.26	2.80	574.8	2,384.6	1.3
1.12	2.85	1,685.0	6,816.0	1.7
2.02	3.51	405.1	1,723.9	1.2
0.93	1.57	874.0	3,319.7	0.8
1.24	1.30	192.3	736.7	0.3
d0.58	1.10	708.0	2,488.1	Nil
2.44	1.48	41.7	171.5	Nil
1.37	1.43	70.5	258.0	Nil
0.67	1.69	196.6	811.0	0.8
1.46	1.64	50.5	201.0	2.1
0.14	1.36	590.1	2,016.8	0.0
0.47	0.17	19.4	103.2	Nil
1.53	1.72	196.1	812.5	0.2

WORKING WITH DECIMAL NUMBERS

3

OBJECTIVES

After mastering the material in this chapter, you will be able to:

1. Read and write decimal numbers. p. 46

2. Round off decimal numbers accurately. p. 48

3. Perform all the fundamental processes on decimal numbers. p. 49

4. Work word problems that require the use of decimal numbers. p. 52

5. Change fractions to decimal numbers. p. 53

6. Understand and use the following terms:
   Place Values
   Rounding Rule
   Decimal Point Placement Rule

SPORT-A-MERICA

Hal notes that nearly every transaction with a customer, the bank, or a supplier involves the use of decimals. Dollars and cents, percents, decimal expressions of fractions and various sales, economic, trade reports, and price quotations all are in use every day. Yesterday, Kerry asked, "What were our sales for last Tuesday?" Hal answered, "$1,083.70 is the amount I deposited in the bank from the cash register." Kerry then asked, "Are there any blue shirts with the tennis racquet logo on the sleeve left?" "We now have them on sale for $29.95, which is a 30% reduction," Hal responded. "Then we must have sold over half of the stock," stated Kerry. "That's the level at which the Retail Sportswear Association says we should begin marking the remaining stock for sale," said Hal, and he smiled. Hal knew that the shirt was a real success. It had only been in the store for a month and it was nearly sold out as of today. He was happy the shirt would make a substantial contribution to the profits of the month.

Selecting the right price and then the right sale price was a function of dealing with decimal numbers. He had Brenda (their sales clerk) mark the sale price on the stock after hours on Saturday. It cost Sport-A-Merica $12.60 per hour to have Brenda mark the sale price. Hal would need to determine how much her 6.5 hours of work would cost. He was sure that the cost would be worth it. The sale the next day was a real success.

Reading and Writing Decimal Numbers
Objective 1

A decimal number is another way of expressing a fraction or portion of a whole unit. The figure below shows the representation of a decimal number as part of a whole. Note that this decimal number can be shown in a way similar to the fraction $\frac{3}{4}$ in the previous chapter. The decimal value .3 can be shown as follows:

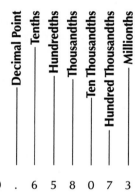

Additional information will be covered in this chapter that will allow you to adapt your knowledge of the fundamental processes to decimal numbers.

place values   To be able to read and write decimal numbers, you must be able to use the **place values** to the right of the decimal point. The following figure will help you identify the place values of decimal numbers.

Decimal Point	Tenths	Hundredths	Thousandths	Ten Thousandths	Hundred Thousandths	Millionths
0 .	6	5	8	0	7	3

Notice that the spelling of the place values to the right of the decimal point has "ths" on the end of the place value name.

The digit in the thousandths place is an 8.

The digit in the tenths place is a 6.

The digit in the ten thousandths place is a 0.

The digit in the millionths place is a 3.

Exercise A

In the number 0.879325

1. The digit in the hundredths place is ____7____.

2. The digit in the tenths place is _____.

3. The digit in the millionths place is _____.

4. The digit in the ten thousandths place is _____.

5. The digit in the hundred thousandths place is _____.

6. The digit in the thousandths place is _____.

Exercise B

In the number .548061

1. The digit in the _____hundred thousandths_____ place is a 6.

2. The digit in the _____ place is a 4.

3. The digit in the _____ place is a 1.

4. The digit in the _____ place is a 0.

5. The digit in the _____ place is a 5.

6. The digit in the _____ place is an 8.

Check your progress by turning to the answers at the back of the book.

Now that you know the place values on both sides of the decimal point, you are able to read and write decimal numbers. The numbers to the right of the decimal place are read somewhat differently than those on the left.

Example 1   The number .507 is read five hundred seven thousandths. (Notice that the number is read, then the label is added on at the end.)

Example 2   The number .9637 is read nine thousand, six hundred thirty-seven ten thousandths.

Exercise C

Read the following decimal numbers to yourself.

1. .6057   2. .9871   3. .500   4. .936   5. .8710   6. .9057

Reading a number that has digits on both sides of the decimal point requires that you use the rules presented in chapter 1 along with those that you have just learned. One additional rule is necessary to make the reading complete. When reading the number, the decimal point is read as "and" or "point."

Example 3   The number 487.903 is read four hundred eighty-seven and nine hundred three thousandths *or* four hundred eighty-seven point-nine-zero-three.

Exercise D

Read the following numbers to yourself.

1. 1.9673   2. 400.5091   3. 87,365.851   4. 987.53   5. 896.57321

Reading values that include dollars and cents requires that you read all of the dollar values followed by the word "dollars" and then read "and _____ cents."

Example 4   Kerry must give a price quotation to a customer over the phone. The Sport-A-Merica price is $581.39. The price is read, "five hundred eighty-one dollars and thirty-nine cents." (Notice that the hyphens and the comma are still used in the writing of the values just as they were in chapter 1.)

Exercise E

Write the following values in words. Place commas and hyphens in the appropriate places.

1. **$823.67**   *Eight hundred twenty-three dollars and sixty-seven cents*

2. **$2,105.46**

3. **$867.28**

4. **$234.87**

5. **$3,869.61**

Exercise F

Write the following values using numbers. Insert commas and dollar signs where necessary.

1. Nine hundred thirty-five dollars and forty-nine cents.
      $935.49

2. Sixty-three dollars and twenty-six cents.

3. Five hundred seventy-two dollars and twenty-one cents.

4. Three thousand, eighteen dollars and seven cents.

5. Fifty-one thousand, three hundred seven dollars and thirty-six cents.

Rounding Off
Objective 2

rounding rule

The rules that you learned in chapter 1 on rounding off will also hold true for decimal numbers.

     The **rounding rule** is the same for decimal numbers as it is for whole numbers. If the instructions for rounding are to "round off your answer to the nearest thousandths," then you must be concerned with the digit in the thousandths place and the digit immediately to the right of it (i.e., the ten thousandths place). In the number .07583, the digit 5 is in the thousandths place. The 5 will be changed to a 6 if the digit immediately to the right of it is a 5 or larger. In the example given, the digit is an 8; therefore, the answer is .076. All of the digits to the right of the number to be rounded are dropped.

Example 5 | Round off .603791 to the nearest hundred thousandths.

Digit in question (to be rounded)

Decision-making digit (5 or larger?)

.603791 = .60379

Exercise G

Round off the following numbers to the place value indicated.

1. .60578 to the nearest tenth ___.6___   2. .4653 to the nearest hundredth _____

3. .0037 to the nearest tenth _____   4. .54621 to the nearest thousandth _____

5. .5467 to the nearest unit _____

Exercise H

Round off the following values to the nearest cent.

1. $605.837 ___$605.84___   2. $57.653 _____

3. $486.938 _____   4. $68,640.273 _____

5. $1,754.114 _____

Exercise I

Round off the following numbers to the place value indicated.

1. 607.8035 to the nearest hundredth ___607.80___   2. 758.9208 to the nearest hundred _____

3. 759.92749 to the nearest tenth _____   4. 547,647.8492 to the nearest ten thousand

_____

5. 14.79849 to the nearest hundredth _____

Addition
Objective 3

The addition of decimal numbers requires the decimal points to be aligned before addition begins.

Example 6 | Add the decimal numbers .03;   6.878;   5.67;   and 181.006.

```
    .03           .03
  6.878         6.878
  5.67          5.67
181.006       +181.006
              193.584
```

(.03 + 6.878 + 5.67 + 181.006 =)

Zeros may be added to the numbers if they will aid in the addition process. They should be added to the right of the last digit, thereby not changing the value of the number. The zeros fill in all the columns and, therefore, reduce errors.

Example 7

Find the sum of the following numbers: 6.53;   80.751;   .7;   9.543;   .00658;   and .801.

With zeros added:

○ *Put those decimal points in line; or make sure you don't forget where the decimal point belongs!*

```
     6.53000
    80.75100
      .70000
     9.54300
      .00658
  + .80100
    98.33158
```

(*Note:* Any whole number has a decimal point to the right of the digit in the units or ones place.)

(6.53 + 80.751 + .7 + 9.543 + .00658 + .801 =)

Exercise J

Find the sum in the following problems.

1. 506.375;   9,831.6;   19.6025;   181;   and
    5,037.62   = 15,576.1975

2. 56.0983;   86.14;   .9573;   658.9401;   and 87

3. .456;   657.98;   76.98;   .86946;   and 8.941

4. 65.9201;   958;   65.937;   and 75.937

5. 15.3;   8.0981;   817.04;   and .05178

Exercise K

Find the sum in the following problems.

1. Eight and forty-three ten thousandths; six hundred twelve thousand; eighteen and three hundredths; and fifty-seven thousandths.

```
        8.0043
  612,000.0000
       18.0300
     + .0570
  612,026.0913
```

2. Three hundred four and six tenths; forty and twenty-seven thousandths; sixty-seven hundredths; and one and four ten thousandths.

3. Fifty-three and six hundred two ten thousandths; four hundred sixteen thousandths; ten and four tenths; and sixteen.

4. Thirty-two and twenty-six thousandths; fifty and ninety-five thousandths; four hundred twenty; and seventy-nine thousandths.

5. One hundred fifty-eight thousand; sixty and twenty-two ten thousandths; four hundred and six thousandths; and three and two hundred four ten thousandths.

Subtraction

Subtraction of decimal numbers also requires the decimal points to be aligned.

Example 8   Subtract 9.853 from 10.1

$$
\begin{array}{r}
10.100 \\
-\ 9.853 \\
\hline
.247
\end{array}
$$

Note that zeros may be added to the minuend if it will aid you in the subtraction process.

(10.1 − 9.853 =)

Exercise L

Find the difference in the following problems.

1. $8.067 - .537 =$
$$
\begin{array}{r}
-\ .537 \\
\hline
7.530
\end{array}
$$

2. $19.6 - 3.167 =$

3. $.9853 - .547 =$

4. $65 - 8.957 =$

5. $86.958 - 45.095 =$

Exercise M

1. $\$863.47 - \$27.98 =$
$$
\begin{array}{r}
-27.98 \\
\hline
\$835.49
\end{array}
$$

2. $\$875.90 - \$345.85 =$

3. $\$65.86 - \$17.69 =$

4. $\$649.09 - \$52.10 =$

5. $\$756.92 - \$478.00 =$

Multiplication

The multiplication of decimal numbers requires the additional step of the placement of the decimal point. When multiplying the whole numbers 37 and 8, we find the answer to be

$$
\begin{array}{r}
37 \\
\times\ 8 \\
\hline
296
\end{array}
\quad
\begin{array}{l}
\text{multiplicand} \\
\text{multiplier} \\
\text{product}
\end{array}
$$

decimal point placement rule   The **decimal point placement rule** states that the answer should have as many decimal places to the right of the decimal point as the multiplier and the multiplicand have in total.

Example 9 $37 \times .8 =$
$$
\begin{array}{r}
37 \\
\times\ .8 \quad (+1) \\
\hline
29.6 \quad\ (1)
\end{array}
$$

(37 × .8 =)

Example 10

Hal and Kerry looked at their cost of labor for their sale at Sport-A-Merica. Brenda had worked 6.5 hours at $12.60 per hour. How much did she earn?

$$6.5 \times \$12.60 =$$

$$
\begin{array}{r}
6.5 \quad (+1) \\
\times \$12.60 \quad (+2) \\
\hline
6300 \\
7560 \\
\hline
\$\ 81.900 \qquad (3)
\end{array}
$$

(6.5 × 12.60 =)

Example 11

$$.0037 \times .00008 =$$

$$
\begin{array}{r}
.0037 \quad (+4) \\
\times \quad .00008 \quad (+5) \\
\hline
.000000296 \quad (9)
\end{array}
$$

Note: Zeros must be added to fill decimal place positions if there are blanks after stepping off the decimal point.

(.0037 × .00008 =)

Exercise N

Find the product in the following problems. Pay particular attention to the placement of the decimal point.

1.
$$
\begin{array}{r}
982 \\
\times\ 8.6 \\
\hline
5892 \\
7856 \\
\hline
8,445.2
\end{array}
$$

2.
$$
\begin{array}{r}
3.4 \\
\times .67
\end{array}
$$

3.
$$
\begin{array}{r}
76.2 \\
\times\ .650
\end{array}
$$

4.
$$
\begin{array}{r}
5.63 \\
\times\ .405
\end{array}
$$

5.
$$
\begin{array}{r}
459 \\
\times\ .006
\end{array}
$$

Division
Objective 4

The division of decimal numbers uses the same rules as the division of whole numbers. The placement of the decimal point is the only additional problem in the division of decimal numbers. As shown in example 12, the division of 42 by 6 requires no decimal point placement, but 42 ÷ .6 does require the decimal point placement.

Example 12

$$
\begin{array}{r}
7 \\
6\overline{)42}
\end{array}
$$

○ *Don't forget to put the decimal point in your answer. It could make a big difference!*

If the problem would have been to divide 42 by .6, the decimal place would need to be moved in the answer the same number of places as the divisor, as shown, in order to correctly place the decimal point.

$$
.6\overline{)42.0} \qquad
\begin{array}{r}
70. \\
.6\overline{)42.0.}
\end{array}
$$

$$+1 \quad +1$$

(43 ÷ .6 =)

Zeros must be added to the end of the dividend in order to fill in the spaces created by moving the decimal point. Additional zeros may be added to the right of the decimal place to aid in division. The number 420 has the same value as 420.0 or 420.000. The decimal point in the answer is placed directly above the decimal point in the dividend after the decimal point has been moved.

Example 13

$$530.1 \div .57 =$$

$$
\begin{array}{r}
930. \\
.57.\overline{)530.10.} \\
\underline{513} \\
171 \\
\underline{171} \\
0
\end{array}
$$

(530.1 ÷ .57 =)

Exercise O

Round off your answers to the nearest tenth in the following problems.

1. $25.0146 \div .7 =$
$$
\begin{array}{r}
35.735 \\
.7.\overline{)25.0.146} = 35.7
\end{array}
$$

2. $.24 \div .6 =$

3. $4.63 \div .015 =$

4. $9.872 \div 5.9 =$

5. $.045 \div 8.2 =$

Changing Fractions to Decimal Numbers
Objective 5

Any fraction can be changed to a decimal number or its equivalent by dividing the numerator by the denominator.

Example 14

○ *A quarter of a dollar is 25¢ or $.25. We interchange fractions and decimals all of the time!*

Common Fraction

$$
\frac{3}{4} = 4\overline{)3.000}^{\,.75}
$$
$$
\begin{array}{r}
2\ 8 \\
\overline{20} \\
\underline{20}
\end{array}
$$

Improper Fraction

$$
\frac{13}{8} = 8\overline{)13.000}^{\,1.625}
$$
$$
\begin{array}{r}
8 \\
\overline{50} \\
\underline{48} \\
20 \\
\underline{16} \\
40 \\
\underline{40}
\end{array}
$$

Mixed Number

$$
4\frac{1}{8} = 4 \text{ plus } 8\overline{)1.000}^{\,.125}
$$
$$
\begin{array}{r}
8 \\
\overline{20} \\
\underline{16} \\
40 \\
\underline{40}
\end{array}
$$

Answer = 4.125

(3 ÷ 4 =)

Exercise P

Change the following fractions into decimal numbers.

1. $\frac{3}{8} = .375$

2. $\frac{9}{16} =$

3. $1\frac{3}{5} =$

4. $\frac{12}{5} =$

5. $2\frac{3}{16} =$

ASSIGNMENT Chapter 3

Name _____ Date _____ Score _____

Skill Problems

In the number 0.471803219

1. The digit in the hundredths place is _____.

2. The digit in the tenths place is _____.

3. The digit in the millionths place is _____.

4. The digit in the ten thousandths place is _____.

5. The digit in the hundred thousandths place is _____.

6. The digit in the thousandths place is _____.

Write the following value in words.

7. $856.45 _____

8. $3,271.99 _____

Round off the following numbers to the place value indicated.

9. .70568 to the nearest tenth _____

10. $705.927 to the nearest cent _____

11. 548,658.4587 to the nearest thousandth _____

12. 64.65738 to the nearest hundredth _____

Write the following using numbers.

13. Four hundred sixty-one thousandths _____

14. Six and three hundredths _____

15. Fifty-two and twenty-seven ten thousandths _____

16. Ten thousand, fifty-one _____

Complete the following Inventory List.

Item Code	Cost	Units in Stock	Value of Inventory
354068	$15.95	647	**17.**
654083	10.65	82	**18.**
654082	11.65	1,264	**19.**
654103	18.23	981	**20.**

Change the following fractions into decimal numbers. Round off to the nearest thousandth.

21. 3/4 _____ **22.** 5/10 _____ **23.** 13/5 _____ **24.** 3/9 _____

Select a letter for the definition.

A. Place Values   B. Rounding Rule   C. Decimal Point Placement Rule

25. _____ The number to the right must be 5 or larger to round up.

26. _____ The answer in multiplication problems should contain as many decimal places as the two values in the problem.

27. 42 ÷ .05 = _____

28. $5\frac{1}{4} \times .62 =$ _____

29. Add the decimal numbers .006; 67.953; 5.005; and 20.

30. Find the sum of the decimal numbers .7; .654; 65.47; and .047.

31. Subtract 9.3 from 12.634.

32. Find the difference between 20.6 and 2.654.

Find the product.

33.   657
      ×9.4

34.   5.4
      ×.24

35.   48.5
      ×.460

36. .0058
      ×2.0

Round off your answers to the nearest hundredth.

37. .032 ÷ 1.6 =

38. 65.4183 ÷ .5 =

38. .56 ÷ .8 =

40. $5.85 ÷ .5 =

Complete the following problems.

41. 3.6 × .042 =

42. 14 ÷ .02 =

43.   4.872
        12.87
      1,095.1037
      +345.81

44.   876.5
      −45.860

45. $7 \frac{1}{4} \times .6 =$

46. .042 ÷ 5 =

47. Round off your answer to the nearest tenth.

      6.032
      ×8.087

48. Round off your answer to the nearest cent.

      $846.13 ÷ 5.6 =

49.   63.08
        9.
         .075
      +   .0039

50. $.3 \times \frac{7}{8} =$

51. Round off your answer to the nearest hundredth.

      .193 × 6.073 =

52. $2 \frac{1}{4} \times .06 =$

53. 1.29
      ×.036

54. $19 ÷ 2.5 =

55. 1,469.
      −6.3807

Business Application Problems

Solve the following problems. Round off to the nearest cent.

1. The manager ordered two cases of copier paper. The price listed was $15.95 per case. Shipping cost is $5.25. What is the total for the two cases of paper?

2. To create a square foot of finished concrete, we require $0.85 for labor. Our customer has $3,000.00 in their budget for outdoor improvements. If the materials cost $1,400, what is the largest amount of square footage we can complete for our customer (rounding off to the whole number)?

3. The employees in the production department increased their output by 1/5 over the standard amount of 420 for the week. Each piece produced is valued at $63.78. What is the value of the output for the week?

4. The Quality Assurance Manager reported to her superior that 1.3 units were defective for each hour worked in the Finish Department. The Finish Department worked 680 hours over a five-day period. How many defective parts were produced?

5. The Procurement Department assigned each employee 58 items to oversee. The average value of each item was $386.45. There are five people in the Procurement Department. What is the value of the items that the department procures?

6. A piece of material that costs $6.83 plus $14 of labor goes into each of the 3,980 light fixtures produced by Jamison, Inc. each month. What is the total cost of the light fixtures produced by the company for the month?

7. A jeweler made a bracelet of gold that weighed 10.7 ounces. The gold is priced at $280 per ounce. The jeweler charged $28 per hour for each of the 8.5 hours on the project. What is the cost of the bracelet?

8. Max has decided that he will lose the 65 pounds that he feels he is overweight over the next year. How much weight will he need to lose each week?

9. The motorcycle that Janice rides can travel 50.4 miles on a gallon of fuel. She is planning a trip of 3,477.6 miles. Fuel will cost $1.22 $^9/_{10}$ per gallon. What will be the cost of the fuel for the trip? Round off your answer to the nearest dollar.

10. Mat will need to decide on which meat to use for the Bar-B-Que for the 11-member team on Saturday. He can either select beef at $4.65 per pound or chicken at $2.69 per pound. He will serve a half pound of beef or 3/4 pound of chicken per player. He will use the lowest total cost. What is the cost of the meat?

11. Dee is the quality assurance manager of a banquet firm. She has been asked to determine the cost of serving a group of twenty-five guests. The price will be $21.90 per pound. Each guest will be served .4 of a pound. What is the cost for the group of twenty-five guests?

12. Fynee is required to check all incoming invoices to make sure that the supplier has extended the totals correctly for each item and totaled the amount. Review her work to be sure that it is correct for the Jules Joyner Designs invoice.

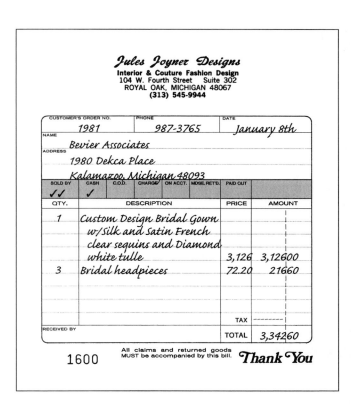

Courtesy of Jules Joyner Designs, Inc.

13. Patricia Barrett worked 7.1 hours at the rate of $15.80 per hour. How much did she earn?

14. The sales staff met to determine the total distance on the route through the city. The distance from the office to the warehouse was 4.6 miles; warehouse to the first stop, 3.2 miles; and sales route to the office, 54.1 miles. What is the total mileage?

15. Allison is paid 8.3 ($.083) cents for each part that she produces. Yesterday she produced 893 parts. To the nearest cent, how much did she earn?

16. Roberto Desin found that five-eighths of the 17.35 tons of chloride imported each year were used for agricultural purposes. To the nearest tenth of a ton, how much was used for this purpose?

17. Al LaFrance paid $10,500 for rent last year for his lock shop. The cost of the rent is $5 per month for each square foot of space. How many square feet of space did he rent?

18. Casoni Manufacturing, Inc. found that the price of needed raw materials had increased by 1.9 times over the past four years. The old price of four years ago was $14.50 per unit. What is the price increase?

19. Marie Kreger sold her second-hand clothing store for $87,300. The balance on the mortgage was $18,546.87. She paid the real estate agent a commission on the sale. The commission was $7,800. She also paid the bank a penalty of $567.12 for early payment on the mortgage. She decided to use the money from the sale to purchase a home that cost $93,900. How much money did she have to borrow to purchase the home?

20. The manager of the shipping department at Belin Fluids discovered that over the past month, 46.8 hours were lost because of accidents, 85 hours because of sickness, and 46 hours because of vacations. The department had scheduled 820 hours for the month. How many hours were actually worked?

21. Joanne Bacher sold pennants at football games. The pennants cost thirty-seven cents each. They were sold for seventy-five cents. Joanne and her sales team sold 167,759 pennants during the season. What were the total sales for the season?

22. Bill Wilson said that he would like to expand his shop by 1.8 times its current area of 1,020 square feet. He expects the addition would cost about $150,000. How many square feet would the new portion of the building have?

23. Julie Smith was offered a job paying $65 a day for eight hours' work. She would like to take the job, but she is not sure if the job pays more than the $8.20 an hour she is earning at her present job. Which is the higher hourly rate that Julie has as an option?

24. The Lennolda Shoppe is taking inventory of the small pieces in the storage area of the shop. The inventory revealed the following: 45 pieces priced at $.54 each, 144 pieces priced at $.19 each, 20 units priced at $1.95 each, a box of 50 pieces priced at $.39 each, and 4 units priced at $.49 each. What is the total value of these small pieces of merchandise?

25. Ernest Seldon bought a bin of bird seed. The bird seed is made up of seven bushels of oats at $5.95 a bushel, ten bushels of cracked corn at $7.95 a bushel, and five bushels of cracked sunflower seeds at $14.95 a bushel. The mixture was sold for $7.50 a half bushel, which included the 10 cents for the bag in which it was packed. How much profit did he make?

26. The lounge chairs of the Marina Bay Condominium are in need of repair. A review of the grounds indicated that there were 27 chairs and 27 loungers that needed to have new webbing installed. The manager received price quotes from area repair shops. The lowest price was $.71 per replaced web. There are 13 webs on each of the chairs and 22 webs on each lounger. How much will it cost to have the 54 units repaired?

27. Mac McLeod projected sales figures for Barcelona, Inc. show the volume of "firm" and "possible" sales by representative. What is the difference between the firm and possible sales figures if each unit sells for $8.45?

Sales Rep.	Firm Sales	Possible Sales
A. Janor	410 units	650 units
B. K. Moore	301 units	800 units
N. Manoto	285 units	1,000 units

28. Write a complete sentence in answer to this question: Fuel for the fleet of trucks operated by Wiise and Company cost $5,689.40 last month. The company is able to purchase fuel at a cost of $1.17 per gallon. To the nearest tenth of a gallon, how many gallons of fuel did the firm purchase last month?

29. Write a complete sentence in answer to this question: Dr. Monroe's office will begin to deliver a small flower arrangement and card to each of the surgery patients as of the first of next month. There are 162 patients scheduled for surgery next month. The flowers will cost $8.50 and the cards will cost $1.25 each. The delivery charge will be $4 on the average. What will be the total cost of this service for the month?

30. Write a complete sentence in answer to this question: The Boxon Corp. has offered to deduct one-eighth of the price of every unit of their product sold after the first 500,000 units are sold, per month. The product sells for $627.50 each. What will be the price of units sold over the 500,000 volume level? Round off your answer to the nearest cent.

★ **Challenge Problem**
The Gayle Company bought a roll of metal 4 feet wide with a length of 500 feet. The metal is to be cut into pieces 4 feet wide by 5.7 feet long. There were 14 pieces that had to be scrapped owing to errors made in cutting. The roll cost $4,678.95. If the firm makes a profit of $14.86 on each piece produced, what will be the total profit made from the sale of the full cut pieces?

✅ MASTERY TEST Chapter 3

Name _____   **Date** _____   **Score** _____

1. Round off the following values as indicated.
   a. 7.9803 to the nearest ten _____
   b. .08712 to the nearest tenth _____
   c. 647.096 to the nearest hundred _____
   d. 64.0968 to the nearest thousandth _____
   e. 43.916 to the nearest tenth _____

2. Write the following value in words: $385.70.
   _____

3. Add 733.042; 8.367; and 13.758.

4. Subtract 9.405 from 57.079.

5. Multiply $1.2 \times .56$ and round off your answer to the nearest unit.

6. Divide $3.72 by 8 and round off your answer to the nearest cent.

7. Change the following to decimal form.
   a. $\dfrac{5}{8}$ = _____   b. $\dfrac{3}{10}$ = _____   c. $4\dfrac{1}{5}$ = _____

8. If a certain industrial product is made from plastic, it costs 6.7 cents ($.067) per unit over a volume of 6,000 pieces. If the same product is made of rubber, it will cost $450 for the same volume. The company has a contract that assures a revenue of $1,560 for the 6,000 pieces made of either material. What is the maximum profit per piece rounded to the nearest cent?

9. AM Corp. has expanded its business 1.6 times over last year's volume of $324,952. What is this year's volume?

10. Anthony James Ureel, the manager of an agribusiness firm, noted that seven-tenths of the 15,479 barrels of herbicide he bought last year were used for soybean production. To the nearest whole barrel, how many were used for soybeans?

(If you missed problem 1, you need to review Rounding Off; problem 2, you need to review Reading and Writing Decimal Numbers; problem 3, you need to review Addition; problem 4, you need to review Subtraction; problems 5, 8, or 9, you need to review Multiplication; problem 6, you need to review Division; problem 7, you need to review Changing Fractions to Decimal Numbers; and problem 10, you need to review Multiplication, Changing Fractions to Decimal Numbers, and Rounding Off.) Record your score on the Record of Performance on Mastery Tests in the back of the book.

BANKING RECORDS

OBJECTIVES

After mastering the material in this chapter, you will be able to:

1. Identify different types of endorsements. p. 66

2. Reconcile a bank statement. p. 67

3. Understand and use the following terms:
   Depositor
   Imprinting Checks
   Automatic Teller Machine (ATM)
   Deposit Slip
   Check Register
   Check Stub
   Check
   Payee
   Blank Endorsement
   Special Endorsement
   Restrictive Endorsement
   Electronic Funds Transfer
   Bank Statement
   Deposits in Transit
   Outstanding Checks
   Service Charges
   Adjusted Balance

SPORT-A-MERICA

Hal receives the bank statement for Sport-A-Merica in the mail on October 3. Recognizing the fact that the company has bills to pay this week, Hal decides to review the bank statement and the company records. He notices that the bank included a $63 service charge for the month. That seems too high.

Kerry and Hal discuss the service charge. Kerry suggests moving the Sport-A-Merica accounts to First National Bank, where she has her personal account. Hal reasons that they would then have to sign new signature cards, transfer the balance, and order new checks. Would all of this be worth the bother? Kerry says she enjoys the ATM after-hours convenience, and in addition, First National Bank has a low minimum balance requirement of $1,000 before service charges are assessed.

Hal doesn't feel ATMs would be of much use to the firm, but understands why Kerry uses them. Furthermore, the Sport-A-Merica account balance is only $690.40 according to the check register, and Hal must pay bills this week. He is not sure of the *actual* balance in the account. Other charges, such as those for overdrafts and collection charges, seem more important from a business point of view.

On the other hand, Kerry knows that $600 will be deposited in *her* account automatically each month from her annuity investment. Her account balance must be adjusted each month. Both Kerry and Hal recognize the need to investigate banking costs to decide what they want to do with the money belonging to Sport-A-Merica.

Most debt payments made by business and by private individuals are made by check. The checking account provides a convenient as well as a safe payment method—a definite improvement over paying cash in person.

When a new checking account is opened, the parties who are authorized to sign checks are required to give a sample signature to the bank. A card containing the signature(s) is kept on file and is available to bank personnel at all times. The procedure of verifying the handwritten signature is used to provide security.

depositor

The new **depositor** is given a set of deposit slips and blank checks. A personalized set of checks is ordered at this time, also. The new, personalized set will contain pertinent information, such as the account holder's name, the account number, the address, the telephone number, and possibly the driver's license number. There is usually a charge for the **imprinting** process. Imprinted checks are easier for the bank to identify and process. The imprinting will also limit use to only the authorized parties whose names are imprinted on them or those authorized to sign for a company or organization.

imprinting checks

automatic teller machine (ATM)

An individual depositor may also request an **automatic teller machine (ATM)** card. This card allows the depositor to make deposits and withdrawals at many locations 24 hours a day by using the card along with a secret code number issued by the bank.

Deposits and Payments

When a depositor wants to deposit money in an account, a **deposit slip** or **deposit ticket** must be completed. The date, the amount of currency, the amount of coin, as well as each check to be deposited, are listed separately on the deposit slip. The deposit slip, along with the monies specified, is given to the bank teller or placed in the ATM machine by the depositor. A duplicate **deposit slip** or some acknowledgment is then given to the depositor. This is retained as proof of deposit, and the deposit is recorded in the check register.

deposit slip

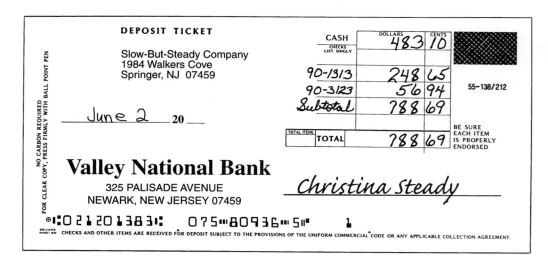

check register

The **check register** or **check stub** provides the form on which the depositor can determine the amount in the account. Deposits are added to the account balance, and issued checks are subtracted from the account balance. It is important to record the deposits and checks *and* to do the necessary addition or subtraction at the time the transaction occurs. This procedure will ensure that sufficient funds are in the account prior to issuing the next check.

Check Register

DATE		CHECK NUMBER	TRANSACTION DESCRIPTION	✓T	(—) AMOUNT OF PAYMENT OR DEBIT		(+OR +) OTHER		(+) AMOUNT OF CREDIT		BALANCE FORWARD	
19 ____			BE SURE TO DEDUCT ANY PER PER ITEM CHARGES, SERVICE CHARGES OR FEES THAT MAY APPLY IN "OTHER" COLUMN.								2,611	80
10	17		TO:								320	00
			FOR:								2,931	80
10	17		TO: *Travel Court Lodge*		156	20					156	20
			FOR: *meeting room*								2,775	60
			TO:									
			FOR:									
			TO:									
			FOR:									
			TO:									
			FOR:									
			TO:									
			FOR:									

check

payee

   When writing a **check,** the depositor must fill in the date; the party to whom the check is payable (the **payee**), which could be a person or an organization; the amount of the check in numbers as well as words; and lastly, authorized signature(s). The check register or check stub should also be filled out at this time. This provides a record that duplicates the information on the check itself.

   Note that a deposit of $320 is added to the $2,611.80 balance brought forward; then check #47 for $156.20 is subtracted to determine the new balance of $2,775.60 for Abol Corporation.

   When the bank receives the completed check, it handles it as a demand by the depositor to pay the payee the specified amount.

Check

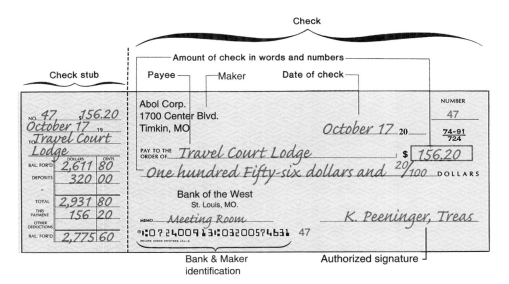

Types of Endorsements
Objective 1

Before the bank will exchange cash for the completed check, the payee must sign the check. This must be done on the back, on the same end that states "Pay to the order of." This signature must be the same name as that specified on the front of the check, such as "John M. Student." The federal government has requested that the endorsement be written in the top $1\frac{1}{2}$ inches of the check. This will allow the bank to use the rest of the area for processing. If the payee simply signs his or her name, the check can be cashed by anyone. The check is, therefore, transferable. This type of endorsement is termed a *blank endorsement* **blank endorsement.** Should a check with a blank endorsement be lost, it could possibly be cashed by any finder. The finder would simply have to sign his or her name below the endorsement of the payee to receive the specified amount of cash, unless the bank requests identification.

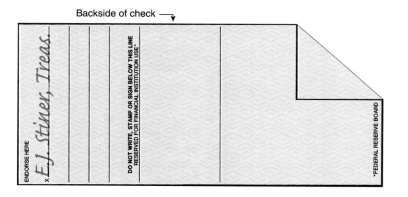

Backside of check

To provide protection to the payee, the depositor could add to the blank endorsement a limitation such as "Payable to J. Hawkins   John M. Student." With J. Hawkin's signature, only J. Hawkins would be allowed to cash the check. This type of endorsement *special endorsement* is termed a **special endorsement.** This endorsement indicates a specific person or organization to which the check is payable.

Another restriction that can be added to the signature specifies exactly what is to be done with the check. "For deposit only   John M. Student" and the payee's signature will restrict the use of the check to only a deposit in the payee's account. This type of *restrictive endorsement* endorsement is termed a **restrictive endorsement.** A restrictive endorsement usually signifies a terminal transfer of the check, that is, the final use of the check.

Exercise A

Identify the endorsements below as blank, special, or restrictive.

restrictive	1. For deposit to the account of S.S.X. Co. Abe Solen
_____	2. Payable to the Green Co. Alice Madison
_____	3. James Amos
_____	4. P. D. Pierson
_____	5. For payment on promissory note #37802 P. Wolver
_____	6. For deposit only J. J. Adams
_____	7. The Big Horn Corp.
_____	8. Payable to J. Swage Alp Smith
_____	9. P. & K. Turin Co.
_____	10. Iverson Co.

Checking services are available not only from banks, but also from savings and loan associations, credit unions, and investment firms. Not all services are available from all types of firms. In addition, many of these services are not available to corporations. It is worthwhile for the individual consumer and the corporate officer to investigate and compare available services.

Many of the "extra" services are available because of computers. An account holder can transfer funds from one account to another; receive interest on checking account balances; make deposits and withdrawals on a 24-hour basis from ATMs; have amounts such as a paycheck or retirement check automatically deposited; have installment payments or utility payments automatically withdrawn; and have purchases deducted through a debit card. The competition will change the services over time.

The account holder may also transfer funds from one account to another and make payments electronically through the use of a personal computer without actually writing a check by hand. Transfers and transactions made electronically, without a paper check, are

electronic funds transfer   termed **electronic funds transfer.**

Reconciliation of a Bank Statement
Objective 2

bank statement

Once a month (or less often by special arrangements) the bank sends a **bank statement** to the depositor. This statement is a record of all of the changes in the account that the bank is aware of since the last statement. It shows the checks that have been paid by the bank; the deposits received by the bank; service charges, if any; as well as other changes in the account balance.

The balance on the statement will probably not agree with the balance in the check register or on the last check stub. This disagreement may not be due to inaccuracy on the part of either the bank or the depositor, because both parties have arrived at their balances by acting on the information available to them at the time. The bank statement and the checkbook records must be *reconciled*. This reconciliation procedure will determine the actual funds available in the account.

When the bank statement is received, the depositor must

1. Organize by check number the cancelled checks returned with the statement. These checks have been received by the bank, and their amounts have been subtracted from the account balance.

outstanding checks

2. Identify what checks have *not* been received by the bank by comparing the returned checks with the check register or check stubs. These checks are termed **outstanding checks.** The total of their amounts must be subtracted from the bank statement balance. They represent a future reduction in the bank's account balance when the check is received by the bank.

3. Review the deposit slips returned with the bank statement or the indication of deposit received by the bank. This will identify the deposits made by the depositor but not received by the bank at the time the statement was prepared. These deposits are

deposits in transit

termed **deposits in transit.** The total deposits in transit must be added to the bank statement balance.

service charges

4. Review the bank statement to determine if any **service charges** have been specified. The total of these charges must be subtracted from the depositor's balance on the check register or check stub. The bank has already reduced the account balance for these charges as shown on the statement.

    Service charges may be levied by the bank for many reasons, such as

    - for each check received
    - for each deposit made
    - for a monthly account use charge
    - for imprinting of checks
    - for an overdraft on the account balance
    - for a collection charge
    - for an ATM transaction

    All these charges are subtracted from the depositor's record (checkbook) as service charges.

5. Review the bank statement for other changes in the account balance. For example,
    An authorized automatic payment made from the account without a check written
        must be subtracted from the checkbook.
    A collection made by the bank and deposited to the account without a deposit slip
        must be added to the checkbook.
    A deposit made to the account without a deposit slip (i.e., payroll checks directly
        deposited to the account through wire transfer by the employer, Social Security
        checks directly deposited into the account by the federal government, or pension
        checks deposited by the employer) must be added to the checkbook.
    Checks returned for insufficient funds must be subtracted from the checkbook. Interest
        earned on the account balance must be added to the checkbook.

adjusted balance   The result of all of these changes in the bank statement balance is the determination of the actual funds available, known as the **adjusted balance.**

○ *The adjusted balance means the amount of money that is available to spend.*

Reconciliation Model

Bank		Checkbook	
Balance (ending)	$ _____	Balance (ending)	$ _____
+ Deposits in Transit	+ _____	+ Credits	+ _____
− Outstanding Checks	− _____	− Service Charges	− _____
		− Debits	− _____
	_____		_____
Adjusted Balance	$ _____	Adjusted Balance	$ _____

Example 1

Hal decided that first things should come first. He needed to find the adjusted balance in the current checking account so that he could pay the bills that are due. He reviewed the Sport-A-Merica, Inc. bank statement that showed an $863.70 balance. The check register showed a $690.40 balance. After organizing the cancelled checks in numerical order, he found that check number 1806 for $423.30 was still outstanding. A deposit for $267, made on the last day of the month, was not shown on the bank statement. Service charges totaled $63. A note for $80 was collected by the bank. The bank statement reconciliation appears here.

Sport-A-Merica, Inc.
Bank Statement Reconciliation
September 30, 20XX

Bank Statement Balance	$863.70	Check register balance	$690.40
Plus: Deposit in transit	+267.00	Less: Service charges	−63.00
	$1130.70		$627.40
Less: Outstanding check			
no. 1806	−423.30	Plus: Note collected	+80.00
Adjusted balance	$707.40	Adjusted balance	$707.40

$(863.70 + 267 - 423.30 = \qquad 690.40 - 63.00 + 80.00 =)$

Exercise B

Reconcile the information below to find the adjusted balance.

1. Statement balance = $877.98
Checkbook balance = $787.44
Service charge = $12.00

Checks outstanding = No. 471	$32.00
477	25.17
478	13.00
479	32.37
	$102.54

Bank bal.	Chbk. bal.
$877.98	$787.44
−102.54	−12.00
$775.44	$775.44

2. Statement balance = $197.40
Checkbook balance = $168.80
Service charge = $9.00

Checks outstanding = No. 213	$14.80
220	75.95
221	21.85

Deposit in transit = $75.00

3. Statement balance = $4,315.17
Checkbook balance = $3,901.07
Service charge = $6.00

Checks outstanding = No. 502	$312.00
503	79.50
504	131.26

Deposits in transit = $62.50, $40.16

4. Statement balance = $1,430.00
Checkbook balance = $1,576.00
Service charge = $13.00

Checks outstanding = No. 32	$429.00
37	362.70
41	336.87
42	133.43

Deposits in transit = $397.12, $65.36, $932.52

Example 2

Kerry showed Hal her bank statement from August. Hal noted the service charge was only $31, which included $10 for new check imprinting and $1 each for the use of the ATM six times. She had a balance of $830 in her checkbook and $1,132 on the bank statement. The month's direct deposit of $600 from her annuity was shown on the bank statement, but not in her checkbook. Check number 107 for $83 was outstanding. A $350 deposit from her paycheck last week was not shown on the statement. Reconcile the balance.

BANK STATEMENT		CHECKBOOK	
Balance	$1,132	Balance	$830
Plus: Deposit in transit	350	Plus: Direct deposit	600
	$1,482		$1,430
Less: Check outstanding	83	Less: Service charges	31
Adjusted balance	$1,399	Adjusted balance	$1,399

$(1,132 + 350 - 83 = \qquad 830 + 600 - 31 =)$

Exercise C

Reconcile the following information to find the adjusted balance. Note that these exercises include ATM, direct payment, and deposit activities.

1. Statement balance = $480
   Checkbook balance = $400
   Service charges = $1 for each ATM use   (11 times this month)
   Direct deposit (by employer) = $400
   Deposits in transit = $309

Bank bal.	Ckbk. bal.
$480	$400
+309	−11
$789	+400
	$789

2. Statement balance = $683
   Checkbook balance = $718
   Note collected by bank = $500 (for the depositor)
   Note collection charge = $10
   Service charges = $18 monthly, $1 for each ATM use (12 times this month)
   Checks outstanding = No. 403 $350
                             411   120
   Deposits in transit = $965

3. Statement balance = $1,300
   Checkbook balance = $1,020
   Direct deposit (pension fund) = $380
   Direct deposit (social security check) = $176
   Checks outstanding = No. 142 $65.75
                             145   50
                             148   250
                             149   10.45
   Deposits in transit = $652.20
   Service charges = $0 (due to high balance)

4. Statement balance = $468
   Checkbook balance = $512
   Service charges = $22
   Direct deposit (by employer) = $238.88
   Checks outstanding = No. 178 $91.50
                             179   50
                             185   91.20
   Deposit in transit = $493.58

5. Statement balance = $1,980
   Checkbook balance = $2,940
   ATM use = 7 times at $1.50 per use
   Service charge = $0 (high balance)
   Direct deposit (by pension fund) = $920
   Deposit in transit = $1,600
   Automatic payment = $309.50
   Checks outstanding = No. 981 $40

ASSIGNMENT Chapter 4

Name _____ **Date** _____ **Score** _____

Skill Problems

Identify the different types of endorsements.

1. _____ Kim Funside
2. _____ Payable to Aree Singleton   Kerri Donald
3. _____ For deposit to account #22-807-5337   Sandra Belson
4. _____ For deposit only   Dennis Urie
5. _____ Tulsa, Inc.
6. _____ Payable to the Reformed Church of Nicaragua   Juan Cortez

Select a letter for each definition.

A. Depositor   B. Imprinting Checks   C. ATM   D. Deposit Slip
E. Check Register F. Check Stub   G. Check   H. Payee
I. Bank Statement J. Deposit in Transit   K. Outstanding Checks
L. Service Charges M. Adjusted Balance

7. _____ The checking account balance from the records of the depositor.

8. _____ Coins, currency, and checks that have not been added to the records of the bank.

9. _____ A machine available 24 hours to accept deposits and provide cash withdrawals from accounts.

10. _____ Issued checks that have not cleared the bank.

Finish the following sentence with the proper term.

11. The _____ was given a set of deposit slips and blank checks.

12. Other charges against your checking account besides checks written may be costs for _____
    and bank _____.

13. A record device to verify amount of money put into an account is a _____.

14. Usually once a month, the bank sends a _____ to the depositor.

Fill in the blanks with the above terms to complete the Reconciliation Model.

Reconciliation Model

Bank		Checkbook	
Balance (ending)	$ 1,282	Balance (ending)	$ 815
15. + _____	+ 200	+ Direct Deposit	+ 600
16. − _____	− 83	17. − _____	− 16
18. _____	$ 1,399		$ 1,399

Reconcile the information below to find the adjusted balance.

19. Statement balance = $987.26   Cks. Out:
    Checkbook balance = $1,182.53   601 $ 10.99
    Service Charge = $16.50   605   6.78
    Deposits in transit = $500.00   606   295.01
                                        609   8.45

Reconcile the information below to find the adjusted balance.

20. Statement balance = $54.36
    Checkbook balance = $42.20
    Service Charge = $7.50 monthly
       6 ATM uses, $1 each
       1 Overdraft, $27.50
    Deposits in transit = $482.04

Cks. Out:

822	$ 60.50
823	54.80
824	399.95
830	19.95

21. Statement balance = $4,631.40
    Checkbook balance = $1,740
    Deposit in transit = $1,287.60
    Checks outstanding = $4,220
    Service charge = $41

22. Statement balance = $4,057.50
    Checkbook balance = $2,418.74
    Service charge = $7.50 monthly; 255 checks, $.25
    each; 20 deposits, $.35 each; Earnings credit on high
    balance, $78.25
    Note collection fee, $6.00
    Notes collected by bank for our company from our customer = $957.64

Cks. Out:

14051 $687.12

23. Statement balance = $2,467.15
    Checkbook balance = $2,607.12
    Health ins. deduction = $917.00
    Deposits in transit = $99.45, $41.05, $65.74, 124.35
    Service charge = $14.35
    Customer's check non-collectable = $12.44

Cks. Out:

1406	$ 375.40
1440	214.00
1402	545.01

Using the information from question 23, complete the following check register.

Number	Date	Description	Amount of Check	✔ T	Amount of Deposit	Balance
	6/8/xx	Balance Forward				$ 3,296.09
1439	6/9/xx	Thuets Products-May	$245.05			24. _____
1440	6/9/xx	PHBS-May	25. _____			26. _____
	6/9/xx	Monday's Receipts			$99.45	27. _____
1441	6/10/xx	Gas & Go	15.50			28. _____
	6/10/xx	Tuesday's Receipts			41.05	29. _____
	6/11/xx	Wednesday's Receipts			65.74	30. _____
1402	6/12/xx	Roger M Thayer-wages	31. _____			32. _____
	6/12/xx	Thursday's Receipts			124.35	33. _____
	6/12/xx	Reconciliation				
		34. _____	_____			35. _____
		36. _____	_____			37. _____
		38. _____	_____			39. _____

40. The answer in number 39 is termed the _____ of the check register, which is the actual funds
available.

Business Application Problems

Find the adjusted balance in the following problems unless otherwise directed.

1. Jamesson Company has a balance in its checking account of $3,587.50 at the end of the month. The statement indicated a balance of $3,662.50. The statement indicated a service charge of $28. Two checks for a total of $564 were not shown on the statement. A deposit of $461 was not shown on the statement. Reconcile the account.

3. Bosditch, Inc. received their bank statement for last month showing a balance of $1,687. The checkbook shows a balance of $1,557.50. The accountant found a check was still outstanding for $226.90 and a deposit for $67.40 was not shown on the bank statement. A service charge of $30 was shown on the bank statement.

2. Barnes, Thorner and Company's check register shows a balance of $586. Its accountant found the following by comparing the company records to the statement.

Checks Outstanding:		Deposits in Transit:
No. 492	$650	$912
495	497.50	261

The bank statement balance was $545.50. If the records of both the bank and the company are correct, what is the amount of the bank service charge?

4. The check register for the Biggs Corporation is shown below. Except for checks 339 and 341, all the checks were shown on the bank statement. The bank statement balance was $5,044.70.

10   BE SURE TO DEDUCT ANY PER ITEM CHARGES, SERVICE CHARGES OR FEES THAT MAY APPLY IN OTHER COLUMN

DATE		CHECK NUMBER	TRANSACTION DESCRIPTION	✓T	(−) AMOUNT OF PAYMENT OR DEBIT		(− OR −) OTHER		(+) AMOUNT OF CREDIT		BALANCE FORWARD	
10	12	335	TO: Jackson, Inc. FOR:		345	60					12,556	20
10	15	336	TO: Marked Charge Co. FOR:		569	-					11,987	20
10	15	337	TO: Al DeRow FOR:		45	-					11,942	20
10	22		TO: FOR:						1,982	50	13,924	70
10	25	338	TO: First Federal Bank FOR:		855	-					13,069	70
10	28	339	TO: Ace Properties FOR:		1,500	-					11,569	70
10	28	340	TO: Top Investments FOR:		8,000	-					3,569	70
10	31	341	TO: I.R.S. FOR:		2,250	-					1,319	70

From the incomplete information available, determine the adjusted bank balance.

5. From the incomplete information in problem 4, determine the service charge for the Biggs Corporation.

6. Jay Marks reviewed The Mathews Inn bank statement for the month which showed a balance of $907.50. The checkbook shows a balance of $869.92. The business manager found that a check was still outstanding for $85.38 and a deposit for $24.90 was not shown on the bank statement. A service charge of $22.90 was shown on the bank statement.

7. Bands Corporation's check register is shown below. Check 511 is the only outstanding check. The deposit made on June 30 is not shown on the bank statement. The bank statement shows a balance of $3,674.95.

10   BE SURE TO DEDUCT ANY PER ITEM CHARGES, SERVICE CHARGES OR FEES THAT MAY APPLY IN OTHER COLUMN

DATE	CHECK NUMBER	TRANSACTION DESCRIPTION	✓T	(−) AMOUNT OF PAYMENT OR DEBIT	(• OR •) OTHER	(+) AMOUNT OF CREDIT	BALANCE FORWARD
6 10	506	TO: Barker, Inc. FOR:		45 -			7,556 20
6 15	507	TO: Charter Bank FOR:		950 -			6,606 20
6 17		TO: FOR:				900 -	7,506 20
6 22	508	TO: Hill Electronics FOR:		1,459 50			6,046 70
6 25	509	TO: First National Bank FOR:		823 75			5,222 95
6 26	510	TO: Ace Properties FOR:		1,500 -			3,722 95
6 27	511	TO: Partical Quant. FOR:		700 -			3,022 95
6 30		TO: FOR:				1,750 -	4,772 95

From the incomplete information available, determine the adjusted bank balance.

8. The bank statement for Toddler Toys, Inc. indicated the following information:
1. The checkbook balance was $3,975.80.
2. There was a service charge on the bank statement for $33.
3. The total of the checks outstanding was $197.50.
4. A deposit made last week for $2,550 was not shown on the bank statement.
5. The bank statement showed a current balance of $1,590.30.

Service Charges = $27.00
Deposits in Transit = $67.33, $109.50, $240.13

9. Margin, Inc. received their bank statement for the month of July showing a balance of $6,063.50. The review of the cash account showed a balance of $7,341. A service charge for $83 was shown on the bank statement. There was also a check outstanding in the amount of $255.50 and two recent deposits totaling $1,450 were not shown on the bank statement.

12. The Bengrin Company received the bank statement for the month of August on the thirty-first. The statement balance was $3,218.40, which compared to $3,176.41 in the cash account of the balance sheet. The $3,176.41 also agreed with the balance in the check register. The accountant reviewed the statement and found that the following points required reconciliation: check No. 493 for $357.18 had not been cashed, the deposits for $195 and $82.19 made on the thirtieth were not shown on the bank statement, and the service charge of ????? was not recorded on the check register.

10. The checkbook for Melger and Associates indicates a balance of $2,528.60. However, the July bank statement showed a balance of $2,972.10. The accounting clerk found that a check was still outstanding for $1,452. The deposit made last week for $952.50 was not shown on the bank statement. A service charge of $56 was also listed on the bank statement.

13. The White Water Corporation bank statement revealed these items:

A balance of $147.30

A check for $31 deposited by the firm that was returned because there were insufficient funds in B. Bennett's account

Check 141 for $101 that was not cashed by the payee

A deposit for $78 that was not received by the bank when the statement was prepared

A service charge of $20 that has already been subtracted from the bank account balance

The firm's check register shows a $175.30 balance.

11. The Sperry Company check register shows a balance of $1,946.80. The bank statement shows a balance of $2,650.47. After a review of the bank statement, the following information is determined.

Checks Outstanding = No.	406	$79.80
	471	20.00
	472	720.50
	481	327.33

14. The Valley Inn has a balance of $5,698.03 on its check register. The bank statement shows a balance of $5,995.90. The following information is determined after a close review of the bank statement and the check register.

Checks Outstanding:	Deposits in Transit:
No. 13 $420.00	$165.00
23 270.50	203.60
31 100.00	103.03

Service charge (monthly account use): $21.00
Reconcile the bank statement.

15. Cyndi found that the bank statement balance and the check register had the same balance of $89. After a review of the statement she found there was a $13 deposit in transit as well as a $29 outstanding check. She also found that she had failed to subtract a $4.50 check in her check register. The service charge amount on the statement was unreadable. What was the amount of the service charge?

16. Scanloni Imports, Inc. received a bank statement on October 10 that revealed a balance of $1,680. The manager, Sally Scanloni, noted that the statement did not include the $150 deposit she made earlier in the week or the two checks she wrote yesterday. The checks totaled $483.75. There was also a service charge for $13. What should be the balance in the account?

17. The Benji Company's check register balance is $73.60. The bank statement balance is $1,002. The following information is determined after a review of the bank statement and the check register:

Checks Outstanding:	Service Charges:
No. 140 $163.80	$18.50 for account use
181 980.00	8.00 for note collection
185 47.23	

Deposits in Transit: $18, $482, and $67.41
What was the value of the note collected by the bank?

18. The check register for the Support Service Company is
shown below. All the checks through No. 841 were
included in the month's statement. The deposit of
October 29 is not shown on the bank statement. The
service charge is $24.

		PLEASE BE SURE TO **DEDUCT** ANY PER CHECK CHARGES OR SERVICE CHARGES THAT MAY APPLY TO YOUR ACCOUNT							
NUMBER	DATE	CHECKS ISSUED TO OR DESCRIPTION OF DEPOSIT	(−) AMOUNT OF CHECK	√ T	(−) CHECK FEE (IF ANY)	(+) AMOUNT OF DEPOSIT		BALANCE 1680.48	
839	10/1	TO/FOR ABERESON CO.	16.00	✓				−16.00	
							BAL	1664.48	
840	10/6	TO/FOR SMITH CORP.	822.20					−822.20	
							BAL	842.28	
841	10/9	TO/FOR O. PETERSON	327.00					−327.00	
							BAL	515.28	
842	10/10	TO/FOR SCHMIDT & SONS	5.00					−5.00	
							BAL	510.28	
843	10/16	TO/FOR SMITH CORP.	200.00					−200.00	
							BAL	310.28	
844	10/23	TO/FOR H.D. HEBNER	29.50					−29.50	
							BAL	280.78	
		TO/FOR DEPOSIT			10/29	583.00		+583.00	
							BAL	863.78	

The balance on the bank statement is $491.28. What
is the adjusted balance?

19. Susan Bigsby noted that her credit union had charged her account $1 for each of the six times she had used the automatic teller machine in the month of August. There were $10 in other service charges. Her statement balance was $265. She had made one deposit since the statement date for $160. The statement indicated an automatic payment from the account in the amount of $32. Her records showed a balance of $473. She had no outstanding checks. What is the adjusted account balance?

20. The November bank statement for the Pondi Company is shown here. The statement does not show a deposit made on November 30 for $587.29, or two checks outstanding—No. 862 for $27 and No. 867 for $368.50. What is the adjusted balance for the month?

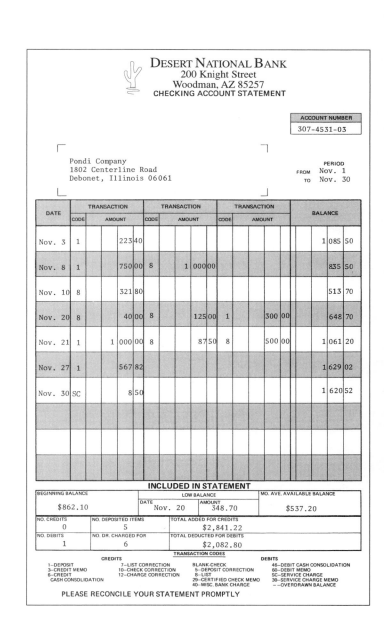

DESERT NATIONAL BANK
200 Knight Street
Woodman, AZ 85257
CHECKING ACCOUNT STATEMENT

ACCOUNT NUMBER
307-4531-03

Pondi Company
1802 Centerline Road
Debonet, Illinois 06061

PERIOD
FROM Nov. 1
TO Nov. 30

DATE	CODE	AMOUNT	CODE	AMOUNT	CODE	AMOUNT	BALANCE
Nov. 3	1	223 40					1 085 50
Nov. 8	1	750 00	8	1 000 00			835 50
Nov. 10	8	321 80					513 70
Nov. 20	8	40 00	8	125 00	1	300 00	648 70
Nov. 21	1	1 000 00	8	87 50	8	500 00	1 061 20
Nov. 27	1	567 82					1 629 02
Nov. 30	SC	8 50					1 620 52

INCLUDED IN STATEMENT

BEGINNING BALANCE	LOW BALANCE		MO. AVE. AVAILABLE BALANCE
$862.10	DATE Nov. 20	AMOUNT 348.70	$537.20

NO. CREDITS 0	NO. DEPOSITED ITEMS 5	TOTAL ADDED FOR CREDITS $2,841.22
NO. DEBITS 1	NO. DR. CHARGED FOR 6	TOTAL DEDUCTED FOR DEBITS $2,082.80

TRANSACTION CODES
CREDITS: 1–DEPOSIT 3–CREDIT MEMO 6–CREDIT CASH CONSOLIDATION 7–LIST CORRECTION 10–CHECK CORRECTION 12–CHARGE CORRECTION
BLANK-CHECK 5–DEPOSIT CORRECTION 8–LIST 29–CERTIFIED CHECK MEMO 40–MISC. BANK CHARGE
DEBITS: 46–DEBIT CASH CONSOLIDATION 60–DEBIT MEMO SC–SERVICE CHARGE 30–SERVICE CHARGE MEMO – –OVERDRAWN BALANCE

PLEASE RECONCILE YOUR STATEMENT PROMPTLY

21. ProJet, Inc. check register shows a $450.40 balance. A computer printer skip on the bank statement conceals the statement balance, but a review of it does reveal the following information.
   a. Service charges: $21.00
   b. Deposits in transit: $87, $600, and $273
   c. Checks outstanding: No. 87 $76.23
                          88     $185
                          98     $632.50

22. Mary Travel, Inc. has a balance of $2,724.63 in its check register. The bank statement shows a balance of $3,027.40. A review of the check register and bank statement shows the following information:

Checks Outstanding:	Deposits in Transit:
No. 328 $421.90	$368.60
345 370.50	103.03

Service charge (monthly account use): $18
Reconcile the bank statement.

23. Write a complete sentence in answer to this question. Able Movers found that their bank statement was not complete with regard to the deposits that they had made yesterday for $3,690. The statement balance was $1,856. There were four checks outstanding that totaled $290. A service charge of $56 for monthly service, transaction charges, and collections was also noted on the statement. The bookkeeper was not available to adjust the balance. What should the balance be according to the available information?

24. Write a complete sentence in answer to this question. John's bank statement showed a charge for use of his automatic teller privileges in the amount of $1.50 per use. John used the auto teller seven times over the course of the month. On his account was also a flat service charge of $12.00. If his checkbook balance is correct at $61.30, what is his adjusted balance?

25. Write a complete sentence in answer to this question. Karen Wallace was notified by her bank that she had not properly endorsed a check for deposit. The check amount was $4,690. The bank indicated over the phone that her balance would be affected by this problem until she could come to the bank and straighten it out. Karen had also transferred $500 from her savings account to the checking account just yesterday. The balance shown in the checkbook is $3,852 before the transfer is recorded. What is her real balance at this point?

★ **Challenge Problem**
The Polk & Stern Company found that a $63.50 check had not been recorded in the check register. The register's uncorrected balance was $3,684.72. The bank statement balance was $1,990.32. A note had been collected by the bank for the depositor for $750 and a charge of $5 was levied against the account for this collection service. A $423.50 check deposited by the company was returned because it was improperly endorsed. New checks were imprinted by the bank for a charge of $25.80. The company had deposited $1,476.60 as the week's receipts since the bank prepared the monthly statement. A $450 monthly interest payment was shown on the bank statement as having been automatically withdrawn from the account. There was no monthly use service charge on the account because of the amount of the balance. What is the adjusted balance?

MASTERY TEST Chapter 4

Name _____ **Date** _____ **Score** _____

1. Using your name as payee, give an example of:
   a. A restrictive endorsement _____
   b. A blank endorsement _____
   c. A special endorsement _____

2. The Baird Corp. check register shows a $1,063.08 balance. The bank statement shows a balance of $1,088. The accountant reviews the bank statement and the check register and finds the following information:

Checks Outstanding:	Deposits in Transit:
No. 184—$161.92	$165.00
187—$180.00	$ 80.00
188—$105.50	$311.97
No service charges.	
Automatic payment from account from bank: $100	Deposit to account without deposit slip: $234.47

   Reconcile the bank statement and find the adjusted balance.

3. Calvin York has a balance of $92.61 in his check register as of April 30. The bank statement received on May 2 indicates a $47 balance. After reviewing both the check register and the bank statement, he finds that the deposit that he made on April 30 for $25 was omitted. He also finds that a check that he had deposited on April 25 was returned because the account holder did not have enough funds to cover the check. The amount of the check is $47.43. There are also two checks outstanding: No. 103 for $12.80 and No. 107 for $25. What is the service charge?

4. Jon Sasinowski has received a message from his bank that his account has been overdrawn. The last statement of balance was $89.70. Since the date of the statement there has only been one check written for $97.10. The statement had included a service charge of $16. There has not been a deposit made to the account since the statement. By how much has he overdrawn his account?

5. From the information given, reconcile the balance. There is a notation that the deposit on June 13 of a check for $1,962.80 was returned to the account holder due to insufficient funds. A service charge of $9 was also noted. Checks No. 474 and 476 were not shown on the statement. There were no outstanding checks from previous months.

PLEASE BE SURE TO **DEDUCT** ANY PER CHECK CHARGES OR SERVICE CHARGES THAT MAY APPLY TO YOUR ACCOUNT

NUMBER	DATE	CHECKS ISSUED TO OR DESCRIPTION OF DEPOSIT	(—) AMOUNT OF CHECK	T	(—) CHECK FEE (IF ANY)	(+) AMOUNT OF DEPOSIT	BALANCE
							940 20
470	6/7	TO/FOR CAMPBELL ASSOC.	50 00	✓			-50 00
							BAL 890 20
471	6/10	TO/FOR BEARDSON CO.	682 40				-682 40
							BAL 207 80
		TO/FOR DEPOSIT			6/13	1962 80	+1962 80
							BAL 2170 60
472	6/20	TO/FOR LITTLE & SONS	423 16				-423 16
							BAL 1747 44
473	6/27	TO/FOR PEARSON CORP.	800 00				-800 00
							BAL 947 44
474	6/27	TO/FOR McTAGGART & CO	12 50				-12 50
							BAL 934 94
475	6/29	TO/FOR R.E. EMERSON	25 60				-25 60
							BAL 909 34
		TO/FOR DEPOSIT			6/29	1827 00	+1827 00
							BAL 2736 34
476	6/30	TO/FOR J. OLSONSITE CORP.	10 00				-10 00
							BAL 2726 34
		TO/FOR					BAL
		TO/FOR					BAL
		TO/FOR					BAL
		TO/FOR					BAL

REMEMBER TO RECORD AUTOMATIC PAYMENTS / DEPOSITS ON DATE AUTHORIZED.

PAYROLL

5

OBJECTIVES

After mastering the material in this chapter, you will be able to:

1. Determine gross earnings for an employee. p. 84

2. Determine net pay for an employee. p. 90

3. Determine the employer's share of payroll taxes. p. 95

4. Understand and use the following terms:
   Pay System
   Gross Earnings
   Time Card
   Payroll Register
   Net Pay
   Gross Sales
   Sales Returns
   Sales Allowances
   Net Sales
   Straight Commission
   Commission Formula
   Deductions
   Exemption
   Form W-4
   Taxable Base
   Tax Rate
   FUTA

SPORT-A-MERICA

The volume of business has been very good the past four months. Kerry is beginning to feel the extra burden of spending so many hours at the store. With the Christmas season about to begin, she is thinking about adding one or two more sales staff to ease the work load. Hiring the extra salespeople now would allow a little time to train them to use the cash register, improve their sales techniques, and inform them about the products.

New people added to the staff would increase the coverage on the sales floor but would also mean higher payroll costs. Kerry would be able to do some additional outside sales and public relations work while the new people handle the floor sales. After all, Kerry's success as an athlete was one of the reasons the business was able to gain momentum so quickly.

Kerry knows additional people will also mean more paperwork, government forms, taxes, and expenses above the simple rate of pay to the employee. On the other hand, Brenda, the current sales clerk, has complained about the overtime hours she has had to work in recent months. Additional employees would help reduce Brenda's overtime hours, reduce Sport-A-Merica's overtime expense, and allow more coverage of the floor as the holiday season begins. Kerry is sure that it costs Sport-A-Merica over $450 a week to have Brenda work all those hours. Hal agrees hiring one or two more employees is a good idea while business is booming.

Kerry would like to hire a sales representative to call on schools if he or she would work on a commission basis. She estimates that Sport-A-Merica is missing over $2,000,000 in business by not employing an outside sales representative. How much could Sport-A-Merica pay such a qualified employee?

Kerry and Hal will need to determine the cost of paying the employees as part of their planning.

Determining Gross Earnings
Objective 1   *pay system*

All organizations that have paid employees must maintain payroll records to determine the amount the employee has earned. The company and the employee enter into an agreement at the time of employment. This agreement determines the type of **pay system** on which the employee will be compensated. There are four basic types of pay systems:

1. A *salary* is paid to employees for a pay period during which the hours or days to be worked may not be limited. This type of pay system is usually part of an open-ended agreement that requires the employee to complete specified tasks. This type of pay system is often provided to managerial employees.

2. An *hourly rate* pay system is used in a variety of situations and is designed to compensate an employee for the amount of time spent on the job. Specified work standards are established to indicate the expected level of output. Under this system, the more hours worked, the more earned. This is the most frequently used pay system. It is almost exclusively found in entry-level jobs and often in jobs involving repetitious work.

○ *The piece-rate system is used in garment manufacturing, some other manufacturing businesses, and agriculture.*

3. A *piece-rate* pay system is used when an operation is repetitious. This system compensates an employee with more pay for more output. The employee is paid a specified amount for each operation, pieces produced, or cycles completed. The more an employee produces, the more he or she earns. Occasionally the employee is actually paid more per piece once a minimum has been produced. This is termed differential piece rate.

4. A *commission* pay system is used in conjunction with the selling of goods or services. This system compensates the salesperson for the amount of sales made, not the number of hours worked. A specified percent of the value of the items sold is paid to the salesperson as a commission. The more items sold, the more earned.

These four basic systems are often combined and are used in conjunction with each other to form a more flexible pay system.

Example 1

1. A combination salary and commission can be used to provide a basic income source for a salesperson in difficult times, as well as to motivate the employee to sell more.
2. A combination hourly and piece-rate system can be used for the same purpose.
3. A combination salary and hourly system can be used to compensate managerial employees for excessive work hours.

These are examples of how pay system combinations can be developed to meet the needs of the organization and its employees.

The frequency of payment is also determined at the time of employment. This is known as the *pay period*.

○ *What the employee earns with each paycheck is determined in part by the number of paychecks issued each year.*

An employee on an annual salary of $20,800 could be paid as follows:

Pay Period	Number of Pay Periods per Year	Amount Paid per Period
Weekly	52	$ 400
Biweekly	26	800
Semimonthly	24	866.67
Monthly	12	1,733.33

Overtime

For employees paid on a hourly rate system, the Fair Labor Standards Act of 1938 requires payment of one and one-half times the regular wage rate on hours in excess of forty worked by an employee in one week. The effect of this is shown in example 2.

Example 2

Brenda, a Sport-A-Merica employee, works 46 hours each week. Her regular wage is $9.00 per hour. Do her gross earnings exceed $450 a week?

40 hours × $9.00/hour = $360.00

6 hours × $13.50/hour $\left(\$9.00 \times 1\frac{1}{2}\right)$ =   +81.00

$450.00? — $441.00

○ *In an hourly system, the more hours worked, the greater the gross earnings.*

(40 × 9.00   6 × 13.50 =)

gross earnings

The total, $441.00, is her **gross earnings** for the pay period. This is not the amount that she will receive on her paycheck, but it is the amount that she has earned. Brenda's hours may be recorded by a time clock. Each time she enters or leaves her place of employment she selects her **time card** from a rack and places it in the time clock. The time clock records the time on the card. The record on the card shows the hours she works and is used to compute her gross pay. Hourly employees are usually paid weekly.

time card

payroll record

net pay

This information—hours worked, hourly rate of pay, and gross earnings—is entered into a **payroll record** (sometimes termed a *payroll register*) by a payroll clerk. This computerized data is used to determine **net pay.**

Exercise A

Redi Transport Company pays its employees weekly. The hours worked in excess of forty per week are paid at the rate of one and one-half times the regular wages as prescribed by law. Determine the gross earnings for the following employees.

Name	Title	Hourly Wage	Hours Worked	Regular Earnings	Overtime Earnings	Total Gross Earnings
1. A. Bennett	Associate	$10.00 (10 × 40 =)	43 (15 × 3 =)	$400.00	$45.00	$445.00
2. J. Dixon	Associate	9.50	42	___	___	___
3. L. Pierce	Associate	8.00	41	___	___	___
4. O. Russell	Sr. Associate	22.00	42	___	___	___
5. B. Smith	Trainee	8.00	40	___	___	___
6. K. Williams	Supervisor	18.00	38	___	___	___
7. J. Wilson	Associate	9.00	41	___	___	___

Exercise B

Williamson Manufacturing pays its employees weekly. The company compensates its employees one and one-half times for all hours worked in excess of eight each day (Monday through Friday), whether or not they work in excess of forty each week. All hours worked on Saturday are one and one-half times regular pay. All Sunday hours are triple time. Determine the gross earnings for the following employees.

Name	M	T	W	Th	F	Sa	Su	Hourly Rate	Regular Earnings	Overtime Earnings	Gross Earnings
1. L. Becker	8	9	7	9	9	8	5	$ 8.00	$312.00	$252.00	$564.00
(8.00 × 39 =)	(12.00 × 11 =)		(24.00 × 5 =)								
2. J. Crimmins	7	6	7	8	8			8.20	_____	_____	_____
3. B. Fortos	6	8	8	9	9	9	8	8.10	_____	_____	_____
4. T. Jones	9	9	8	7	8			12.50	_____	_____	_____
5. B. Lemis	5	7	8	10	8			16.00	_____	_____	_____
6. K. Mason	10	9	6	8	8	8	5	9.50	_____	_____	_____
7. A. Rust	8	10	7	8	9	9	3	8.50	_____	_____	_____
8. P. Zist	8	9	7	9	9	8		10.00	_____	_____	_____

○ The more pieces made on a piece-rate system, the greater the gross earnings.

If an employee is paid on a piece-rate basis, the weekly wage is determined by multiplying the number of pieces produced by the amount paid to produce each piece. If the employee produces 1,463 pieces in a week and will be paid 32¢ for each piece produced, the employee's gross earnings for that week will be $468.16 (1,463 × .32). If 1,289 pieces are produced the next week, the gross earnings will be $412.48 (1,289 × .32).

gross sales

For an employee who is paid on a commission system, the pay is based on the value of the item sold. The total value of the sales is termed **gross sales.** The product of price per unit and the number of units sold will determine gross sales. Occasionally the amount of the sale is changed because of occurrences after the sale is completed. Should a customer find that the product is not what was expected—the wrong color, size, or quantity, or other such circumstances—the merchandise may be returned to the seller. The seller gives the buyer credit for the return by reducing the total invoice amount. This action also reduces

sales returns

the sale amount for the salesperson. This reduction in sales is termed **sales returns.**

sales allowances

Another factor that may reduce sales is termed **sales allowances.** Instead of asking the buyer to return the merchandise for credit, the seller may choose to offer a price

○ The more items sold on a commission pay system, the greater the gross earnings.

reduction to the buyer. If the buyer keeps (and sells) merchandise that is damaged or is the wrong color, size, etc., the buyer is given a sales allowance.

Sales returns and sales allowances reduce the amount of actual merchandise sold and, therefore, the amount of commission earned. By subtracting the value of sales returns and *net sales* sales allowances from gross sales, **net sales** can be determined. This is the amount on which a commission is calculated.

> gross sales
> −sales returns
> −sales allowances
> net sales

Example 2

Sharon Pierson sold 4,072 units of her firm's product in one week. The selling price per item was $4.95. The selling firm had twenty-eight units returned and made a 20 percent (.20) price allowance on another forty units. What were the net sales for Sharon for the week?

4,072	= Units sold
×$4.95	= Price per unit
$20,156.40	= Gross sales
−138.60	= Less sales returns (28 units × $4.95 per unit)
−39.60	= Less sales allowances (40 units × $4.95 × 20%)
$19,978.20	= Net sales for the week

(4,072 × 4.95 =   −138.60 − 39.60 =)

Exercise C

Determine the net sales for each employee.

Employee	Units Sold	Price/Unit	Units Returned	Sales Allowance Units	Sales Allowance Percent	Net Sales
1. L. Potter (963 − 12 = 951 951 × 60 =)	963	$60.00	12	2	•10% (2 × 60 × .10 =)	$57,048.00
2. M. Wierick	900	2.20	420	10	18	_____
3. T. Bowden	2,500	8.00	0	9	7	_____
4. M. Maxwell	9,500	41.75	2	0	—	_____
5. N. Hildredth	700	23.80	35	80	50	_____

---

straight commission A salesperson compensated by a specified percent of net sales is paid on a **straight commission** basis. Net sales are multiplied by a percent agreed upon by the firm and the salesperson. The product of net sales and the percent rate of commission determine gross earnings.

Example 3

Dale Kirkby is a manufacturer's representative for Patio Playtime Furniture. He earns a commission of 3 percent on all net sales. Net sales for the month ended June 30 were $230,000. What is his commission for the period?

$230,000	= Net sales
×.03	= Straight commission rate
$6,900	= Gross earnings

(230,000 × .03 =)

Exercise D

Determine the gross earnings in the following exercise.

Gross Sales	Sales Returns	Sales Allowances	Net Sales	Straight Commission Rate	Gross Earnings
1. $84,600	$400	$ 710	$83,490	3% (.03)	$2,504.70
2. 37,000	700	1,950	_____	6 (.06)	_____
3. 38,000	260	0	_____	7 (.07)	_____
4. 97,000	300	1,250	_____	4 (.04)	_____
5. 98,520	850	13,570	_____	3.5 (.035)	_____

Commission Formula

Extra sales beyond a certain level may be more difficult to achieve. A salesperson may not feel that the extra effort is really worth the extra compensation. Additionally, the selling firm may enjoy lower cost of selling as sales increase for a sales representative. In these cases, an additional incentive may be necessary in order to gain the extra sales. With a **commission formula,** the additional incentive is a salary or a commission rate that increases as sales increase. This makes every extra sale worth even more to the salesperson.

Example 4

○ *A commission formula can motivate an employee to sell even more items.*

Kerry found a person she felt could do a good job representing Sport-A-Merica to colleges and other schools in the area. She explained that the position would be a commissioned job, and therefore, gross pay would be determined by performance. The job would pay 1 percent commission on all net sales, and an additional 3 percent on net sales above $140,000 per month. Kerry felt that the additional 3 percent would provide a real incentive. She told the new employee that once some groundwork had been completed, she expected net sales to be about $185,000 per month. She explained how the commission would be determined (as shown here).

$$\begin{array}{rl} \$185,000 & = \text{Net sales per month} \\ \underline{\times .01} & = \text{Commission rate on all net sales} \\ \$1,850 & = \text{Commission} \end{array}$$

$$\begin{array}{rl} \$185,000 & \\ \underline{-140,000} & \\ \$45,000 & = \text{Net sales above } \$140,000 \\ \underline{\times .03} & = \text{Commission rate on net sales above } \$140,000 \\ \$1,350 & = \text{Commission} \end{array}$$

$$\begin{array}{rl} \$1,850 & = (1\% \text{ commission}) \\ \underline{+1,350} & = (3\% \text{ commission}) \\ \$3,200 & = \text{Gross earnings per month} \end{array}$$

(185,000 × .01 =   45,000 × .03 =)

Example 5 | Lori Driscol has been the top sales associate for Piedmont & Co. for the past six years. She is paid a commission of 1 percent on net sales up to $100,000, 2.5 percent on net sales between $100,000 and $150,000, and 4 percent on net sales above $150,000. Lori's net sales for the month totaled $164,000. What are her gross earnings for this month?

$100,000 = (First $100,000 of net sales)
×.01 = (1% commission rate)
$1,000 = Commission

$50,000 = (Net sales between $100,000 and $150,000)
×.025 = (2.5% commission rate)
$1,250 = Commission

$14,000 = (Net sales above $150,000) ($164,000 − 150,000 =)
×.04 = (4% commission rate)
$560 = Commission

$1,000 = (1% commission)
1,250 = (2.5% commission)
+560 = (4% commission)
$2,810 = Gross earnings

(100,000 × .01 =   50,000 × .025 =   14,000 × .04 =)

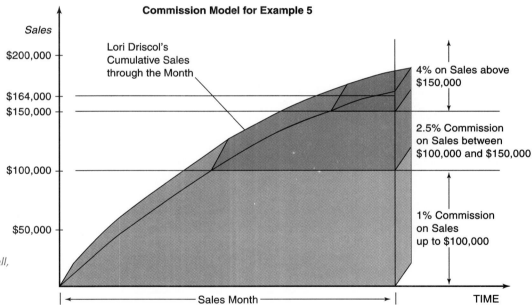

Commission Model for Example 5

○ *Watch for the key words all, above, and between in commission formula problems.*

Exercise E

Determine the gross earnings in these exercises.

Net Sales	Commission Formula	Gross Earnings
1. $86,000	1% (.01) on net sales and 2% (.02) on sales over $60,000 (86,000 × .01 = 26,000 × .02 =)	$1,380.00
2. 127,000	2% (.02) on net sales above $35,000	_____
3. 134,900	2% (.02) on net sales and 3.5% (.035) on sales over $120,000	_____
4. 208,500	2% (.02) on net sales between $150,000 and $200,000, 3% (.03) on net sales over $200,000	_____
5. 168,300	1.5% (.015) on net sales above $150,000 and 2% (.02) on net sales above $200,000	_____

Determining Net Pay
Objective 2

deductions
net pay

The amount of money that an employee earns through any of the pay systems is not the amount received on payday. From the total earnings, or gross pay, a variety of **deductions** must be subtracted. After these have been subtracted, the remainder, known as **net pay,** is payable to the employee.

Deductions are of two types: mandatory and voluntary. Payroll taxes—local and state (where applicable) and federal—are mandatory. They are mandatory for the employee to pay, as well as for the employer to collect as an agent for the taxing authority. The second type is voluntary. Voluntary deductions are discussed on page 93.

Mandatory Deductions
Federal Income Tax

The laws of the United States require that the **federal income tax** be paid as wages are earned. The employer is required to collect or withhold from the employee's paycheck a specified amount each pay period. This amount and the amounts withheld from other employees' paychecks are then periodically sent to the federal government's Internal Revenue Service (IRS). A similar procedure is used for state and local income taxes.

To determine the amount of income tax to be withheld, most employers use a set of charts provided by the IRS. The charts are for weekly, monthly, and biweekly pay periods. Income tax is based on an employee's ability to pay (i.e., the more money earned, the greater the amount of tax). The tax is computed on the amount of gross earnings and the number of exemptions an employee claims. The employee is entitled to claim one **exemption** for himself or herself and one for each dependent. The employee makes this claim on **Form W-4,** which is shown below.

exemption
Form W-4

Form W-4 has been completed for employee Jacqulyn S. Retzloff. She has claimed herself, her spouse, and two children—a total of four. This information with the gross earnings is used to determine the amount of income tax from the charts shown on page 91.

Form **W-4**				
Department of the Treasury Internal Revenue Service	**Employee's Withholding Allowance Certificate** ▶ For Privacy Act and Paperwork Reduction Act Notice, see page 2.			OMB No. 1545-0010 **2000**

1 Type or print your first name and middle initial *Jacqulyn Sue* Last name *Retzloff* **2** Your social security number 378 : 58 : 7122

Home address (number and street or rural route) *300 Agler* **3** ☐ Single ☒ Married ☐ Married, but withhold at higher Single rate. Note: *If married, but legally separated, or spouse is a nonresident alien, check the Single box.*

City or town, state, and ZIP code *Your City, State*   *01010* **4** If your last name differs from that on your social security card, check here. You must call 1-800-772-1213 for a new card . . . ▶ ☐

5 Total number of allowances you are claiming (from line **H** above **OR** from the applicable worksheet on page 2) **5** *4*

6 Additional amount, if any, you want withheld from each paycheck . . . . **6** $ *0*

7 I claim exemption from withholding for 2000, and I certify that I meet **BOTH** of the following conditions for exemption:
● Last year I had a right to a refund of **ALL** Federal income tax withheld because I had **NO** tax liability **AND**
● This year I expect a refund of **ALL** Federal income tax withheld because I expect to have **NO** tax liability.
If you meet both conditions, write "EXEMPT" here . . . . ▶ **7**

Under penalties of perjury, I certify that I am entitled to the number of withholding allowances claimed on this certificate, or I am entitled to claim exempt status.
Employee's signature (Form is not valid unless you sign it) ▶ *Jacqulyn Sue Retzloff*   **Date** ▶ *November 16, 2000*

8 Employer's name and address (Employer: Complete lines 8 and 10 only if sending to the IRS.) *BEST Wages, Inc.* *100 Main Street   Big City, MO   010101* **9** Office code (optional) **10** Employer identification number 42 : 064 1173

Example 6

If Ms. Retzloff's earnings for the week are $508, the amount of the income tax would be $25. The amount of tax is found in the "Married Persons—Weekly" chart. By determining the correct range in which her $508 income for the week would fall and by moving across until the column for the number of exemptions claimed is a 4, to correspond with the W-4 certificate, the value of $25 is then found.

MARRIED Persons—WEEKLY Payroll Period

(For Wages Paid in 2000)

| If the wages are– | | And the number of withholding allowances claimed is— | | | | | | | | | | |
At least	But less than	0	1	2	3	4	5	6	7	8	9	10
		The amount of income tax to be withheld is—										
400	410	42	34	26	18	10	2	0	0	0	0	0
410	420	44	36	27	19	11	3	0	0	0	0	0
420	430	45	37	29	21	13	5	0	0	0	0	0
430	440	47	39	30	22	14	6	0	0	0	0	0
440	450	48	40	32	24	16	8	0	0	0	0	0
450	460	50	42	33	25	17	9	1	0	0	0	0
460	470	51	43	35	27	19	11	3	0	0	0	0
470	480	53	45	36	28	20	12	4	0	0	0	0
480	490	54	46	38	30	22	14	6	0	0	0	0
490	500	56	48	39	31	23	15	7	0	0	0	0
500	510	57	49	41	33	25	17	9	1	0	0	0
510	520	59	51	42	34	26	18	10	2	0	0	0
520	530	60	52	44	36	28	20	12	4	0	0	0
530	540	62	54	45	37	29	21	13	5	0	0	0
540	550	63	55	47	39	31	23	15	7	0	0	0
550	560	65	57	48	40	32	24	16	8	0	0	0
560	570	66	58	50	42	34	26	18	10	2	0	0
570	580	68	60	51	43	35	27	19	11	3	0	0
580	590	69	61	53	45	37	29	21	13	5	0	0
590	600	71	63	54	46	38	30	22	14	6	0	0
600	610	72	64	56	48	40	32	24	16	8	0	0
610	620	74	66	57	49	41	33	25	17	9	1	0

SINGLE Persons—WEEKLY Payroll Period

(For Wages Paid in 2000)

| If the wages are– | | And the number of withholding allowances claimed is— | | | | | | | | | | |
At least	But less than	0	1	2	3	4	5	6	7	8	9	10
		The amount of income tax to be withheld is—										
300	310	38	30	22	14	6	0	0	0	0	0	0
310	320	40	32	23	15	7	0	0	0	0	0	0
320	330	41	33	25	17	9	1	0	0	0	0	0
330	340	43	35	26	18	10	2	0	0	0	0	0
340	350	44	36	28	20	12	4	0	0	0	0	0
350	360	46	38	29	21	13	5	0	0	0	0	0
360	370	47	39	31	23	15	7	0	0	0	0	0
370	380	49	41	32	24	16	8	0	0	0	0	0
380	390	50	42	34	26	18	10	2	0	0	0	0
390	400	52	44	35	27	19	11	3	0	0	0	0
400	410	53	45	37	29	21	13	5	0	0	0	0
410	420	55	47	38	30	22	14	6	0	0	0	0
420	430	56	48	40	32	24	16	8	0	0	0	0
430	440	58	50	41	33	25	17	9	1	0	0	0
440	450	59	51	43	35	27	19	11	3	0	0	0
450	460	61	53	44	36	28	20	12	4	0	0	0
460	470	62	54	46	38	30	22	14	6	0	0	0
470	480	64	56	47	39	31	23	15	7	0	0	0
480	490	65	57	49	41	33	25	17	9	0	0	0
490	500	67	59	50	42	34	26	18	10	2	0	0
500	510	68	60	52	44	36	28	20	12	3	0	0
510	520	70	62	53	45	37	29	21	13	5	0	0
520	530	71	63	55	47	39	31	23	15	6	0	0

SINGLE Persons—MONTHLY Payroll Period

(For Wages Paid in 2000)

| If the wages are– | | And the number of withholding allowances claimed is— | | | | | | | | | | |
At least	But less than	0	1	2	3	4	5	6	7	8	9	10
		The amount of income tax to be withheld is—										
3,240	3,280	578	513	447	382	317	281	246	211	176	141	106
3,280	3,320	589	524	459	393	328	287	252	217	182	147	112
3,320	3,360	600	535	470	404	339	293	258	223	188	153	118
3,360	3,400	612	546	481	416	350	299	264	229	194	159	124
3,400	3,440	623	557	492	427	361	305	270	235	200	165	130
3,440	3,480	634	569	503	438	373	311	276	241	206	171	136
3,480	3,520	645	580	515	449	384	319	282	247	212	177	142
3,520	3,560	656	591	526	460	395	330	288	253	218	183	148
3,560	3,600	668	602	537	472	406	341	294	259	224	189	154
3,600	3,640	679	613	548	483	417	352	300	265	230	195	160
3,640	3,680	690	625	559	494	429	363	306	271	236	201	166
3,680	3,720	701	636	571	505	440	375	312	277	242	207	172
3,720	3,760	712	647	582	516	451	386	320	283	248	213	178
3,760	3,800	724	658	593	528	462	397	332	289	254	219	184
3,800	3,840	735	669	604	539	473	408	343	295	260	225	190
3,840	3,880	746	681	615	550	485	419	354	301	266	231	196
3,880	3,920	757	692	627	561	496	431	365	307	272	237	202
3,920	3,960	768	703	638	572	507	442	376	313	278	243	208

Social Security Tax

The Social Security tax is the second major mandatory deduction. The Social Security tax is levied on the gross earnings of the employee. Again, the employer (with a few exceptions) is obligated to withhold a specified amount from the employee's gross earnings. Payment is later made to the federal government by the employer. The Social Security tax differs from the federal income tax in that it is not based on an ability to pay, and there are no exemptions to consider. Deductions are made based on the first $76,200 that the employee earns from each employer each year.

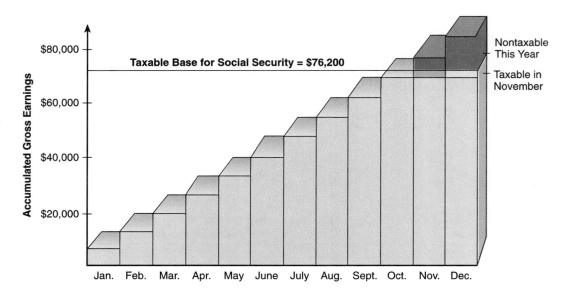

At year's end, the employee pays tax on only $76,200, even if several employers withheld Social Security tax totaling more than the tax based on an income of $76,200. The $76,200 represents the **taxable base.** Earnings beyond this amount are not subject to Social Security for that employer in that year.

taxable base

Example 7

If Ms. Retzloff's year-to-date earnings before this pay period were $75,900 from this employer, then only the amount necessary to bring the accumulated taxable gross earnings to the $76,200 maximum will be taxed this pay period. Once again using $508 as her gross earnings, the taxable amount must first be found.

$76,200
−75,900
$ 300   taxable this pay period

Under these circumstances, the Social Security tax for this period would be $18.60 ($300 × .062). For the next pay period, Social Security will not be deducted, because the first $76,200 of earnings have already been taxed for the year.

Ms. Retzloff's Medicare tax deduction would be unchanged. This deduction would amount to $7.37 ($508 × .0145).

(300 × .062 =)

tax rate

The $76,200 is taxable at a **tax rate** of 6.2 percent as it is earned. Charts are also available for determining the amount of this deduction. The taxable base ($76,200) and the tax rate (6.2 percent) are subject to change by Congress.

Medicare Tax

The Medicare tax rate is 1.45 percent on all earnings. There is not a taxable base maximum and there are no deductions or exemptions considered. The Medicare tax is levied on the gross earnings of the employee.

Example 8

Using the previous example, if Ms. Retzloff's gross earnings for the week were $508, her Social Security tax would be $31.50. This amount is found by multiplying the gross earnings by the tax rate, that is, $508 × .062. Her Medicare tax would be $7.37 ($508 × .0145).

(508 × .062 =)   *(508 × .0145 =)*

State and Local Income Taxes

Most states and many cities have an income tax. The structure of the tax specifications differs widely, but is similar to that of the federal income tax, Social Security, or Medicare tax. The tax may be a graduated tax (*progressive,* in the sense that the tax rate increases as the gross income increases), a flat rate taxed on all gross income, or a rate taxed on wages up to a maximum taxable base (similar to Social Security).

Example 9

Sarah has weekly earnings of $830. She lives in a city that has an income tax of 1 percent of all earnings. The city is in a state that has a progressive income tax. The state tax is 2 percent on the first $750 of earnings a week, $2\frac{1}{2}$ percent on the next $200 of earnings a week, and 3 percent on earnings above $1,000 a week. What is the amount of city and state income tax that Sarah will have deducted from her paycheck? (*Note:* 1% = .01, 2% = .02, $2\frac{1}{2}$% = .025, and 3% = .03)

$830 ×   .01 = $8.30 city income tax

$$\begin{array}{r} \$830 \\ -750 \times\ .02 = \quad 15.00 \\ \hline \$80 \times .025 = \quad +2.00 \\ \hline \$17.00 \text{ state income tax} \end{array}$$

(830 × .01 =   750 × .02 =   80 × .025 =)

Example 10

Veronica Smathers lives in a state that has an income tax that is a flat 3.7 percent of the first $15,000 earned per year. Veronica has earned $14,849 so far this year. This week's paycheck will be based on gross earnings of $650. What will be the deduction for state income tax for this week? (*Note:* 3.7% = .037)

$$\begin{array}{r} \$15,000 \\ -14,849 \\ \hline \$151 \quad \text{taxable this pay period} \\ \times.037 \\ \hline \$5.59 \quad \text{state income tax} \end{array}$$

(151 × .037 =)

Voluntary Deductions

The second type of deduction is voluntary deductions authorized by the employee. A payroll deduction authorization slip is signed by the employee and given to the payroll clerk. This okays deductions from gross pay for any one of a variety of purposes specified by the employee.

Some examples of voluntary deductions are union dues, uniform rental, credit union payments, savings contributions, savings bonds, retirement programs, profit-sharing plans, Individual Retirement Act (IRA) plans, and direct deposits to bank accounts. Some of these "voluntary" deductions may be required by union contract, by a court action or settlement, the result of a financial contract (credit union payment, etc.), but they are not necessarily mandated *by law*.

Once all of the deductions have been calculated, they are subtracted from gross earnings to determine net pay. This amount is made payable to the employee on the paycheck and is the actual amount that the employee has available to spend.

Example 11

Tom has been a Sport-A-Merica employee for a short time. He is paid a salary of $500 a week. Tom has a wife and three children, and claims all of them as exemptions. His earnings prior to this paycheck were $3,500 for the year. He has deductions of $40 per paycheck for a credit union payment, and $10 for the purchase of U.S. savings bonds. What is his net pay?

$500.00 = Gross earnings
− 17.00 = Federal income tax (five exemptions)
−31.00 = Social Security tax
− 7.25 = Medicare tax
−40.00 = Credit union payment
−10.00 = U.S. savings bonds
$394.75 = Net pay

(500 − 17.00 − 31.00 − 7.25 − 40.00 − 10.00 =)

Exercise F

Determine the net pay of each employee at Burson Techniques, Inc. for the first week of the year from the information below.

Employee	Gross Earnings	Federal Income Tax	State Income Tax	Medicare Tax	Social Security Tax	Union Dues	IRA Deduction	Net Pay
1. Johnny Biddy	$ 600.00	$ 53	$28.00	8.70	$37.20	$6	$ 0	$467.10
2. James Carson	1,263.40	196	61.73			0	40	
3. J. Grason	898.00	96	42.88			6	120	
4. Kay Lasher	756.83	75	37.16			6	0	
5. Sharron Orr	694.20	66	34.72			6	25	

Exercise G

Determine net pay for the week from the incomplete payroll register below. All the employees are married.

Payroll Register Employee	Gross Earnings	Exemptions Claimed	Federal Income Tax	Gross Earnings Before Pay Period	Medicare Tax	Social Security Tax	Net Pay
1. Sarah Blair	$435.00	2	$	$31,000	$	$	$
2. Jock Lieber	521.00	2		29,795			
3. Fran Smith	487.00	1		47,900			
4. John Ulrick	507.00	5		24,958			
5. Kay Wilson	496.00	0		41,800			
6. Bill Puhl	462.00	1		42,000			
7. A. Lingle	595.00	2		77,850			
8. Jim Seal	460.00	1		47,940			
9. Jean King	427.00	2		38,900			
10. Ed Ely	479.00	4		76,891			

Employer's Payroll Taxes
Objective 3

The employer, like the employee, is obligated to pay taxes based on the employees' gross earnings. The taxes are an extra expense to the employer, increasing the cost of labor.

The most significant of these taxes is the Social Security tax. The employer is required to pay an amount matching the employee's contribution. Six point two percent (6.2 percent or .062) of the first $76,200 each employee earns for each year must be paid to the federal government. When the employer makes the payment, the employer is really forwarding an amount equal to 12.4 percent of the employee's earnings. The 12.4 percent results from a 6.2 percent deduction from the employee plus the 6.2 percent additional tax from the employer.

The employer will also pay Medicare tax at 1.45 percent (2.9 percent with the employee share). The Medicare tax is based on all earnings for each employee. That is, there is not a taxable base maximum as with Social Security. As a result, very high-income individuals will pay more in Medicare tax than in Social Security tax in a year.

A third employer payroll tax is the two-part unemployment tax. The two parts are the Federal Unemployment Tax (FUTA) and the state unemployment tax (may have different names by state). Both taxes are based on the first $7,000 of gross earnings of each employee. This is an employer-only tax.

The FUTA rate is 6.2 percent (.062). A credit of 5.4 percent may be granted against the FUTA rate for the unemployment tax levied by the state. This means that the FUTA tax rate is really .8 percent (6.2 percent − 5.4 percent). This tax is remitted to the federal government quarterly for amounts due of $100 or more and annually for amounts due of less than $100.

The state may vary the 5.4 percent tax rate based on the unemployment history within the state and/or the unemployment history of the individual employer. New companies will initially pay the highest rate. This tax is remitted to the state government.

These payroll taxes levied against the employer are, in effect, increases in the cost of labor, even though these taxes do not appear as part of the employee's gross earnings.

Example 12

The accountant has provided Hal with the payroll information for the current period. The amount of gross earnings for Sport-A-Merica employees for the period equals $4,620. The company must remit the Social Security tax, Medicare tax, and state and federal unemployment taxes for the period to the federal and state governments. The state unemployment tax rate is 5.4% (.054). The federal unemployment tax is .8% (.008). All wages are to be taxed. How much must be remitted?

$\begin{array}{ll} \$4,620 & \text{taxable earnings this pay period} \\ \times\ .124 & (6.2\% \times 2) \\ \hline \$572.88 & \text{(for Social Security tax)} \end{array}$

$\begin{array}{ll} \$4,620 & \\ \times\ .029 & (1.45\% \times 2) \\ \hline \$133.98 & \text{for Medicare tax} \end{array}$ $\begin{array}{ll} \$4,620 & \\ \times\ .054 & \\ \hline \$249.48 & \text{for state unemployment tax} \end{array}$

$\begin{array}{ll} \$4,620 & \\ \times\ .008 & \\ \hline \$\ 36.96 & \text{for federal unemployment tax} \end{array}$

$\begin{array}{l} \$572.88 \\ 133.98 \\ +36.96 \\ \hline \$743.82 = \text{amount to be remitted to federal government} \\ \$249.48 = \text{amount to be remitted to state government} \end{array}$

(4,620 × .124 =)   (4,620 × .029 =)   (4,620 × .008 =)   (4,620 × .054 =)

Exercise H

Determine the amount due the federal government for the Social Security tax, Medicare tax, and F.U.T.A. and to the state government for state unemployment tax. The amount should include both the employee and employer share when appropriate.

Gross Earnings of Employees for the Pay Period	Social Security Tax	Medicare Tax Due	F.U.T.A. Due (.8%)	State Unemployment Tax Due (5.4%)
$4,890	1. $ 606.36	6. $ 141.81	11. $ 39.12	16. $ 264.06
12,850	2. _____	7. _____	12. _____	17. _____
3,478	3. _____	8. _____	13. _____	18. _____
23,600	4. _____	9. _____	14. _____	19. _____
8,920	5. _____	10. _____	15. _____	20. _____

ASSIGNMENT Chapter 5

Name _____ **Date** _____ **Score** _____

Skill Problems

Select the best term/phrase for the statements below.

A. Pay System   B. Gross Earnings   C. Time Card   D. Payroll Register
E. Net Pay   F. Gross Sales   G. Sales Returns   H. Sales Allowances
I. Net Sales   J. Deductions   K. Exemptions   L. Form W-4
M. Tax Rate   N. Taxable Base   O. Straight Commission P. Commission Formula

1. _____ An item of control that allows a payroll department to determine the number of hours worked by employees.

2. _____ The amount due a person after deductions have been subtracted from gross pay.

3. _____ The dollar value of sales after sales returns and sales allowances have been subtracted from gross sales.

4. _____ The amount earned by the employee in a pay period.

5. _____ A form showing gross earnings, deductions, and net pay for all employees.

6. _____ Legal dependents and the employee; can be claimed by the employee in order to reduce the amount of income tax withheld.

7. _____ Amounts subtracted from gross pay for such items as taxes, contributions, and savings.

8. _____ All the sales of a firm for a time period.

9. _____ A combination of ever-increasing rates of pay for sales designed to motivate an employee to sell more.

10. _____ Merchandise returned to the seller because of damage, errors in ordering, etc.

11. _____ Adjustments made in the price of the merchandise in order to compensate the buyer for keeping damaged or incorrectly shipped merchandise.

12. _____ The maximum amount of earnings that can be taxed for Social Security and Medicare tax purposes for that year.

Johson Haber, Associates pays its employees weekly. The hours worked in excess of forty per week are paid at the rate of one and one-half times the regular wages as prescribed by law. Determine the gross earnings for the following employees.

Name	Hourly Wage	Hours Worked	Regular Earnings	Overtime Earnings	Total Gross Earnings
13. J. Yale	11.50	46			
14. R. Ott	9.75	44.25			
15. P. Birkson	10.25	28			
16. L. Parker	8.85	39			
17. K. Mathews	9.40	46.5			

Determine the net sales in the following exercise.

	Units Sold	Price Per Unit	Units Returned	Sales Allowance Units	Sales Allowance Percent	Net Sales
18.	845	$65.20	5	3	20%	
19.	201	65.20	9	8	30%	
20.	540	69.50	0	0	20%	
21.	942	60.10	9	7	15%	
22.	150	75.00	2	0	20%	

Determine the net sales and gross earnings in the following exercise.

	Gross Sales	Sales Returns	Sales Allowances	Net Sales	Straight Commission Rate	Gross Earnings
23.	$64,500	$ 540	$ 780		4%	
24.	21,450	280	85		2	
25.	5,410	1,450	0		25	
26.	200,460	4,870	2,401		.5	

Determine the gross earnings in these exercises.

	Net Sales	Commission Formula	Gross Earnings
27.	$910,140	2% on net sales up to $800,000, 3% on sales over $800,000	
28.	4,845	10% on first $4,000 of net sales, 25% on sales over $4,000	
29.	245,670	2% on net sales above $95,000	
30.	2,780	10% on net sales and additional 2% on sales over $3,000	
31.	5,641	15% on net sales and additional 10% on sales over $4,500	

Determine the net pay for each employee for the first week of the year from the information below. All employees are married and claim two exemptions.

Employee	Gross Earnings	Federal Income Tax	Social Security Tax	Medicare Tax	Credit Union	Union Dues	Net Pay
32. H. Barker	$431	$ _____	$ _____	$ _____	$ 0	$3.50	$ _____
33. L. Lowery	587	_____	_____	_____	40.00	4.20	_____
34. P. Norris	412	_____	_____	_____	17.00	3.50	_____
35. F. Tunks	490	_____	_____	_____	78.00	3.50	_____
36. B. Willard	525	_____	_____	_____	25.00	4.00	_____

Determine the net pay for each salaried employee for the month of March from the information below. All employees are single.

Employee	Gross Earnings	Exemptions Claimed	Federal Income Tax	Medicare Tax	Social Security Tax	Net Pay
37. J. Cousins	$3,868.00	5	$ _____	$ _____	$ _____	$ _____
38. K. Dale	3,751.00	1	_____	_____	_____	_____
39. J. Kerns	3,460.00	0	_____	_____	_____	_____
40. M. Kurt	3,421.00	3	_____	_____	_____	_____
41. P. Wisfal	3,672.00	3	_____	_____	_____	_____

Complete the following payroll ledger. All of the employees are single.

Employee	Gross Earnings (Week)	Exemptions Claimed	Federal Income Tax	Gross Earnings Before Pay Period	Medicare Tax	Social Security Tax	Net Pay
Tait, J.	$356.00	4	_____	$ 5,612.00	_____	_____	_____
Willis, C	391.00	2	_____	65,035.00	_____	_____	_____
Woods, D	357.50	5	_____	4,652.50	_____	_____	_____
Yale, M.	384.25	0	_____	85,421.80	_____	_____	_____
Totals	42. _____	****	43. _____	****	44. _____	45. _____	46. _____

47. Using the ledger above, what is the total amount due to the federal government for Social Security and Medicare tax only? The amount should include employee and employer share.

48. Using the ledger above, what is the total amount due to the federal government for federal unemployment? The rate for this company is .8% of the first $7,000.00.

49. Using the ledger above, what is the total amount due to the state government for unemployment? The rate for this company is 2.7% of the first $7,000.

50. Using the ledger above, what is the total amount due to the federal government?

Business Application Problems

1. Susie earns a salary of $380 a week on her job as a data entry clerk. She is single and claims no withholding allowances. What is the deduction for federal withholding tax?

2. An employee mentioned to a friend that she had $34.32 deducted from her paycheck for Social Security tax in the first week of January. What were her gross earnings?

3. Gerrard has a salary of $2,000 per month as the manager of a used car lot. He also earns a commission of 5 percent (.05) on the sales made by the lot. Last month the lot sold 12 vehicles totaling $56,470. What was Gerrard's gross earnings for the month?

4. Karen has earned a salary of $2,488 a week for the first 32 weeks of this year. What is the Social Security deduction for this week?

5. A commission of 4.5 percent on all sales as well as a salary of $200 per week is paid to all sales associates at World Tickets. Last week Manuel sold $17,600 in travel packages. What is his gross income for the week?

6. Alex worked 52 hours last week. He worked 58 hours this week. His employer pays time and a half for hours in excess of 40 per week. Alex earns $9.80 per hour. How much more did Alex earn this week?

7. The employees at Bedford Inc. had earnings of $14,860 for the week. The Social Security and Medicare tax amount on the earnings must be noted in the accounting records. What is the total amount for these taxes combined for the employer and employee?

8. A combination piece rate and hourly rate is earned by Beth. She worked 38 hours last week and produced 648 pieces. Her hourly rate is $9. She earns $.32 for each finished piece. What are her gross earnings for the week?

9. First National employs Suzanne as a mortgage broker. She had a record month in June with sales of $365,000 in new mortgages. She earns a commission of 1.3 percent on all sales above $50,000. What is her gross earnings amount?

10. Dale earns his income from Daytime Productions through a commission formula of 1 percent on all sales up to $60,000, 2.5 percent on all sales between $60,000 and $100,000, and 4 percent on all sales above $100,000. What are his gross earnings for the month if his sales were $120,800 for the period?

11. Brad is reviewing Gwen's paycheck authorization. She earned $385 last week. Her W-4 form indicates that she claims two exemptions and that she is single. She has also authorized a deduction for $40 toward retirement. Her gross earnings to date are $35,670. What is the net pay amount for this check?

12. Determine the amount of gross earnings for an employee from the time card information shown below.

Days	M	T	W	Th	F
	10	8	9	7	10

Notes: 1. Pay rate = $9 per hour
          2. Overtime paid at one and one-half times the regular pay rate for all hours in excess of 40 per week.

13. The Bridgeton Manufacturing Company pays a commission of 3 percent on all sales, and an extra $1\frac{1}{2}$ percent on sales above $250,000. The top sales associate sold $267,970 last month. There were thirty-five units returned during the month. Each unit is priced at $659.00. What was the commission earned by the top salesperson this month?

14. Kay Smite was paid a salary of $1,490 a week. She worked each week of the year. What was the deduction for Social Security on the last (52nd) week of the year?

15. What was the total Social Security tax deduction from Ms. Smite's earnings during the year?

16. Kolan Industries, Inc., employs Kerry as a telemarketer. Her gross earnings this week are made up of a salary of $320, plus $138 in commissions she earned on sales. She is married, has one child, and claims one exemption. What is her net pay if her gross earnings to date are $25,837?

17. Rob is paid a salary of $600 a month, plus a 2 percent commission on net sales over $40,000 and an additional 1 percent on net sales over $100,000. Rob sold $164,800 of merchandise this month, with only $3,200 of returns. What were Rob's gross earnings?

18. Sun Labs, Inc. employs Sam on a piece-rate basis at $.50 per piece. His W-4 form shows that he claims a total of two exemptions and that he is single. His earnings to date are $82,412. What is his net pay?

Output Record

Monday	Tuesday	Wednesday	Thursday	Friday
193	189	109	128	141

19. Darcy had a gross income of $549.35 for the past week. He lives in a city that has an income tax of 1 percent. The state also has an income tax of 2 percent on the first $30,000. Darcy has an accumulated earnings of $29,288 prior to this week. What is Darcy's net income for the week if the deduction for federal income tax is $87.00?

20. What would be the net income for Darcy for the following week if his gross income was the same as problem 19?

21. Vee has determined that the company must pay the Social Security tax due the federal government. The amount withheld from employees totals $4,680. What is the amount of the check payable to the federal government for Social Security tax (employee plus employer)?

22. Deductions by the Payroll Office of $18 for union dues, $27.50 for uniform rental, $50 for a savings program, and $10 for tool rental are authorized by Peter Wilson. The payroll clerk also notes that he claims one exemption even though he is married. He has earned $14,900 prior to this pay period. His hourly rate is $13. He worked forty-two hours this week. The company pays double the regular rate for hours in excess of forty. What is his net pay?

23. The earnings for Dixie Brown total $9,867 for the year to date. Dixie is married but claims no exemptions. Her gross earnings for this week are $410. She has $50 deducted from her check each week for savings. She also has a $15 deduction for parking each week. The city in which Dixie works has a 1 percent tax for residents and a $\frac{1}{2}$ percent tax for nonresidents on the first $10,000 of gross earnings. Dixie is not a resident of the city. What is her net income for the week?

24. Perie has a full-time job that pays $8 per hour at Adain Products, Inc. He worked 43 hours last week. The firm pays time and one-half for all hours in excess of 40 per week. Perie is single. He had worked double shifts and a great deal of overtime earlier in the year, bringing his year-to-date gross earnings to $41,800. He has also agreed to have $50 deducted from his check each week for savings. What is his net pay?

25. Write a complete sentence in answer to this question. Janice Noteri is considering changing the number of exemptions she claims in order to increase her take-home pay. Janice is a single mother and has a salary of $350 per week. She now only claims herself, but she is considering claiming three exemptions. What will be the difference in her take-home pay if she makes the change?

26. Write a complete sentence in answer to this question. Al Derton receives a salary of $2,365 per month. He claims one exemption and is single. This paycheck included a bonus of $1,500 in addition to his salary. How much will Al receive in this monthly paycheck if his earnings to date total $23,000?

27. Write a complete sentence in answer to this question. If an employee is now earning an annual salary of $41,800 and receives a raise of $5,000, what will be the additional take-home pay for the year? The W-4 form indicates that the employee claims three exemptions and is single. She receives a monthly paycheck.

★ **Challenge Problem**

Catty is employed with the Marks Corporation. She earns a salary of $200 a week, as well as a commission of $1\frac{1}{2}$ percent of sales. Her commission last month was $840. She will receive one-fourth of that amount this week in her gross earnings. She is married, has five children, but only claims three exemptions. She also earns $40 a week by doing artwork on the side out of her own home. She collects this income in cash. Catty has earned $14,900 to date from her sales for the Marks Corporation and $3,880 for her artwork. She has $47 taken out of her paycheck each week for savings and $100 is directly deposited into her bank account. What is her paycheck amount for the week?

Name _____ **Date** _____ **Score** _____

True or False

T F **1.** Gross earnings + deductions = net pay.

T F **2.** A W-4 form is used to determine the amount of withholding allowances for an employee.

T F **3.** Overtime pay is not subject to F.I.C.A. tax because it is $1\frac{1}{2}$ times the usual rate of pay.

T F **4.** Net pay is determined from reading the numbers on the time card.

T F **5.** FUTA is deducted from each employee's wages.

T F **6.** Gross earnings is the amount that is used to determine federal income tax.

7. Compute the employee's gross earning's from the piece-rate information below.

Output Record

Days	M	T	W	Th	F
Units of output	74	83	91	59	87
Hours worked	8	6	8	7.5	8.1

Piece-rate = $1.15 per unit

8. Determine the amount of federal income tax to be withheld if the gross earnings are $400 for the week, the number of exemptions claimed is one, and the employee is single.

9. The payroll clerk noted that $911.57 was withheld from the employees of the company for Social Security tax for the week. What is the total amount that the employer must send the federal goverment for Social Security tax for the week?

10. Peter earns an hourly rate of $9.50. He is paid weekly and worked forty-seven hours last week. He is paid overtime for hours in excess of forty at the rate of one and one-half times. His earnings to date are $26,280. He is married and claims two exemptions. What is his net pay?

11. Jeff Hazel earns $9 per hour. He worked 42 hours last week. He is paid double time for all hours worked in excess of 40 each week. His earnings to date are $7,947. He is single and claims one exemption. What is his gross pay for the week?

12. Compute the gross earnings for an employee who is paid on a combination hourly and piece-rate system.

Day	M	T	W	T	F
Hours worked	9	7.6	8	7.2	8
Units of output	93	112	141	86	152

Hourly rate = $7.00
Piece rate = 7.8¢ per unit

BUSINESS MEASUREMENTS

6

OBJECTIVES

After mastering the material in this chapter, you will be able to:

1. Perform all of the fundamental processes in problems dealing with measurements. p. 108
2. Find the solution to problems dealing with linear, surface, and volume measure (using the traditional measurement system). p. 108
3. Find the solution to problems dealing with linear, surface, and volume measure (using the metric system). p. 112
4. Convert measurements to the metric system or to the traditional measurement system. p. 113
5. Use currency exchange rates in business transactions. p. 115
6. Understand and use the following terms:
   Linear Measure
   Surface Measure
   Volume Measure
   Currency Exchange

SPORT-A-MERICA

The addition of two new sales clerks helps relieve some of the pressure for Kerry, but the relief doesn't last long. As the holiday buying season begins to build, it is apparent that Sport-A-Merica has established itself in the retail community. The customer count is growing at such a rate that it is clear the store will soon be bulging at the seams. The stock on the floor is interfering with the traffic flow of customers throughout the store. The sales clerks, Kerry, and Hal are almost on top of one another. The merchandise processing room needs to be at least 20 ft. by 20 ft. Hal would also like to bring in a line of Italian bicycles. He has no room to show them in the present store. With the favorable exchange rate he feels that Sport-A-Merica is losing potential income.

Hal realizes that as soon as the holiday season is over, a new store location must be found. Sport-A-Merica must expand before competition comes along and takes business away. Though the need to move is apparent, a new and bigger store in a good location will be difficult to find. The right location will mean the possibility of even greater growth in the future. This decision is not only important for the immediate future, but will possibly have an effect on the next growth move.

Realizing that the moving decision cannot wait until after the holidays, Kerry and Hal scour the classified ads, speak with realtors, and spend weekends looking for a suitable, affordable store. They did not realize the full amount of the considerations that come into play on a business expansion decision. The square footage of floor space is a prime issue. The length of exposure the store has to the street, the volume of cubic feet of air to be heated or cooled (Hal has budgeted for $7,000 a year), the cost of improvements, the existence of nearby parking—all are items that must be given serious attention. Though time consuming, the prospect of enlarging the operation, as well as the profits, keeps the search under way. Persistence will pay off!

Understanding dimensional measurements is useful to a business student because this can be applied in the areas of building decoration and landscaping (carpeting, painting, sodding, and seeding), construction (determining the amount of earth removal, insulation required, and the area to be heated or cooled), and establishing retail and wholesale space allocations (determining overhead expenses by area, shelf allocations, and sales by floor space).

Objective 1

The first part of this chapter includes using some basic measuring principles. The second part introduces the metric measurement system and problems relating to metric conversion as well as to metric measurement. You will be able to make use of your knowledge in working with whole numbers, fractions, decimals, and the fundamental processes in this chapter. Do your work carefully to be sure to:

1. perform accurately the basic fundamentals on measurement information
2. select the correct conversion factor from the measurement table
3. reduce or borrow units from larger or smaller units

The following is a table of measurements. This table will be useful to you in working the problems in the first part of the chapter.

Measurements Table

Linear or Straight Measure

12 inches	=	1 foot
3 feet	=	1 yard
$5\frac{1}{2}$ yards	=	1 rod
$16\frac{1}{2}$ feet	=	1 rod
66 feet	=	1 chain
320 rods	=	1 mile
5,280 feet	=	1 mile
1,760 yards	=	1 mile

Cubic or Volume Measure

1,728 cubic inches	= 1 cubic foot
27 cubic feet	= 1 cubic yard
128 cubic feet	= 1 cord of wood

Liquid Measure

4 gills	= 1 pint
2 pints	= 1 quart
4 quarts	= 1 gallon
$31\frac{1}{2}$ gallons	= 1 barrel
$7\frac{1}{2}$ gallons	= 1 cubic foot
32 ounces	= 1 quart

Dry Measure

2 pints	= 1 quart
8 quarts	= 1 peck
4 pecks	= 1 bushel

Surface or Square Measure

144 square inches	= 1 square foot
9 square feet	= 1 square yard
$30\frac{1}{4}$ square yards	= 1 square rod
160 square rods	= 1 acre
640 acres	= 1 square mile or section

Counting

12 units	= 1 dozen
12 dozen	= 1 gross

Avoirdupois Weight

16 ounces	= 1 pound
100 pounds	= hundredweight
20 hundredweights	= 1 ton
2,000 pounds	= 1 ton
2,240 pounds	= 1 long ton

Time

60 seconds	= 1 minute	100 years	= 1 century
60 minutes	= 1 hour	$4\frac{1}{3}$ weeks	= 1 month
24 hours	= 1 day	30 days	= 1 month
7 days	= 1 week	52 weeks	= 1 year
365 days	= 1 year	13 weeks	= 1 quarter (year)
366 days	= 1 leap year	12 months	= 1 year

Linear Measure
Objective 2

Linear measure considers only one dimension: length.

linear measure

|←———————————————— Length ——————————————→|

To perform the basic fundamentals on measurement information you must begin in the same manner as you would any other whole-number problem. The first step is to set up the problem; then perform the necessary function.

Example 1

4 yd, 1 ft, 5 in. × 2 = 4 yd, 1 ft, 5 in.
$$\underline{\times 2}$$
8 yd, 2 ft, 10 in.

(4 × 2 =   1 × 2 =   5 × 2 =)

Example 2

4 hr, 37 min, 14 sec + 1 hr, 40 sec =   4 hr, 37 min, 14 sec
$$\underline{+1 \text{ hr} \qquad\quad 40 \text{ sec}}$$
5 hr, 37 min, 54 sec

○ *How much time in a day remains after subtracting your time in class and studying?*

(4 + 1 =   14 + 40 =)

The examples so far have not required borrowing or reducing in order to complete the problem. In the following examples, you must either borrow or reduce in order to find the correct answer.

When reducing measurement problems, always begin with the smallest measurement (in size) and work up to the largest.

Example 3

17 yd, 2 ft, 9 in. × 3 =   17 yd, 2 ft, 9 in.
$$\underline{\times 3}$$
51 yd, 6 ft, 27 in.

This answer must be reduced to lowest terms.

2 ft, 3 in.   51 yd, 6 ft
12)27 in. $$\underline{+2 \text{ ft, 3 in.}}$$
                      51 yd, 8 ft, 3 in.

2 yd, 2 ft   51 yd   3 in.
3)8 ft $$\underline{+2 \text{ yd, 2 ft}}$$
                      53 yd, 2 ft, 3 in.

(17 × 3 =   2 × 3 =   9 × 3 =)

Example 4

7 yd, 1 ft, 3 in. − 5 yd, 2 ft, 10 in.

○ *How far is it from your desk to the chalkboard? How wide is your desk?*

In this example, ten inches cannot be subtracted from three inches; so, a foot's worth of inches must be borrowed and added to the three inches. The sum, fifteen inches, is enough to allow the subtraction. The next step is to subtract the feet. A yard's worth of feet must be borrowed and added to the zero feet. The sum, three feet, allows the subtraction of two feet. The last step is to subtract the yards.

7 yd, 1 ft, 3 in.   7 yd, 0 ft, 15 in.   6 yd, 3 ft, 15 in.
$$\underline{-5 \text{ yd, 2 ft, 10 in.}}$$ $$\underline{-5 \text{ yd, 2 ft, 10 in.}}$$ $$\underline{-5 \text{ yd, 2 ft, 10 in.}}$$
                                                    1 yd, 1 ft, 5 in.

(6 − 5 =   3 − 2 =   15 − 10 =)

Exercise A

Solve the following problems. Remember to reduce your answers.

1. 13 yd, 2 ft, 7 in. × 4 = 52 yd, 8 ft, 28 in.
   = <u>55</u> yd, <u>1</u> ft, <u>4</u> in.

2. 8 lb, 6 oz ÷ 6 = _____ lb _____ oz

3. 1 hr, 6 min, 10 sec ÷ 2 = _____ min, _____ sec

4. 60 lb, 11 oz × 7 = _____ lb, _____ oz

5. 5 hr, 17 min − 3 hr, 50 min, 13 sec = _____ hr, _____ min, _____ sec

Surface Measure

surface measure

Surface measure (or square measure) considers two dimensions: length and width. Answers are labeled as square feet, square yards, etc. Surface measure is the result of finding the product of the length and width (i.e., L × W).

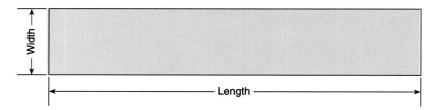

Example 5

The area Sport-A-Merica currently has includes a merchandise processing room that measures 19 ft long and 11 ft wide. How many square feet are there in the room?

```
  19 ft
 ×11 ft
   19
   19
 209 sq. ft
```

(19 × 11 =)

Example 6

Surface measure problems also need to be reduced. In the previous example, the answer should be reduced to square yards and square feet. The measurements table indicates there are nine square feet in a square yard.

○ *Can you estimate the size of a room by the number of floor or ceiling tiles?*

```
      23 sq. yd, 2 sq. ft
  9)209 sq. ft
     18
     29
     27
      2
```

(209 ÷ 9 =)

Exercise B

Complete the problems below and reduce your answers.

1. 4 sq. yd, 2 sq. ft + 22 sq. yd,
   8 sq. ft, 7 sq. in. = 26 sq. yd, 10 sq. ft,
   7 sq. in. = _27_ sq. yd, _1_ sq. ft., _7_ sq. in.

2. 71 sq. yd ÷ 3 = _____ sq. yd, _____ sq. ft

3. $9\frac{1}{2}$ ft × $6\frac{1}{2}$ ft = _____ sq. yd, _____ sq. ft

4. 3.2 mi × 6.8 mi = _____ sq. mi

5. 5 sq. yd − 3 sq. yd, 2 sq. ft = _____ sq. yd, _____ sq. ft

---

Volume Measure

volume measure

Volume measure (or cubic measure) considers the three dimensions of length, width, and depth. Answers are labeled as cubic feet, cubic yards, etc. Cubic measure is the result of finding the product of the length, width, and depth of an object. In some problems, height may be substituted for the depth factor (i.e., L × W × D).

Example 7

Hal needs to estimate the cost of the heating/cooling bill for the new space Sport-A-America is considering. How many cubic feet of air are there in a one-story building measuring 80 feet long, 32 feet wide, with a ceiling 8 feet high?

L	×	W	×	D	=
80 ft	×	32 ft	×	8 ft	= 20,480 cu. ft

Hal has budgeted $7,000 for heating/cooling the new space for one year. The utility company estimates the annual cost of heating/cooling this type of building at $.32 per cubic foot. What is the estimated cost of heating/cooling for the year?

20,480 cubic ft × $.32 = $6,553.60

Volume measure problems, as well as surface measure problems, may need to be reduced; in the previous problem, the answer could be reduced to cubic yards and cubic feet. After consulting the measurements table, you will find that there are 27 cubic feet in a cubic yard.

$27)\overline{20,480}$ = 758 cu. yd 14 cu. ft

(80 × 32 × 8 =   20,480 × .32 =)

Exercise C

Complete the problems below and reduce your answers.

1. 6 yd × 3 ft × 2.5 ft = 135 cu. ft = __5__ cu. yd
2. 8 ft × 10 ft × 3 ft = _____ cu. yd, _____ cu. ft
3. 42 sq. ft by 7 ft high = _____ cu. yd, _____ cu. ft
4. 16 cu. yd, 19 cu. ft + 3 cu. yd, 12 cu. ft = _____ cu. yd, _____ cu. ft
5. A pool is 40 feet long, 20 feet wide, and averages 5 feet deep. Find the total volume to the nearest cubic foot.
   _____ cu. ft

The Metric System
Objective 3

Most packaged goods sold in the United States are currently labeled in our traditional measurement system, as well as in the metric system. All other industrialized nations use the metric system. Our government has made a commitment to integrate the metric system into our business activities. Since exports and imports are ever greater factors in our economy, it is imperative that we fully learn this international language of measure.

First we will look at the metric system and see how it is structured. Then we will discuss how to convert traditional measurements to the metric system. Finally, we include some practice problems in measurement conversion.

The liter, the meter, and the gram are the basic units of measure. For a rough comparison:

1 liter = approximately 1 quart

1 meter = approximately 1 yard

1 kilogram = approximately 2 pounds

The Metric Measurement System

Linear Measure

10 millimeters (mm)	= 1 centimeter (cm)
10 centimeters	= 1 decimeter (dm)
10 decimeters	= 1 meter (m)
10 meters	= 1 decameter (dcm)
10 decameters	= 1 hectometer (hm)
10 hectometers	= 1 kilometer (km)
10 kilometers	= 1 myriameter (mym)

Square Measure

100 square millimeters (sq. mm)	= 1 square centimeter
100 square centimeters (sq. cm)	= 1 square decimeter
100 square decimeters (sq. dm)	= 1 square meter
100 square meters (sq. m)	= 1 square decameter
100 square decameters (sq. dcm)	= 1 square hectometer
100 square hectometers (sq. hm)	= 1 square kilometer
100 square kilometers (sq. km)	= 1 square myriameter (sq. mym)

Cubic Measure

1,000 cubic millimeters (cu. mm)	= 1 cubic centimeter
1,000 cubic centimeters (cu. cm, cc)	= 1 cubic decimeter
1,000 cubic decimeters (cu. dm)	= 1 cubic meter
1,000 cubic meters (cu. m)	= 1 cubic decameter
1,000 cubic decameters (cu. dcm)	= 1 cubic hectometer
1,000 cubic hectometers (cu. hm)	= 1 cubic kilometer
1,000 cubic kilometers (cu. km)	= 1 cubic myriameter (cu. mym)

Liquid and Dry Measure

10 milliliters (ml)	= 1 centiliter (cl)
10 centiliters	= 1 deciliter (dl)
10 deciliters	= 1 liter (L)
10 liters	= 1 decaliter (dcl)
10 decaliters	= 1 hectoliter (hl)
10 hectoliters	= 1 kiloliter (kl)
10 kiloliters	= 1 myrialiter (myl)

Weight Measure

10 milligrams (mg)	= 1 centigram (cg)
10 centigrams	= 1 decigram (dg)
10 decigrams	= 1 gram (g)
10 grams	= 1 decagram (dcg)
10 decagrams	= 1 hectogram (hg)
10 hectograms	= 1 kilogram (kg)
10 kilograms	= 1 myriagram (myg)
10 myriagrams	= 1 quintal (q)
10 quintals	= 1 tonneau (T)

Notice that in all the metric units of measure, the prefix is the same (i.e., milli, centi, deci, etc.).

kilo-	hecto-	deca-	basic unit	deci-	centi-	milli-
(larger)			of measure			(smaller)

Factors

Notice that all the metric system measurements have a base of 10 (i.e., by dividing 1 meter by 10, the result is 1 decimeter, 10 hectometers equals 1 kilometer, and so forth). Once you become familiar with the prefixes of this system, it is much easier than the 12 inches to a foot, 3 feet to a yard, 1,760 yards to a mile, etc., that we have been using.

The square measure unit is found by the same method as it is in the traditional measurement system (i.e., length times (×) width). If 1 meter is equal to 10 decimeters, then 1 square meter must be equal to 100 square decimeters ($10 \times 10 = 100$).

The same holds true for volume or cubic measure. Length × width × depth is the method used to determine cubic measure. This is the reason that 1 cubic meter is equal to 1,000 cubic decimeters ($10 \times 10 \times 10 = 1,000$).

The previous explanations should give you a basic understanding of the manner in which the measurement units are found.

The base of 10 allows for easy conversion from meters to, say, decameters; 53.6 meters equals 5.36 decameters, since there are 10 meters to a decameter ($10)\overline{53.6} = 5.36$).

7.6 square kilometers = 760 square hectometers (100 square hectometers equals one square kilometer (i.e., $100 \times 7.6 = 760$)).

53,065.2 cubic millimeters = 53.0652 cubic centimeters ($53,065.2 \div 1,000 = 53.0652$).

Exercise D

Convert the following measurements to the base indicated.
1. 963.2 centimeters to meters _____ 9.632 _____
2. 856 square hectometers to square kilometers _____
3. 806,329 cubic centimeters to cubic meters _____
4. .76 liters to centiliters _____
5. 47,357 grams to decigrams _____

Objective 4

It may be necessary for you to convert from the traditional measurement system to the metric system, or vice versa. The conversion table will aid in that process.

Example 8

A hole 47 ft. long, 18 in. wide, and 42 in. deep equals _____ cubic meters.

47 ft. × 1.5 ft. × 3.5 ft. = 246.75 cu. ft.

246.75 cu. ft.
×.02832 (1 cu. ft. = .02832 cu. m) (from conversion table)
6.98796 cu. m

(47 × 1.5 × 3.5 =   246.75 × 0.2832 =)

Example 9 | 437 kilometers equals _____ miles.

437 kilometers
×.6214 (1 km = .6214 mi) (from table)
271.5518 miles

(437 × .6214 =)

Conversion Table

Linear Measure

1 in. = 2.54 cm	1 mm = .03937 in.
1 ft = .3048 m	1 cm = .3937 in.
1 yd = .9144 m	1 dm = .3281 ft
1 rd = 5.029 m	1 m = 39.37 in.
1 mi = 1.6093 km	1 m = 3.281 ft
	1 m = 1.0936 yd
	1 dcm = 1.9884 rd
	1 km = .6214 mi

Square Measure

1 sq. in. = 6.452 sq. cm	1 sq. mm = .00155 sq. in.
1 sq. ft = .0929 sq. m	1 sq. dm = .1076 sq. ft
1 sq. yd = .8361 sq. m	1 sq. km = .3861 sq. mi
1 sq. mi = 259 ha or 2.589 sq. km	1 sq. cm = .155 sq. in.
1 sq. rd = 25.293 sq. m	1 sq. m = 1.196 sq. yd
	1 ha = 2.471 acre

Cubic or Volume Measure

1 cu. in. = 16.3872 cu. cm (cc)	1 cu. cm (cc) = .06102 cu. in.
1 cu. ft = 28.317 cu. dm or .02832 cu. m	1 cu. dm = .0353 cu. ft
1 cu. yd = .7646 cu. m	1 cu. m = 1.308 cu. yd
1 cd = 3.624 steres (st)	1 st = .2759 cd

Liquid and Dry Measure

1 dry qt = 1.101 L	1 L = .908 dry qt
1 liquid qt = .9463 L	1 L = 1.0567 liquid qt
1 liquid gal = .3785 dcl or 3.785 L	1 dcl = 2.6417 liquid gal
1 pk = .881 dcl or 8.81 L	1 dcl = 1.135 pk
1 bu = .3524 hl	1 hl = 2.8377 bu

Weight Measure

1 gr troy = .0648 g	1 g = 15.432 gr troy
1 oz troy = 31.104 g	1 g = .03215 oz troy
1 oz avoir. = 28.35 g	1 g = .03527 oz avoir.
1 lb troy = .3732 kg	1 kg = 2.679 lb troy
1 lb avoir. = .4536 kg	1 kg = 2.2046 lb avoir.
1 T (short) = .9072 met. t	1 met. t = 1.1023 T (short)

Exercise E

Fill in the missing values. *Note:* Your answers may vary from the answers in the back of the book because of your choice of the rounded conversion factors from the table.

1. 4 yd, 7 in. equals __3.8354__ m (meters). (4 × .9144 =   7 × 2.54 =)
2. 4 yd, 2 ft, 6 in. equals _____ m (meters).
3. 36.1 m equals _____ yd (yards).
4. 16 liquid qt equals _____ L (liters).
5. 91 sq. ft equals _____ sq. m (square meters).
6. 7 sq. mi equals _____ sq. km (square kilometers).
7. 2.6 sq. m equals _____ sq. yd (square yards).
8. 4 cu. ft, 18 cu. in. equals _____ cu. m (cubic meters).
9. 3 lb, 14 oz avoir. equals _____ kg (kilograms).
10. 7.6 metric tons equals _____ tons.

Exchange Rates
Objective 5

The United States is one of the largest economic markets in the world in terms of its population and its spending power. Recent increases in international trade have made it very important for us to provide product information in the metric system. The exchange of currency has also become an important aspect of trade to understand as international transactions increase. Economies of the various nations of the world are constantly changing, and thus the demand for their currencies is also changing. The rate of exchange

EXCHANGE RATES		
Country	U.S. $ equiv.	Currency per U.S. $
Argentina (Peso)	1.0005	.9995
Australia (Dollar)	.6524	1.5328
Austria (Schilling)	.07573	13.204
Bahrain (Dinar)	2.6525	.3770
Belgium (Franc).	.0258	38.7100
Brazil (Real)	.5811	1.7210
Britain (Pound)	1.6013	.6245
Canada (Dollar)	.6784	1.4740
Chile (Peso)	.002025	493.85
China (Renminbi)	.1208	8.2780
Columbia (Peso)	.0005934	1685.33
Czech. Rep. (Koruna).		
Commercial rate.	.02768	36.130
Denmark (Krone)	.1401	7.1390
Ecuador (Sucre)		
Floating rate	.00010753	9300.00
Finland (Markka)	.1753	5.7055
France (Franc)	.1589	6.2946
Germany (Mark)	.5328	1.8768
Greece (Drachma)	.003209	311.59
Hong Kong (Dollar)	.1290	7.7549
Hungary (Forint)	.004185	238.94
India (Rupee)	.02330	42.915
Indonesia (Rupiah)	.0001231	8125.00
Ireland (Punt)	1.3233	.7557
Israel (Shekel).	.2427	4.1195
Italy (Lira)	.0005382	1858.05
Japan (Yen)	.008229	121.52
Jordan (Dinar).	1.4104	.7090
Kuwait (Dinar)	3.2616	.3066
Lebanon (Pound)	.0006631	1508.00
Malaysia (Ringgit)	.2632	3.8000
Malta (Lira)	2.4777	.4036
Mexico (Peso)		
Floating rate	.1030	9.7100
Netherland (Guilder)	.4729	2.1147
New Zealand (Dollar)	.5356	1.8671
Norway (Krone)	.1261	7.9273
Pakistan (Rupee)	.01935	51.670
Peru (new Sol)	.3007	3.3253
Philippines (Peso)	.02630	38.020
Poland (Zloty)	.2504	3.9942
Portugal (Escudo)	.005198	192.38
Russia (Ruble) (a)	.04092	24.440
Saudi Arabia (Riyal)	.2666	3.7505
Singapore (Dollar)	.5796	1.7252
Slovak Rep. (Koruna).	.02291	43.643
South Africa (Rand)	.1600	6.2510
South Korea (Won)	.0008432	1186.00
Spain (Peseta)	.006263	159.66
Sweden (Krona)	.1161	8.6118
Switzerland (Franc)	.6547	1.5275
Taiwan (Dollar)	.03054	32.743
Thailand (Baht).	.02696	37.090
Turkey (Lira)	.00000248	403522.00
United Arab (Dirham)	.2723	3.6728
Uruguay (New Peso)		
Financial	.08938	11.118
Venezuela (Bolivar)	.001670	598.75

of one currency for another currency is of daily concern to business firms. The rate of currency exchange is printed in the business section of most newspapers and business periodicals daily.

The member nations of the European Union (E.U.) have agreed to use a single currency, the Euro. Beginning in the year 2001, the Euro will be used for common exchange. Currently, all economic data, major transactions, and economic forecasts use Euros as the basis of valuation. This is an important issue as the E.U. is the largest economic unit in the world in terms of population and a very sizable force in economic buying power. The Euro, like other currencies, changes value. The Euro is valued at approximately one U.S. dollar.

The table of exchange rates shows the value of the monetary units of other countries compared to that of the U.S. dollar. Notice that some currencies are worth more than the U.S. dollar (Kuwait's, for example) and others are worth less (Japan's).

Example 10

Mota International, Inc., sold 400 units of their product to a firm in Kuwait at a price of $58 per unit. How many Kuwait dinars would it take to pay the invoice?

400 × $58 = $23,200 total invoice price in U.S. dollars
$23,200 × .3066 = 7,113.12 Kuwait dinars

(23,200 × .3066 =)

Example 11

If the same order were sold to a customer in Japan, how many yen would it take to pay the invoice?
$23,200 × 121.52 = 2,819,264 Japanese yen

(23,000 × 121.52 =)

Exercise F

Determine the amount of foreign currency needed to make the following purchases.

Customer Country	Invoice Amount (U.S.)	Currency Needed
1. Australia	$4,600 (4,600 × 1.5328 =)	7,050.88 dollars (Australian)
2. Norway	15,900	_____ krone
3. Chile	7,000	_____ pesos (Chile)
4. Canada	230,000	_____ dollars (Canadian)
5. Mexico	450	_____ pesos (Mexico)

The currency exchange rate is equally important for U.S. firms that buy from producers in other countries. The U.S. firm will need to make a similar type of calculation to determine the amount of U.S. dollars needed to pay the invoice.

Example 12

Hal is quoted a price of 835,000 lira for each of the 20 bicycles that Sport-A-Merica will purchase from a firm in Italy. What is the amount of the invoice in U.S. dollars?

20 × 835,000 lira = 16,700,000 lira
16,700,000 lira × .0005382 = 8,987.94 U.S. dollars

(16,700,000 × .0005382 =)

Exercise G

Determine the number of U.S. dollars necessary to make the following purchases.

Purchase from	Invoice Amount	U.S. Dollars Needed
1. South Africa	600 rands (600 × .16 =)	$96.00
2. India	7,000 rupees	_____
3. Finland	500 markkas	_____
4. South Korea	9,200,000 wons	_____
5. Philippines	50,000 pesos	_____

The exchange tables are printed daily because the exchange rates may vary from one day to the next. When a firm agrees to a purchase, they must think about the possibility of exchange rates changing between the date of the purchase and the date that the invoice is to be paid.

Example 13   Wiedmont Associates had agreed to purchase an item for 161,000 Peruvian sols. The exchange rate on the date of purchase was .3007. The exchange rate on the date that the invoice was due is .3019. How much more in U.S. dollars did it take to pay the invoice when it was due than it would have taken when the purchase was made?

```
  .3019
 −.3007
  .0012    .0012 × 161,000 sols = $193.20
```

(.3019 − .3007 =   .0012 × 161,000 =)

ASSIGNMENT Chapter 6

Name _____ **Date** _____ **Score** _____

Skill Problems

Solve the following problems. Remember to reduce your answers.

1. 4 yd, 2 ft, 10 in. − 1 yd, 11 in. =

2. 14 lb, 8 oz ÷ 4 =

3. 4 hr, 30 min, 25 sec ÷ 5 =

4. 20 lb, 8 oz × 3 =

5. 3 hr, 23 min − 1 hr, 40 min, 24 sec

6. 8 sq. yd, 2 sq. ft + 15 sq. yd, 7 sq. ft, 4 sq. in. =

7. 46 sq. yd ÷ 3 =

8. 5 1/2 ft × 8 1/2 ft =

9. 4.8 mi × 8.3 mi =

10. 7 hr, 40 min, 30 sec ÷ 6 =

11. 18 yd × 2 ft × 1.5 ft =

12. 9 ft × 11 ft × 5 ft =

13. 14 cu. yd, 21 cu. ft + 4 cu. yd, 12 cu. ft =

14. 50 ft long, 12 ft wide, and 6 in. deep equals how many cubic feet (nearest cubic foot)?

15. 3 yd, 2 ft, 8 in. × 2 = _____ yd, _____ ft., _____ in.

16. 36 lb, 8 oz ÷ 4 = _____ lb, _____ oz

17. 7 min, 20 sec × 83 = _____ hr, _____ min, _____ sec

18. 2 sq. yd, 7 sq. ft × 7 = _____ sq. yd, _____ sq. ft

19. 5 yd, 2 ft × 4 ft × 7 yd, 1 ft = _____ cu. yd, _____ cu. ft

20. 2.3 sq. ft × 5 ft = _____ cu. ft

Convert the following measurements to the base indicated

21. 4,362 dcm = _____ meters.

22. 18 L equals _____ cl

32. 9 cu. ft, 16 cu. in. = _____ cu. m (cubic meters).

23. 786.5 cm to meters:

33. 47 m = _____ yd (yards).

24. 642 cubic hectometers to cubic decameters:

34. 8.4 sq. m = _____ sq. yd (square yards).

25. 807,604 cu. cm to cubic meters:

35. 47 liquid qt = _____ L (liters).

26. .46 L to centileters:

36. 42 sq. mi = _____ sq. km (square kilometers).

27. 37,480 g to decigrams:

Fill in the missing values.

28. 870 tons = _____ metric tons.

37. 95 sq. ft = _____ sq. m (square meters).

29. 8 yd, 9 in. = _____ m (meters).

38. 781 cm = _____ in. (inches).

30. 7 lb, 12 oz avoir. = _____ kg (kilograms).

31. 5 yd, 2 ft = _____ meters.

Customer Country	Invoice Amount (in customers' currency)	Currency Needed
39. United States	$5,971	_____ marks (Germany)
40. Canada	10,943	_____ U.S. dollars
41. Greece	98,740	_____ U.S. dollars
42. Kuwait	345,158	_____ U.S. dollars
43. Taiwan	324,482	_____ U.S. dollars
44. South Africa	654,871	_____ U.S. dollars
45. United States	421	_____ dollars (Singapore)
46. Mexico	25,480	_____ U.S. dollars
47. Brazil	7,804	_____ U.S. dollars
48. Israel	16,300	_____ U.S. dollars
49. United States	5,000	_____ yen (Japan)
50. Germany	6,850	_____ U.S. dollars

Business Application Problems

1. A piece of land sold as a building site for $27,000. The piece of land was 1.89 acres. What was the cost per acre?

2. The owner of the land in problem 1 will build a single-story house with 3,500 square feet and a shop with 2,000 square feet on the site. How many square feet of land will not be covered by buildings?

3. The owner of the land in problem 1 will spray the land not covered with buildings with an insecticide. The spray costs $13 per liter. Each liter will cover 5,000 square feet. What is the cost of the spray material? Only full liters are available for purchase.

4. A contract valued at 42,000,000 Indian rupees has an expected delivery of one year. At the end of the completed contract, the Indian rupee had dropped in value by .00011. What is the amount of the savings to the buyer in dollars, if the old rate was .02386 Indian rupees to one U.S. dollar?

5. An Idaho firm contracted to sell goods to a firm in Argentina. The contract is valued at 50,000 Argentine pesos. What is the value of the deal in U.S. dollars?

6. Kariva will paint the four walls of her store, which measures 28 feet by 40 feet. The walls are 8 feet high. The paint she is considering will cover 500 square feet per gallon. Each gallon costs $19.95. Ignoring the area of windows and doors, what will be the cost of the paint? (*Note:* She must purchase full gallons.)

7. Taxson, Inc. purchased electronic parts from a firm in Japan last year. The price of each piece was 81.4 yen. Taxson purchased 48,000 pieces at intervals of 4,000 each month over the year. The contract called for payment at the end of each month. The average exchange rate of the yen was .0094 U.S. dollars. What was the total cost of the purchase in U.S. dollars?

8. A company will purchase 5,000 pounds of a chemical material known as "Simcon-2" from one of two suppliers. The American supplier's price is $3.95 per pound. The Venezuelan supplier's price is 4,840 bolivars per kilogram (avoir). Both prices include delivery. What is the lowest total price for the 5,000 pounds?

9. Kem-Kleene, Inc. purchases soap in a 55-gallon container. The soap is packaged into 6-ounce packages for use in dispensers for their commercial customers. How many 6-ounce packages can be developed from one 55-gallon container?

10. If Kem-Kleene, Inc. in problem 9 were to enter the European market, the packages would be .5 liter in size and would sell for $1.95 each. What is the value of a 55-gallon container of soap packaged for the European market?

11. Carol Huntington was assigned a new task by her supervisor. The task is expected to take her 1 hour and 53 minutes. She must complete the task two times a day. How much time will be left out of an 8-hour day for other activities?

12. Moore and Associates imported raw materials from Thailand at a quoted price of 57,000 baht. How many U.S. dollars will be required to pay the invoice?

13. The shipping department must deliver 500 boxes that are $1\frac{1}{2}$ cubic feet in size. The boxes will be packed into a container that is able to accommodate 75 cubic feet. How many containers will be necessary to ship the 500 boxes?

14. Bayburn & Jackson purchased machinery priced at 500,000 South Korean won. When the invoice was due, the exchange rate had changed from .000841 to .000832. How much money did Bayburn & Jackson make or lose by delaying payment?

15. The Shaw family is considering a new pool for their summer home. They contact the Cool Pool Company, a local distributor. The pool company representative stated that the first job would be to remove the sod from an area 30 ft × 60 ft. Next, a hole would be dug 25 ft × 55 ft × an average of 6 ft deep. The pool would then be built in the hole. The finished pool would measure 20 ft × 50 ft × 5 ft deep (average). How many square yards of sod must be removed from the lawn before the digging can begin?

16. How much would it cost the Shaw family for earth removal at $46 a cubic yard (nearest cubic yard)?

17. The Shaws are concerned about the amount of time it would take to fill the pool if their well will only pump 225 gallons per hour. How many hours will it take to fill the pool to within one foot of the top?

18. George and Delores will mix one cubic meter of chocolate, 400 cubic decimeters of nuts, and 700 cubic decimeters of coconut together in their candy shop. From this mixture they will make Mother's Day hearts that will be 50 cubic centimeters in size. How many hearts will they be able to make?

19. Write a complete sentence in answer to this question. The Dive Shop will have a new sign on the top of their building measuring 10 feet high and 27 feet long. The material that will be used costs $41 per square meter. What will be the cost of the material?

20. Write a complete sentence in answer to this question. The mayor's office must be recarpeted. The room is 30 ft wide and 32 ft long. An 8-ft by 4-ft area near the desk in the corner of the room is to be left uncarpeted. The carpet store has suggested a nylon beige tweed at $37 a square yard. The carpet can only be sold in full square yards. How much will it cost to carpet the portion of the room as described?

★ **Challenge Problem**

The J. B. Colson Company will build a new building in July. In order to determine the potential savings in utilities expense gained by installing insulation, an estimation of costs and savings must be developed. The building will be 75 ft wide by 260 ft long. The side walls will be 10 ft high in the single-story structure. All of the ceilings will be insulated to a depth of 7 in., except that of the Shipping and Receiving Department, which has an area of 40 ft × 65 ft. The local utility company has stated there will be a $60 per month savings on heating and cooling expenses if the building is insulated to the 7-in. depth in the areas described. The insulating material will cost $2.78 for each cubic-meter bag. How many months of fuel savings will it take to repay the cost of the insulation?

1. The Harry Cornell Co. has just received a contract to construct a bicycle path across a part of the county. The company had three groups of men to work on the project, which was 4 miles, 895 yards long. The company has decided to divide the distance equally among the three groups. To the nearest foot, how long of a distance will each group be responsible?

2. If Clara's job operation requires 37 seconds to complete, how many complete operations can she do in a period of 5 hours?

3. Bell Korten is about to remodel his insurance office. The room is 30 ft × 42 ft with 8-ft side walls. He will paint the walls and ceiling all one color. The paint that he picked out will cover 500 sq ft per gallon. How many gallons must be purchased to do the job if he does not allow for doors, windows, etc.? (Partially used gallons can not be returned.)

4. A building must have an addition that measures a total of 400 feet (for the three sides). The purchasing agent must determine the amount of cement to be used to fill a footing trench $2\frac{1}{2}$ ft deep by 18 in. wide. Determine, to the nearest cubic yard, how many cubic yards of cement will be needed to fill the trench.

5. 53 sq. m ÷ 17 equals _____ sq. ft.

6. A dump truck hauls 7.6 cubic yards of gravel per load. How many cubic meters would the truck haul?

7. Ninety pounds of candy is to be sold at $.50 for each 70-gram box. How many boxes can be made from the 90 pounds?
(Use avoir.)

8. 8 yd, 2 ft, 9 in. − 5 yd, 1 ft, 11 in. = _____?

9. Convert 143 square decameters to squre meters:
_____.

10. What is the U.S. dollar value of an invoice from China for 42,300 renminbi?

11. The owner of a store contacted on air conditioning firm for an estimate. The owner indicated that the dimensions of the store were 60 ft by 120 ft, with 12-ft-high side walls. The air conditioning firm stated that their Model X-1,000 could handle 75,000 cubic feet. Would the Model X-1,000 be large enough for the store?

12. Bolton, Inc., sold a machine priced at $3,900 to a firm in Ecuador. What amount of sucre will Bolton, Inc., expect in payment for the sale?

PERCENT

OBJECTIVES

After mastering the material in this chapter, you will be able to:

1. Define what *percent* means. p. 128
2. Change percent to a fraction. p. 128
   percent to a decimal. p. 128
   decimal to a fraction. p. 128
   decimal to a percent. p. 128
3. Write all the above with fractional percent. p. 129
4. Identify the base, rate, and amount in word problems. p. 130
5. State the formulas to be used to find the base, the rate, and the amount. p. 130
6. Solve for the unknown using one of three methods. p. 131
7. Work word problems that apply percent to business situations. p. 134
8. Understand and use the following terms and concepts:
   Percent
   Base
   Rate
   Amount

SPORT-A-MERICA

 Kerry and Hal have found two possible new store locations for Sport-A-Merica. One store, on Washington Avenue, is a perfect location, but it is a little smaller than Hal had wanted. The space is only 3,000 square feet. Current store space is 1,000 square feet. The second location has 3,700 square feet and is in a good area of town, on Cedar Street, a busy highway area. Both stores have the space divided into several rooms, which would provide more display space on the walls.

Hal realizes that now is the time to take some serious measurements. Will the new space allow for more merchandise? Should the store add more lines of merchandise? The store currently handles 890 items. Hal wants to add 47 new items. What percent increase will that be? Should they advertise the percent increase? Will there be room to park the new delivery vehicle they purchased last month? Utilities will be a big consideration. Hal will need to investigate the cost of electricity for a space that is 200 to 270 percent larger than the present store. Should the cost of utilities be divided based on the area of floor space or based on the type of merchandise in a department? The additional utilities will add to operating costs. This would mean that the price of the merchandise would have to be increased, unless Sport-A-Merica can increase sales by at least as much as the increase in costs.

Making the move to a larger store is clearly necessary, as Hal sees it. There is a very good indication that more business would be available if they could expand. To make the right decision, Hal must compare the current space with the new possibilities and with new product opportunities. Hal and Kerry know that the amount of space, costs, and profits will not stay the same.

What *Percent* Means
Objective 1

Percent is a way of expressing the relationship between two numbers. It is shown in hundredths (i.e., 1 percent is the same as $\frac{1}{100}$). In business applications, percent is used to compute payroll deductions, cash and trade discounts, depreciation, interest, returns, markup, and taxes—to name just a few. The importance of percent cannot be emphasized too much.

percent

The sign % signifies percent and, therefore, hundredths. You can also express hundredths as a fraction by showing a denominator of 100 (i.e., $\frac{?}{100}$). In chapter 3, we saw that the second position to the right of the decimal point is the hundredths position. One hundredth is written .01, five hundredths is .05, and forty-three hundredths is .43, and so on.

Percent to a Fraction
Objective 2

To change a number expressed as a percent to a fraction with a denominator of 100: (1) drop the percent sign, and (2) add a denominator of 100.

Example 1

1. $47\% = \dfrac{?}{100}$

2. $47\% = \dfrac{47}{100}$

Percent to a Decimal

In chapter 3, we found that any fraction can be changed into a decimal number by dividing the numerator by the denominator. In the previous example, 47 percent was changed into a fraction of $\frac{47}{100}$. This fraction can be changed into a decimal number by dividing 47 by 100, which gives an answer of .47 (i.e., forty-seven hundredths).

Example 2

$7\% = .07$
$135\% = 1.35$

(7 ÷ 100 =   135 ÷ 100 =)

Decimal to a Fraction

If a fraction can be changed to a decimal number, we should be able to reverse the process and change a decimal to a fraction: (1) drop the decimal point, and (2) put the number over 100.

Example 3

1. $.76 = \dfrac{?}{100}$

2. .76 is read as seventy-six hundredths, which can be expressed as $\dfrac{76}{100}$

Example 4

$.08 = \dfrac{8}{100}$ $1.53 = \dfrac{153}{100}$

(.08 × 100 =   1.53 × 100 =)

Decimal to a Percent

A decimal can similarly be changed to a percent. Again, recalling that *percent* is hundredths, .67 can be changed to 67 percent; .67 is read "sixty-seven hundredths" or "sixty-seven percent." To complete the process: (1) drop the decimal point, and (2) add the percent sign.

Example 5

$.06 = 6\%$
$1.15 = 115\%$

(.06 × 100 =   1.15 × 100 =)

You should now be able to interchange percents, fractions, and decimals in any order. Use the previous page of this chapter as a guide to do the first few problems of exercises A and B.

Exercise A

Express each of the following as a decimal.

1. $17\% = \dfrac{17}{100} = \underline{\quad .17 \quad}$ 2. $56\% = \dfrac{56}{100} = \underline{\qquad}$ 3. $11\% = \dfrac{11}{100} = \underline{\qquad}$

4. $9\% = \dfrac{9}{100} = \underline{\qquad}$ 5. $156\% = \dfrac{156}{100} = \underline{\qquad}$

Exercise B

Fill in the blanks in the following problems.

1. $\underline{\quad 41 \quad}\% = \dfrac{41}{100} = \underline{\quad .41 \quad}$ 2. $\underline{\qquad}\% = \dfrac{}{100} = .62$ 3. $\underline{\qquad}\% = \dfrac{}{100} = 1.27$

4. $\underline{\qquad}\% = \dfrac{}{100} = .1$ 5. $21\% = \dfrac{}{100} = \underline{\qquad}$

Fractional Percent
Objective 3

○ *A fractional percent is very small. It is less than $\frac{1}{100}$.*

Quite often, a percent is expressed as a mixed number, such as $8\frac{1}{2}$ percent or $20\frac{1}{4}$ percent. In order to use these percents in calculations, you must be able to change the mixed number, or *fractional percent,* to a decimal. The process requires these steps: (1) change the fraction to a decimal, (2) place the decimal behind the whole number, (3) drop the percent sign, and (4) divide by 100.

Example 6

$$9\frac{3}{4}\% = 9.75\% = \frac{9.75}{100} = .0975$$

$$\frac{1}{5}\% = .2\% = \frac{.2}{100} = .002$$

(9.75 ÷ 100 =   .2 ÷ 100 =)

When you read a fractional percent that does not include a whole number, such as $\frac{1}{4}$ percent, the percent can be read as "one-fourth percent" or, more clearly, as "one-fourth of one percent." One-fourth percent is realistically a fraction of 1 percent. It should not be confused with 25 percent.

Exercise C

Fill in the blanks in the following problems.

1. $8\frac{1}{5}\% = \dfrac{8.2}{100} = \underline{\quad .082 \quad}$ 2. $1\frac{3}{8}\% = \dfrac{}{100} = \underline{\quad}$ 3. $\underline{\qquad}\% = \dfrac{.6}{100} = \underline{\quad}$

4. $52.9\% = \dfrac{}{100} = \underline{\quad}$ 5. $\underline{\qquad}\% = \dfrac{}{100} = .295$

Percent Problems
Objective 4

In this portion of the chapter, you will learn how to put the percent skills you have learned to work for you. By using the percent skills and the relationships to other information, you will be able to solve business problems. Before we proceed, you must become very familiar with the descriptions of the following three terms. Understanding these descriptions will make your efforts in problem solving begin correctly.

base   The **base** is: always the whole.

always 100%.

always what something else is compared to.

the number we take some percent "of."

the number that follows the word "of" in a problem sentence.

defined as "the factor of which so many hundredths is taken."

not always the largest number.

The base would be the number of students in your class, for example. In a business situation, it might be the total number of employees.

rate   The **rate** is: a percent.

usually expressed in a percent form, but can also be expressed as a fraction or a decimal (i.e., 25%, or $\frac{1}{4}$, or .25).

defined as "a certain number of hundredths."

For example, the rate would be the percent of students in your class over 5 feet 8 inches tall. In the business situation noted previously, it could be the percent of females in the firm.

amount   The **amount** is: related to the base because of the size of the rate.

not always the smallest number (may be larger than the base!).

sometimes termed *part, portion,* or *percentage.*

defined as "the result of taking a certain number of hundredths of the base."

○ *Basic algebra substitution can also be used to solve for the unknown.*

The amount would be the *actual number* of students over 5 feet 8 inches tall. In the business situation noted earlier, it could be the 176 females in the company.

Take time now to develop an understanding of the three terms and their meanings.

There are three ways to begin solving a percent problem. You could memorize the formulas shown here, you could use the model as a memory aid, or you could learn to solve for the unknown. In order to provide you with as many options as possible, all three ways will now be explored.

The Formulas
Objective 5

In working percent problems, two of the factors will always be known or can be easily found. You must find the third unknown factor. The formulas are as follows:

Amount = Base × Rate   A = B × R

Base = $\dfrac{\text{Amount}}{\text{Rate}}$ B = A ÷ R

Rate = $\dfrac{\text{Amount}}{\text{Base}}$ R = A ÷ B

The Model

By putting your finger on the factor that you wish to find, the remaining uncovered portion of the model reveals the math operations that are used to find the unknown factor. The model is as follows:

Example 7

If you know Amount and Rate in a problem but need to find the Base, put your finger over B.

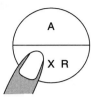

Base is found by dividing A by R.

Solving for the Unknown
Objective 6

We often make statements like, "You have $15 and I have $8. Together we have $23." This could be expressed in another way by saying, "How much money do we have together if you have $15 and I have $8?" This is an expression. We could alter the expression a little more by changing the words to "M is equal to the amount of money you have ($15) and I have ($8)." By making the statement in mathematical terms only, we could write M = $15 + $8. This is known as a *mathematical expression*. What it says is that the values (letters or numbers or a combination of both) on one side of the equals sign are equal to the values on the other side. It is as simple as that!

It is also possible to restate the same facts in other ways to make other expressions. For example, "Between the two of us we have $23. You have $15. How much do I have?" That expression could be written as $23 = $15 + I. We know that if we were to try to find what value I is, we would need to determine what must be added to $15 to make $23. The answer is $8. This could be solved in yet another way: by subtraction. You may do any fundamental operation (add, subtract, multiply, or divide) to an expression as long as you do the operation *on both sides of the expression*. Using the last expression, $23 = $15 + I, $15 can be subtracted from both sides of the expression to determine the answer.

$$
\begin{array}{ccc}
\$23 & = & \$15 + I \\
-\$15 & & -\$15 \\
\hline
\$8 & = & \$0 + I \\
\end{array}
$$
or I = $8

Using the same information, another expression can be developed: "The total amount of money we have between us, less $10, is equal to $13. What is the amount of money we have between us?" The mathematical expression can be written as T − $10 = $13. The solution can be found by addition.

$$
\begin{array}{ccc}
T - \$10 & = & \$13 \\
+ \$10 & & + \$10 \\
\hline
T + \$0 & = & \$23 \\
\end{array}
$$
or T = $23

A few ground rules should be covered at this point in order to create some consistency and understanding regarding this type of math. The letters that are used are termed *variables*. The variable is usually shown on the left side of the expression. The multiplication and division of variables and numbers can be expressed in several ways. These expressions mean the same thing.

4S, 4(S), 4 × S, and 4 · S

(Note that both the parentheses and the dot mean "multiply.")

$\frac{1}{5}$R, $\frac{1}{5}$ × R, $\frac{R}{5}$, and R ÷ 5

Expressions can include multiplication and division, as well as addition and subtraction. An expression can be written as 5A = 40. To find A, you need to divide both sides of the expression by 5.

$$\frac{5A}{5} = \frac{40}{5}$$ The answer is A = 8.

The multiplication function can be used in much the same fashion. $\frac{H}{2}$ = 90 can be solved by multiplying both sides of the expression by 2.

$$2 \times \frac{H}{2} = 90 \times 2$$ The answer is H = 180.

Exercise D

In the following expressions, solve for the variable.

1. J + 20 = 56
   $\frac{-\ 20\ -\ 20}{J\qquad =\ 36}$

2. E − 2 = 9
   $\frac{+\qquad +}{=}$

3. M − 3 = 56
   $\frac{+\qquad +}{=}$

4. $\dfrac{P \times 7}{} = \dfrac{42}{}$

5. $\dfrac{L}{4} = 80$

6. $\dfrac{8K}{} = \dfrac{10 + 6}{}$

7. $\dfrac{7(A)}{} = \dfrac{42}{}$

8. T · 4 + 10 = 50
   $\dfrac{+\qquad =}{\underline{\qquad -\qquad -\qquad}}$
   $=$

9. H + 5 = 30
   $\dfrac{-\qquad -}{=}$

10. $G \times \dfrac{1}{2} = 32$

When solving a problem with a variable, the first step is to put the problem into a statement that is an expression. This statement can, in turn, be made into a mathematical expression. You can then solve for the unknown in that expression.

Some additional hints in working toward an expression: The word "of" indicates that the multiplication function is to be used. As an example, one-fourth of the students is equal to 20. What is the total? The mathematical expression for the two previous sentences is $\frac{1}{4}S$ = 20. This statement says that $\frac{1}{4}$ of S is equal to 20. The word "of" prior to the word "students" indicates that $\frac{1}{4}$ should be multiplied by S (for students) in order to find the total.

In the percent problems that follow, there will always be two factors known or easily found, and one unknown. You now know how to solve for the unknown using an expression. Some example problems will help show how the percent problems are solved. Use the formulas, the model, or solving for the unknown to follow along in the solutions.

Example 8

Base = 400	A = B × R
Rate = 20%	A = 400 × .20
Amount = Unknown	A = 80

(400 × .20 =)

Example 9

$B = 60$ $A = B \times R$

$R = 140\%$ $A = 60 \times 1.40$

$A = $ Unknown $A = 84$

(60 × 1.40 =)

Example 10

$A = 918$ $B = \dfrac{A}{R}$

$R = 153\%$ $B = \dfrac{918}{1.53}$

$B = $ Unknown $B = 600$

(918 ÷ 1.53 =)

Example 11

The current electric bill is about $450 per month. The new building for Sport-A-Merica will cost $90 more per month for electricity. What is the percent of increase?

$B = \$450$ current bill $R = \dfrac{A}{B}$

$A = \$90$ increase $R = \dfrac{\$90}{\$450}$

$R = $ Unknown $R = 20\%$

(90 ÷ 450 =)

Four-Step Process

Notice that a four-step process can always be used:

1. Identify the factors (base, rate, and amount).
2. Select the formula for the unknown.
3. Substitute the values for the factors.
4. Complete the fundamental process called for in the problem.

In the following problems, find the unknown factor. Round off your answer to the nearest unit or nearest percent.

Exercise E

1. $B = 93$
   $R = 107\%$
   $A = \underline{\ 100\ }$ (93 × 1.07 =)

2. $B = 116$
   $R = \underline{\hspace{1cm}}$
   $A = 100$

3. $B = 700$
   $R = 3\dfrac{3}{4}\%$
   $A = \underline{\hspace{1cm}}$

4. $B = 492$
   $R = 14\%$
   $A = \underline{\hspace{1cm}}$

5. $B = 92$
   $R = \underline{\hspace{1cm}}$
   $A = 10$

6. $B = \underline{\hspace{1cm}}$
   $R = 36\%$
   $A = 180$

Exercise F

1. B = $17   2. B = _____   3. B = $860
   R = 42%   R = 30%   R = _____
   A = <u>$7.14</u> (17 × .42 =)   A = $54   A = $4,400

4. B = _____   5. B = $48   6. B = $47
   R = 250%   R = $14\frac{3}{4}$%   R = _____
   A = $58   A = _____   A = $20

Objective 7

You will find it much easier to use this four-step process when working the word problems at the end of this chapter. The next several examples include an expression that will aid you in solving for the unknown.

Example 12

Forty percent of the students at Bisvern College are in the Vocational Division. There are 6,800 students at the college. How many are in the Vocational Division?

40% of the students are in the Vocational Division.
(R)   (B)
40% of 6,800 = A
40% × 6,800 = A
     2,720 = A

$(.40 × 6,800 =)$

If the same information were used in the problem with a different unknown value, the approach would be the same regarding the use of the four-step process.

Example 13

Of the 6,800 students at Bisvern College, 2,720 are in the Vocational Division. What is the percent of Vocational Division students?

2,720 students of 6,800 students is _____ percent?
(A)   (B)
2,720 ÷ 6,800 is equal to _____ percent?
2,720 ÷ 6,800 = R
      40% = R

$(2,720 ÷ 6,800 =)$

Again, the same information can be phrased differently to ask for a different unknown value. The four-step process would still be used.

Example 14

The 2,720 Vocational Division students represent 40 percent of the total students at Bisvern College. What is the total number of students at Bisvern College?

2,720 students represent 40 percent of the total.
2,720 students represent 40 percent of B.
2,720 = 40% × B
6,800 = B

$(2,720 ÷ .40 =)$

Increases and Decreases

The same formula can be used to find increases and decreases in an amount. The way the problem is worded is very important to the solution. In most cases, you will begin by finding the difference in the previous figure and the current figure.

Example 15

Mark is the manager of the produce section of the local grocery store. On Monday he had 14 boxes of lettuce. On Wednesday he had only 5 boxes left. What percent was sold? Nine of the total (14) were sold.

14	B = 14 R = A ÷ B
−5	A = 9 R = 9 ÷ 14
9 were sold (difference)	R = ? R = 64%

(9 ÷ 14 =)

Example 16

If the question in example 15 were to read "What percent was left?", the solution would be a little different. Five of the total (14) were left.

B = 14   R = A ÷ B

A = 5 (left)   R = 5 ÷ 14

R = ?   R = 36%

(5 ÷ 14 =)

Example 17

Because of increased competition, the sales of the Morten Company decreased from $459,098 to $398,013 over the period of one year. What is the percent of decrease? $61,085 is what percent of $459,098?

$459,098	B = $459,098 R = A ÷ B
−398,013	A = $61,085 R = $61,085 ÷ $459,098
$61,085 (difference or decrease)	R = ? R = 13%

(459,098 − 398,013 =   61,085 ÷ 459,098 =)

The same approach is used to find increases.

Example 18

Ellen had 147 credit customers at the beginning of the quarter. At the quarter's end, she had 159. What is the percent of increase in credit customers? Twelve is what percent of 147?

159	B = 147 R = A ÷ B
−147	A = 12 R = 12 ÷ 147
12 (increase or difference)	R = ? R = 8%

(12 ÷ 147 =)

Example 19

There were 890 items in stock at Sport-A-Merica at the beginning of the year. During the year the company added 47 new items. What is the percent of increase in inventory items? Forty-seven is what percent of 890?

B = 890   R = A ÷ B

A = 47   R = 47 ÷ 890

R = ?   R = 5%

(47 ÷ 890 =)

Example 20   Jason was earning an hourly rate of $8.00. He asked his boss for a raise. In his next paycheck he noticed his new rate of $9.60. What is the rate of the increase? $1.60 is what percent of $8?

$9.60	B = $8	R = A ÷ B
−8.00	A = $1.60	R = $1.60 ÷ $8
$1.60 (increase or raise)	R = ?	R = <u>20%</u>

(1.60 ÷ 8 =)

Exercise G

In the following problems, find the unknown factor. Round off your answer to the nearest unit, cent, or percent.

1. Student enrollment grew from 8,000 to 10,000 or a ____<u>25</u>____ percent increase. (2,000 ÷ 8,000 =)

2. Sales decreased from $670,000 to $600,000 or a _____ percent decrease.

3. Employment increased from 93 to 106 employees or a _____ percent increase.

4. The state budget increased by $16 million from $633 million or a _____ percent increase.

5. The cost of a computer printer decreased by $21 from $189 or a _____ percent decrease.

6. Serious accidents increased by 10 percent from the 850 from last year or to _____ accidents this year.

7. Jim sold 20 percent less this month than the $500,000 that he sold last month or _____ sales for this month.

8. Advertising increased by 30 percent this month to a level of $150,000. The advertising was _____ last month.

9. The number of clients grew to sixty from fifty last month or a _____ percent increase.

10. The price of a product decreased from $5.50 to $5 or a _____ percent decrease.

ASSIGNMENT Chapter 7

Name _____ Date _____ Score _____

Skill Problems

Select the letter for the definition.

A. Amount   B. Base   C. Rate

1. _____ The basis of comparison; the number of which so many hundredths is taken.

2. _____ A means of expressing some part of 100.

Fill in the blanks in the following problems.

3. $18\% = \dfrac{\rule{1cm}{0.4pt}}{100}$

4. $47.5\% = \underline{\hspace{2cm}}$

5. $.41 = \dfrac{\rule{1cm}{0.4pt}}{100}$

6. $2.5 = \underline{\hspace{2cm}} \%$

7. $\dfrac{7}{10} = \underline{\hspace{2cm}}$

8. $\dfrac{1}{5}\% = \underline{\hspace{2cm}}$

9. $1.759 = \underline{\hspace{2cm}} \%$

10. $\dfrac{4.6}{100} = \underline{\hspace{2cm}} \%$

11. $458\% = \underline{\hspace{2cm}}$

12. $\dfrac{1}{4}\% = \underline{\hspace{2cm}}$

13. $\underline{\hspace{1cm}} \% = \dfrac{30}{100} = .30$ **14.** $12\% = \dfrac{\rule{0.8cm}{0.4pt}}{100} = .12$ **15.** $47\% = \dfrac{47}{100} = \underline{\hspace{1cm}}$ **16.** $.5\% = \dfrac{.5}{100} = \underline{\hspace{1cm}}$

17. $17\% = \dfrac{17}{100} = \underline{\hspace{1cm}}$ **18.** $\underline{\hspace{1cm}} \% = \dfrac{134}{100} = 1.34$ **19.** $\underline{\hspace{1cm}} \% = \dfrac{26}{100} = .26$ **20.** $81\% = \dfrac{\rule{0.8cm}{0.4pt}}{100} = .81$

21. $10\ 1/5\% = \dfrac{10.2}{100} = \underline{\hspace{2cm}}$

22. $\underline{\hspace{1cm}} \% = \dfrac{1.75}{100} = .0175$

23. $3.6\% = \dfrac{\rule{1cm}{0.4pt}}{100} = .036$

24. $30\ 1/2\% = \dfrac{\rule{1cm}{0.4pt}}{100} = .305$

Solve for the unknown factor. Round off your answer to the nearest unit, cent, or percent.

25. B = 62
R = 410%
A = _____

26. B = _____
R = $6\dfrac{1}{2}\%$
A = 852

27. B = $17,500
R = _____ %
A = $16,000

28. B = $56
R = _____ %
A = $800

29. B = _____
R = 125%
A = $81.00

30. B = $420
R = 140%
A = _____

31. B = $16
R = 30%
A = _____

32. B = _____
R = 40%
A = $48

33. B = $8,000
R = _____
A = $12,000

34. B = _____
R = 350%
A = $71

35. B = $70
R = 15 3/4%
A = _____

36. B = $82
R = _____
A = $15

37. B = 80
R = 105%
A = _____

38. B = 600
R = _____
A = 100

39. B = 440
R = 2 1/2%
A = _____

40. B = 40
R = 18%
A = _____

41. B = 87
R = _____
A = 12

42. B = _____
R = 24%
A = 185

In the following expressions, solve for the variable.

43. A + 21 =   56

44. B − 8 = 54

45. $\dfrac{C \times 6}{} = \dfrac{36}{}$

46. $\dfrac{D}{5} = 60$

47. $\dfrac{9E}{} = \dfrac{10 + 8}{}$

48. $\dfrac{3(F)}{} = \dfrac{21}{}$

49. G*8 + 6 = 22

50. H × 1/3 = 45

In the following problems, find the unknown factor. Round off your answer to the nearest unit, cent or percent.

51. Sales increased from $125,000 to $260,000 or a _____ percent increase.

52. The price of a product decreased from $8.50 to $6.80 or a _____ percent decrease.

53. Complaints increased by 10 percent from 20 from last year or to _____ complaints this year.

54. The budget must be cut by 9% from last year's $3,000,000 or a _____ cut.

55. Joe received a 25% commission of $2,046. How much did Joe sell? _____

Business Application Problems

Round your answer to the nearest unit, cent, or percent.

1. A U.S. manufacturer has 23 percent of their parts produced outside of the country. If their product costs $18,470, what is the value of the foreign-made parts?

2. Martin and Associates purchased $562,960 worth of products from Allied Corp. last year. Total purchases from all sources for the year were $3,628,950. What percent of total purchases were made from Allied Corp.?

3. The average price of used homes went from $108,900 to $127,200 in one year. What is the percent of increase?

4. The average employee of Maxon Incorporated earns $644 per week. The average deduction is $136. What is the percent of take-home pay for the average employee?

5. Of the 6,000 shareholders of Tarmix Corporation, 420 attended the annual meeting and voted in person. Another 200 voted through proxy ballots. What percent voted?

6. A tourist site analyzed their customers to determine how many were from out of state. In the past three months that number totaled 5,890. The out-of-state customers represented 68 percent of the total. How many customers did the firm have in total?

7. The Book Nook sales for the month were $82,645. Book sales accounted for 62 percent of the total sales. What is the dollar value of the book sales for the month?

8. The number of default loan customers for the Grand Tower Bank dropped to 84, or 1.5 percent, for the month of March. How many loan customers does the bank have?

9. Tomison Technical College enrollment includes 480 students who are being retrained as a result of downsizing in the area last year. The balance of the student population of 916 has recently graduated from high school. What is the percent of recent high school graduates?

10. The Needam Nuggets minor league baseball team attendance was 4,756 last night. The stadium was 82 percent sold out. What is the capacity of the stadium?

11. The Eye Care Center reviewed their customer list and found that 20 percent of their 340 customers used a credit card to make their purchases. How many used a credit card?

12. There were 14,600 paid visitors to the theme park by noon on Saturday. The manager estimated that those arriving before noon represented 40 percent of those that would visit the park that day. How many will visit the park?

13. An ad indicated that a new computer priced at $600 could be purchased by putting $102 down. What percent is the down payment required?

14. The price of a product increased 17 percent. The increase brought the price up to $1,000. How much did the product originally sell for?

15. Mr. Albert E. Hodge found that 70 percent of his assets were in land and buildings. A real estate broker indicated that he could sell the land and buildings for $140,000. How much are Mr. Hodge's assets worth?

16. Bill Otis noted that his total mileage for the month was 2,400. He realized that only 600 miles were for family reasons. What percent were for family reasons?

17. A company wished to buy a building priced at $30,000. They had a down payment of 20 percent. How large of a loan will they have to make?

18. The Ace Security Company has decided to give Fred Jantz a 5 percent raise over his current rate of $4.80 per hour. What will be his new rate?

19. Nancy Pryor noticed that her paycheck stated that she had earned $247 for last week's work. She also took note of the fact that $32.50 was deducted from her check for taxes. What percent was deducted?

20. Spot Deliveries purchased an item by putting $5,000 down and financing the balance. The down payment was 20 percent of the cost of the item. What was the cost of the item?

21. Sammy is a salesperson at a men's store. Yesterday he sold $865 of merchandise. Sammy stated that this was 110 percent of the amount he sold on Tuesday. How much did Sammy sell on Tuesday?

22. The Deli section of the grocery store enjoyed sales of $890 last Tuesday. The store had sales of $39,740 on the same day. What percent of the sales came from the Deli?

23. The sales staff of Reilly and Associates made up 80 percent of the total employees. The sales staff and the legal staff together made up 90 percent of the total employees. There are 12 members of the legal staff. How many employees are there in total?

24. A corporate budget was increased by 2.4 percent over the previous year. The previous budget called for expenditures of $45,970,000. What is the new budget amount?

25. The average expenditure for food in the Malory family is $80 per week. The family spends 16% of gross income on food. What is the family gross income?

26. The Home Mortgage Company found that 164 of their clients were in default for the month of July. Those who defaulted represented 2 percent of the total mortgages held by the company. How many mortgages does the company hold?

27. Carrie spends $480 for rent each month. She has totaled her expenditures on food and has found that she spends $260, or 16 percent of her gross income, on food. What percent of her gross income does she spend on rent?

28. Write a complete sentence in answer to this question. If the unemployment rate changed from 7.3 percent to 7.5 percent, how many additional workers would be unemployed in a work force of 90,000,000 people?

29. Write a complete sentence in answer to this question. All of the sales staff of the Donaldson Corp. made the quota for the month. The quota was a sales volume of $260,000. Seven of the ten sales representatives went over the quota by 15 percent. One additional sales representative exceeded the quota by 20 percent. What was the excess of quota achievement using the figures given?

30. Write a complete sentence in answer to this question. Grain and Company will increase their sales this year to $3,500,000. The increase will come from $200,000 received from sales of their "Carl" product, plus a 10 percent increase over last year in sales of their other product lines. What were sales last year?

★ **Challenge Problem**

The Groome Corp. has determined to give 5 percent of their employees a 5 percent raise. The total of the salaries paid to employees is $3,000,000. The average salary is $25,000. What will be the total cost of salaries after the raise?

Name _____ **Date** _____ **Score** _____

Fill in the blanks in the following problems.

1. 9 % = _____

2. $\frac{7}{8}$ = _____ %

3. .21 = _____ %

4. $9\frac{1}{2}$% = $\frac{}{100}$

5. **T F**   A = R × B

6. **T F**   R = B ÷ A

7. **T F**   250 = 50% of 50

8. B = $600
R = 25%
A =

9. The Sales Department of Jacklen Marketing Co. reported that gross sales were $642,800 for the first quarter of the year. Sales returns were $12,856 for the same period. What percent were the returns to gross sales (nearest percent)?

10. The cost of materials for the boat manufacturing industry increased by $8\frac{1}{2}$ percent over last year. An item that sells for $24 wholesale this year would have been priced lower last year by what amount?

11. Food is about 16 percent of the average family budget. What will be the expenditures for food by a family that has an income of $1,500 a month?

12. The work force in Vitar County is approximately 85,000 people. The unemployment rate recently decreased .7 percent. How many additional people were put to work?

13. An apartment building increased in appraised value by $7,000 over a period of one year to $86,000. What was the percent increase in appraised value?

14. A bank had 40 percent of its holdings in stocks and bonds. The amount invested totaled $68 million in all investments. How much was invested in stocks and bonds?

15. The bank in problem 14 added $816,000 to total investments the following year. What is the percent increase in investments?

INSURANCE

OBJECTIVES

After mastering the material in this chapter, you will be able to:

1. State the purpose of insurance. p. 144
2. Calculate premiums in fire, life, and workers compensation insurance problems. p. 144
3. Calculate the amount of payment on a fire loss covered by coinsurance. p. 144
4. Identify different types of life insurance coverage. p. 146
5. Understand and use the following terms and concepts:
   Premiums
   Policy
   Coinsurance
   Beneficiary
   Term Life Policy
   Straight Life Policy
   Cash Surrender Value
   Limited Pay Life Policy
   Endowment Life Policy
   Universal Life Policy
   Variable Life Policy
   Workers' Compensation

SPORT-A-MERICA

Hal met Mark Barink, a State Mutual Insurance agent, yesterday at two o'clock. "Mark," Hal said, "We have made some rather important changes at Sport-A-Merica over the past few months. You have seen us grow since we started out as a small three-person store to what we are today. We expect to grow even bigger over the next few years. We now have a larger store we'll soon be moving into, a new delivery vehicle on the road, and a greater dependence on management, in addition to other insurance considerations we'd like to discuss with you."

Mark replied, "As you continue to grow, your insurance needs will grow with you."

"That is why I thought we should meet now," said Hal. As the conversation continued, the more questions Hal asked, the more questions he developed. Hal recognizes the need for insurance, which spreads the risk and reduces potential losses for Sport-A-Merica, and he also realizes he needs to learn some basic insurance information by relying on the expertise of an insurance representative to cover the details for the business. He especially needs to know the cost of the fire insurance for the new store. And, what would it cost for a life insurance policy on him? After all, he is now 47 years old! What about Sport-A-Merica employees getting injured on the job? When Mark mentioned that a discount would apply if Sport-A-Merica had two kinds of insurance, Hal began thinking about the future needs of the firm even more.

Objective 1

The purpose of insurance is to minimize the cost of unforeseen losses. If a firm or person foresees a chance of high risk, some types of insurance coverage may be warranted. There are many types of insurance—health, fire, casualty, life, crop, automobile, marine, workers' compensation, and even rain insurance. This financial protection can be purchased for groups of people, for individuals, and even for another person. Some insurance may be purchased electronically using a computer at home. In some instances, it can be purchased from the federal government or from privately owned firms. Private firms are either stock companies (owned by stockholders of that firm) or mutual companies (owned by policyholders of that firm). An insurance policy is usually sold through an insurance agent who is a local representative of an insurance company. Due to the vastness of this topic, only a brief coverage of fire insurance, life insurance, and workers' compensation insurance will be included.*

premiums

Insurance protection is purchased by making fixed payment amounts at periodic intervals. The payment amounts are termed **premiums.** The premium amount is established by the insurance company, the insurer, by first determining the chance of loss through statistical analysis, and then adding on the cost of doing business. Premiums are usually paid quarterly, semiannually, or annually. There is an economic savings to the insured who pays annually because this decreases the handling costs of the insurance company. A

policy

policy, or a written contract, is then given to the insurance owner; this policy identifies the areas and amounts of coverage.

Property Insurance

Property damage or loss is a common type of insurance. There are many types of property insurance, including that which covers damage due to wind, fire, flood, vandalism, and breakage. Fire is the greatest damage threat and is included in most property coverage. Fire insurance rates vary by the amount of fire protection available nearby, the age of the dwelling, its construction materials, and the presence of fire alarm system or extinguishers on the premises, among other factors. The higher the risk, the higher the premium. Premium rates are stated in terms of $100 of coverage. Rates for coverage over $100 are multiplied by the number of $100s of coverage desired.

Objective 2

Example 1

Determine the fire insurance premium on the new space for Sport-A-Merica for $110,000 of coverage. The rate per $100 of coverage is $.325 on this type of building.

Number of 100s × Premium per 100 = Annual premium

$110,000 =   1,100 hundreds
              ×$.325 per $100 of coverage
              $357.50 annual premium

(1,100 × .325 =)

Property insurance, including fire protection, is often purchased for periods longer than one year. Policies for three to five years offer a reduction in the amount of the premium because of the reduced cost of doing business.

Objective 3   *coinsurance*

Fire Insurance

A common clause in most fire insurance policies is the **coinsurance** clause. This clause states that the policyholder must provide a minimum percent of coverage on the property. A common coinsurance percent is 80 percent. A building that is valued at $41,000 must be insured for $32,800 ($41,000 × .80; insurance required) in order to receive maximum payment if a loss occurs. If the property was insured for only $24,600 (60 percent of $41,000; insurance carried) and was damaged by $14,000, the insurance

*Further information is available from your local insurance agency or by writing the Consumer Information Center, Pueblo, Colorado 81009, for these free booklets: Insurance for Renters and Homeowners 018D, Insurance for Your Health, Car, and Life 019D, Questions and Answers about No-Fault Auto Insurance 256D.

company would not pay for the full amount of the loss. A partial payoff of $10,500 would be made. This amount is a prorated payment. It is found by

$$\frac{\text{insurance carried}}{\text{insurance required}} \times \text{amount of loss}$$

$$\frac{\$24,600}{\$32,800} \times \$14,000 = \$10,500 \text{ (amount of payoff)}$$

Example 2

James Garison's warehouse is covered by an insurance policy that has an 85 percent coinsurance clause. The building is valued at $120,000. It is insured for $78,000. Fire did $40,000 worth of damage to the building. What is the amount of the settlement from the insurance company?

building value × coinsurance clause = insurance required
$120,000 × .85 = $102,000

$$\frac{\text{insurance carried}}{\text{insurance required}} \times \text{amount of loss} = \text{payoff}$$

$$\frac{\$78,000}{\$102,000} \times \$40,000 = \$30,588.24$$

$(78,000 \div 102,000 = \quad \times 40,000 =)$

Though the property owner may carry more insurance than the coinsurance clause specifies, the payoff will never exceed the face value of the policy. Also, the payoff will never exceed the value of the loss. The maximum payoff on property insured to the coinsurance limit will always be the face value of the policy or the amount of the loss, whichever is lower. Keep this in mind when completing the assigned problems.

Exercise A

Calculate the amount of the annual premium in the problems below.

	Value of Property	Amount of Coverage	Premiums Per $100 of Coverage/Yr.	Total Premium Per Year
1.	$46,000	$36,000	$.310 (360 × .310 =)	$111.60
2.	31,900	24,000	.396	
3.	54,000	42,000	.462	
4.	67,000	51,000	.332	
5.	21,300	16,500	.405	
6.	87,000	70,000	.358	
7.	36,000	27,000	.490	
8.	61,900	50,000	.318	
9.	54,800	42,500	.325	
10.	21,700	17,000	.431	

Exercise B

Calculate the amount of payoff in these problems. Do not round off the insurance carried to the insurance required factor before multiplying the amount of loss. This will provide the most accurate answer.

	Value of Property	Amount of Coverage	Coinsurance Clause Specification	Amount of Loss	Amount of Payoff
1.	$46,000	$36,000	80% (36,000/36,800 ×	$12,000 =)	$11,739.13
2.	65,000	52,000	80	20,000	

(Continued on next page)

Exercise B Continued

Calculate the amount of payoff in these problems. Do not round off the insurance carried to the insurance required factor before multiplying the amount of loss. This will provide the most accurate answer.

	Value of Property	Amount of Coverage	Coinsurance Clause Specification	Amount of Loss	Amount of Payoff
3.	$41,500	$ 30,000	75	$ 6,000	_____
4.	60,000	40,000	80	40,000	_____
5.	260,000	180,000	75	30,000	_____
6.	95,000	76,000	80	10,000	_____
7.	120,000	60,000	80	80,000	_____
8.	80,000	48,000	75	15,000	_____
9.	60,000	45,000	75	22,500	_____
10.	300,000	200,000	80	28,500	_____

Life Insurance

Life insurance has many purposes, the most common of which is to provide some financial security or protection for the survivors of the insured. The survivors that are to receive the financial benefits are named by the insured in the policy. These named persons or

beneficiary   organizations are termed **beneficiaries.**

Life insurance can be purchased in many forms by making monthly, quarterly, semiannual, or annual premium payments. There is an economic savings earned by paying annually. Some of the many types of life insurance are discussed here.

Objective 4

Term Life

The **term life policy** is often offered by employers as part of a fringe benefit package. It provides protection only as long as the premiums are paid. So long as an employee remains with the employer, the term life insurance policy will remain in effect. Should the employee be terminated, the term life insurance is then terminated. It is important, therefore, to have separate insurance coverage not provided by the employer. This type of policy is the most inexpensive for the amount of protection.

Term life premium rates are determined by measuring life expectancy and then adding the cost of doing business and an amount of profit. The premium cost rises each year due to the reduced life expectancy. A review of the Annual Premium table on page 148 shows how rates change for each year of age.

Straight Life

The **straight life policy** also termed whole life, is similar to the term life policy. The insured is protected only as long as premiums are made. Straight life insurance premiums are fixed per year and are higher than term life insurance premiums because part of the premium is set aside by the company for the future. The surplus portion of the premium above the cost of a term life policy in the early years is retained to cover the cost of the insurance in later years. The straight life policy allows the insured to keep the same premium rate over the life of the policy.

Cash Surrender Value

The surplus premium over the cost of term insurance is termed **cash surrender value** or **loan value**. The surplus is paid to the insured by the insurer if the policy is cancelled. They can also be borrowed if the insured needs cash and is willing to pay an interest charge on the amount borrowed. The cash surrender value increases with the length of time that the insurance is in force. As you will note from the following table, the type of insurance purchased also determines the amount of cash surrender value that the insured's policy is worth. This table is predetermined in the policy.

Cash Surrender Value/Loan Value, Age of Insured 25, per $1,000 of Coverage	Years Policy in Force	Straight Life	Limited (20) Pay Life	Endowment Life (20)
	3	$ 15	$ 45	$ 91
	5	31	97	170
	10	57	235	402
	15	105	360	673
	20	194	518	990

Example 3

If Guy, a 25-year-old, purchased a $50,000 Limited Pay Life policy 10 years ago, how much can he borrow on the policy?

$$50 = \text{Number of \$1,000 of coverage}$$
$$\underline{\times \$235} = \text{Loan value per \$1,000 of coverage for 25-year-old}$$
$$\$11,750 = \text{Amount Guy can borrow on policy}$$

(50 × 235 =)

Limited Pay Life

The **limited pay life policy** is similar to the straight life policy. Part of the premium is invested for the insured, but the insured is required to make only a limited number of premium payments such as 20 years' worth. This reduces the financial pressure on the insured when earnings decline, such as after retirement. Because of the limited number of premium payments, the premium is higher than the premium of term or straight life insurance.

Endowment Life

The **endowment life policy** is similar to the limited pay life policy. It has a limited number of premium payments and has a savings factor. The savings factor is larger in this policy than in limited pay life. Endowment life is used to prepare financially for retirement, college, a trip, or a business venture. Because of the greater savings factor, the premium payments are the highest of all the types of life insurance discussed.

Premiums are stated per $1,000 of life insurance coverage. The premium amount varies from company to company due to the efficiency of the firm, the quality of investments made, or type of insurance company. A representative schedule of premiums follows.

Annual Premium per $1,000 of Insurance

Age of Insured*	Term Life	Straight Life	Limited Pay Life	Endowment Life
18	$ 4.42	$14.98	$21.35	$35.45
19	4.43	15.09	21.68	35.48
20	4.45	15.11	22.02	35.50
21	4.47	15.46	22.40	35.52
22	4.49	15.80	22.85	35.55
23	4.53	16.17	23.30	35.58
24	4.58	16.55	23.75	35.62
25	4.61	16.96	24.25	35.67
26	4.67	17.40	24.76	35.74
27	4.72	17.82	25.26	35.81
28	4.80	18.27	25.80	35.87
29	4.89	18.75	26.35	35.95
30	4.98	19.26	26.92	36.05
31	5.08	19.75	27.50	36.17
32	5.22	20.33	28.15	36.28
33	5.39	20.95	28.80	36.43
34	5.60	21.55	29.40	36.60
35	5.77	22.20	30.11	36.79
36	6.07	22.91	30.82	36.99
37	6.40	23.62	31.55	37.25
38	6.76	24.40	32.30	37.52
39	7.17	25.20	33.15	37.81
40	7.60	26.05	33.99	38.15
41	8.09	26.95	34.85	38.55
42	8.63	27.85	35.75	38.95
43	9.23	28.84	36.70	39.40
44	9.86	29.92	37.80	39.90
45	10.55	30.99	38.81	40.38
46	11.35	32.10	39.90	40.95
47	12.20	33.35	40.89	41.55
48	13.20	35.70	42.10	42.25
49	14.21	36.00	43.35	43.00
50	15.39	37.50	44.70	43.80

*Premiums for ages two weeks through eighty years old and beyond are also available.

The annual premium for a $15,000 limited pay life insurance policy would be $342.75 if the insured were twenty-two years old at the time the policy was purchased.

$15,000 =   15 = $1,000 of coverage
              ×$22.85 = Annual premium for a twenty-two-year-old's limited pay life
                          insurance policy
           $342.75 = Annual premium

Example 4

What will be the annual premium on Hal at Sport-A-Merica for a term life policy of $8,000 if he is forty-seven years old?

Number of 1,000s × Premium per 1,000 = Premium

          8 = $1,000 of coverage
    ×$12.20 = Annual premium for forty-seven-year-old term life
    $97.60 = Annual premium

(8 × 12.20 =)

Exercise C

Calculate the annual premium in the following problems.

	Amount of Insurance Coverage	Type of Insurance	Age of Insured	Annual Premium
1.	$ 20,000	Straight	18 (20 × 14.98 =)	$299.60
2.	55,000	Term	23	
3.	100,000	Limited pay	20	
4.	20,000	Term	28	
5.	60,000	Straight	32	
6.	5,000	Endowment	24	
7.	20,000	Limited pay	30	
8.	25,000	Straight	45	
9.	20,000	Term	26	
10.	120,000	Endowment	35	

Exercise D

Calculate the cash surrender value for a 25-year-old in the following problems.

Amount of Coverage	Type of Insurance	Years in Force	Cash Surrender Value	
1. $64,000	Limited pay	5	$ 6,208	(64 × 97 =)
2. 125,000	Endowment	20		
3. 300,000	Straight	3		
4. 150,000	Term	10		
5. 200,000	Endowment	15		

The life insurance industry has developed additional insurance products that are very flexible. It is possible to design a life insurance coverage program that is tailored to one single person. Within a flexible insurance program would undoubtedly be found a universal life policy or a variable life policy, in addition to a personally funded term policy. While the term policy provides the best amount of protection in the form of the amount of the policy, universal or variable life will provide the best amount of flexibility.

Universal Life

Universal life policies allow for changes in premium amounts and, thereby, changes in the amount of insurance coverage, all under the same policy. As the income of the insured changes, the premium amount can be changed. This works especially well for those starting their career. As income and responsibility increases, the amount of insurance can be increased. The surplus portion of the premium for a universal life policy is usually invested in fixed-rate-return vehicles, such as bonds or U.S. Treasury Bills.

Variable Life

Variable life policies are very much like universal life policies in the sense that the premium and coverage amount can be changed over the life of the policy based upon affordability or need. The major difference between these two types of flexible policies is that variable life policy premium surpluses are invested in stocks and mutual funds. The return from these investments is not fixed. In periods of prosperity, the returns will be very high. When the economy slumps, the returns are apt to be poor.

In addition, it is possible to make only one large premium payment. The surplus is invested and then used for future premium payments. A policyholder can also indicate that no more premiums will be paid. A paid-up policy can then be issued. Or, the insured can simply withdraw the cash value of the policy. These possibilities—and many others—require that each person and each business stay in close contact with a qualified insurance agent.

Workers' Compensation Insurance

Employers in all fifty states are required to have some type of **workers' compensation** insurance. This type of insurance provides compensation to employees who have been injured on the job. It pays for medical costs for injuries and work-related disease. Workers are also compensated for lost time from work due to injuries or disease. This type of insurance is a major cost consideration for employers in the mechanized and technical part of our economy. The current total national premium for workers' compensation insurance is in excess of $40 billion per year. In some industries, the cost of workers' compensation insurance premiums may be as much as 60 percent of employee wages. In other industries, the cost could be less than 1 percent.

The amount of the premium is determined by the accident experience of the industry or the individual firm. A firm operating in an industry that is associated with a low level of employee injury or disease will pay a low premium. However, if the firm or industry has a record of high risk to employees, indicating that injury, disease, or death is likely to occur, the premium will be higher.

Example 5

What is the semiannual premium for workers' compensation insurance for a group of 15 employees at Gerrard Grate and Company? The insurance agent has quoted a premium of $400 per employee per year.

Premium per employee × Number of employees = Annual premium
$400 × 15 = $6,000
Annual premium ÷ 2 = Semiannual premium
$6,000 ÷ 2 = $3,000

(400 × 15 =   6,000 ÷ 2 =)

A business manager will need to consider the cost of workers' compensation insurance when determining the total cost of labor.

Example 6

LeFlure Corporation employs 73 workers at their Wisconsin installation. The insurance agent has quoted a premium of $7,300 per employee for workers' compensation insurance coverage. What is the amount of the insurance premium per hour for each employee, if each employee works 2,000 hours per year?

Annual premium ÷ Hours worked annually = Cost per hour
$7,300 ÷ 2,000 = $3.65 per hour

(7,300 ÷ 2,000 =)

Exercise E

Calculate the total annual premium in the following problems.

	Number of Employees	Annual Premium per Employee	Total Annual Premium for Firm
1.	43	$3,650 (43 × 3,650 =)	$ 156,950
2.	67	1,200	
3.	1,258	2,360	
4.	259	18,900	
5.	372	9,374	

Exercise F

Calculate the hourly cost of the premium for each employee in exercise E. Assume that each employee works 2,000 hours per year.

1. _____$1.83_____ (3,650 ÷ 2,000 =)

2. _____

3. _____

4. _____

5. _____

ASSIGNMENT Chapter 8

Name _____ **Date** _____ **Score** _____

Skill Problems

Select the best answer for the statements below.

A. Premiums
B. Policy
C. Coinsurance
D. Beneficiary
E. Term Life Policy
F. Universal Life Policy
G. Limited Pay Life Policy
H. Endowment Life Policy
I. Workers' Compensation

1. _____ Provides protection only for as long as premiums are paid. No surplus or savings factor built into premium.

2. _____ Provides protection for injuries or illness because of occupational activities.

3. _____ The policy holder and the insurance company share in the financial loss of property damage.

4. _____ A large portion of this premium is directed toward a savings plan that may be used for college, retirement, etc.

5. _____ Payments for insurance protection.

Calculate the amount of the annual premium in the problems below. Round off to the nearest cent.

Fire Insurance

Value of Property	Amount of Coverage	Premium Per $100 of Coverage/Yr.	Total Premium Per Year
$200,000	$170,000	$.412	**6.** _____
130,000	100,000	.410	**7.** _____
80,000	60,000	.359	**8.** _____
350,000	280,000	.375	**9.** _____
62,500	53,000	.415	**10.** _____
65,000	54,000	.246	**11.** _____
45,000	40,000	.310	**12.** _____
52,500	60,000	.332	**13.** _____
7,650	5,000	.490	**14.** _____
101,650	100,000	.325	**15.** _____
34,050	30,000	.294	**16.** _____
25,470	24,000	.805	**17.** _____
90,650	70,000	.758	**18.** _____
250,860	200,000	.651	**19.** _____

Life Insurance

	Amount of Coverage	Type of Insurance	Age of Insured	Annual Premium
20.	$150,000	Straight	43	_____
21.	50,000	Term	23	_____
22.	20,000	Endowment	22	_____
23.	250,000	Straight	50	_____

Amount of Coverage	Type of Insurance	Age of Insured		Annual Premium
24. $ 75,000	Term	23		_____
25. 40,000	Limited pay	27		_____

			Years in Force	Cash/Loan Value
26. 150,000	Endowment	25	10	_____
27. 20,000	Straight	25	3	_____
28. 20,000	Term	25	20	_____

Workers' Compensation Insurance

Number of Employees	Annual Premium per Employee	Total Annual Premium for Firm
29. 357	$ 650	_____
30. 730	200	_____
31. 258	5,600	_____
32. 90	750	_____
33. 25	980	_____
34. 26	1,340	_____
35. 1,546	4,860	_____
36. 544	16,470	_____
37. 130	9,405	_____

Calculate the amount of payoff in the problems below. For the most accurate answer, do not round off insurance carried to insurance required before multiplying.

Value of Property	Amount of Coverage	Coinsurance Clause Specification	Amount Required by Clause	Amount of Loss	Amount of Payoff
$ 65,000	$50,000	80%	**38.** $_____	$14,000	**39.** $_____
45,000	30,000	80	**40.** _____	42,000	**41.** _____
52,500	30,000	70	**42.** _____	10,000	**43.** _____
29,000	5,500	75	**44.** _____	4,000	**45.** _____
101,650	50,000	80	**46.** _____	80,000	**47.** _____
40,000	30,000	85	**48.** _____	23,000	**49.** _____
31,500	24,000	90	**50.** _____	12,000	**51.** _____
90,650	70,000	85	**52.** _____	4,500	**53.** _____
250,860	200,000	95	**54.** _____	16,000	**55.** _____

	Value of Property	Amount of Coverage	Coinsurance Clause	Amount of Loss	Amount of Payoff
56.	$24,000	$17,250	80%	$10,000	_____
57.	80,000	50,000	80	40,000	_____
58.	62,000	46,500	75	6,500	_____
59.	100,000	80,000	85	90,000	_____
60.	20,000	14,000	80	9,500	_____

Business Application Problems

1. Ken purchased a $300,000 straight life insurance policy at the age of 30. What is the amount of annual premium?

2. What is the amount of premium for a 46-year-old male who will purchase a $350,000 endowment life policy?

3. An office building was purchased for $260,000. The new owner will purchase $200,000 of fire insurance though the fire insurance company has an 80 percent coinsurance clause. The rate per $100 is $.395. What is the annual premium?

4. If a fire caused $148,500 of damage in problem 3, what is the amount of payoff?

5. If there had been $160,000 of fire insurance coverage in problem 3, what would have been the payoff if the building was a complete loss?

6. Baxter Associates has a workers' compensation premium that is 73 percent of wages. Gene is paid $16 per hour as a skilled worker. What is the cost of the workers' compensation insurance per hour for Gene?

7. Adam paid the first premium on his $56,000 limited pay life insurance policy when he was 32 years old. What is the annual premium amount?

8. Adam, in problem 7, is now 9 years older. How much has he paid in premiums to date?

9. What would be the additional premium if Adam, in problem 7, were to purchase another $56,000 of limited pay life insurance at his current age (41)?

10. A new firm in the warehouse business will be charged a rate of $6.10 per hour worked for each employee for workers' compensation insurance. The firm will begin operations with 23 employees. How much should the accountant plan to pay for workers' compensation insurance in the first year if employees average 42 hours a week?

11. The Burton Corp. wishes to purchase a term life insurance policy on their president, Ann Pierson. The company officials feel that if she died, there would be a decline in earnings because of the absence of her leadership. The company will be named the beneficiary. What will be the cost of a $125,000 term life policy on Ann Pierson if her age is fifty?

12. Phil Buckner has decided that he can afford $12 now each month for life insurance protection. He is 28 years old, married, and has two children. He is not sure which type of policy to purchase, but has narrowed the decision down to term life or straight life. Once selected, he will maintain that type of policy regardless of premium changes in the future. If he dies at age 35, how much more will his family receive if he decides to purchase term life?

13. The Rectan Corp. allows their employees to select from a group of insurance options (dental, optical, life, medical, etc.). The firm will pay up to $120 per month per employee for insurance. Eddie, age 47, has decided to put $80 of his allowance per month into term life insurance. How much insurance will the firm be able to purchase for Eddie?

14. Personnel Plus has purchased an office building for $160,000. Their insurance agent indicated a rate of $.405 per $100 of fire insurance coverage. The insurer has a 75 percent coinsurance clause on fire protection. What will be the annual premium if the firm follows the 75 percent clause stipulation?

15. Polisar Corp. incurred a $35,000 fire loss on an empty warehouse. The building is valued at $100,000. The insurance company specified a 75 percent coinsurance clause in the fire protection policy. What amount of insurance coverage was necessary to receive the maximum payment on the loss?

16. Lambtree Corporation will hire 42 additional employees for summer work. The monthly premium for workers' compensation is $80 per employee. What is the amount of additional premium per month for this workers' compensation insurance for the new employees?

17. The Lawson home was totally destroyed by fire at an estimated loss of $74,800. The insurance policy contained an 80 percent coinsurance clause. The face value of the policy was $38,000. What was the amount of the payoff to the Lawson family?

18. Ben Wolfson will hire two employees for the men's sportswear department. The cost of workers' compensation insurance will be $2.70 per hour for each employee. What is the annual premium for both employees if each employee works 2,000 hours per year?

19. A total loss due to fire was experienced by the Burkholdt Corp. The building had a market value of $450,000 and was insured for $360,000. The insurance policy stipulated a coinsurance clause of 80 percent. What is the payoff?

20. A company has added to the staff over the last month. Each new employee will receive a term life insurance policy in the amount of $10,000. The premiums are paid by the company. The employees are listed with their respective ages. What is the total cost of the insurance coverage to the company?

Bob Akers, 24 years old
Andy Watts, 37 years old
Carla White, 45 years old

21. A new building was insured for $120,000 for fire protection. The structure was appraised at $160,000. A fire in the second story caused $42,000 in damage. What was the payoff to the owner if the insurance policy specified an 80 percent coinsurance clause?

22. An employee can be hired at $8.00 per hour plus fringe benefits that cost $3.00 per hour. An additional cost of an annual (2,000 work hours) workers' compensation insurance premium of $3,500 must be considered. A labor service offers to provide a worker for $12 per hour. Which is less expensive?

23. A new commercial building for Beck and Company will be completed on the first of the month. The insurance agent has indicated that the property insurance policy will specify a 75 percent coinsurance clause. The building will have a market value of $580,000. What is the amount of insurance coverage that Beck and Company will need to carry? Round off your answer to the nearest ten thousand dollars.

24. Write a complete sentence in answer to this question. Carol Reddy, a 43-year-old, has $200 per month to use for additional insurance coverage. She is considering term and straight life policies. She is interested in the difference in the amount of coverage that she can obtain from the same investment. What is the difference in coverage that she can purchase? Round off your answer to the nearest $1,000 of coverage.

25. Write a complete sentence in answer to this question. What is the hourly cost for workers' compensation insurance for a group of 15 employees? The annual premium per person is $800. All employees work 2,000 hours per year.

★ **Challenge Problem**

The Denberton Company pays an annual fire insurance premium of $185.50, at $.35 per $100 of coverage of fire insurance. The building is valued at $85,000. Last week they had a $50,000 fire. There is a 75 percent coinsurance clause. What is the amount of the loss to the Denberton Company not covered by insurance?

Name _____ **Date** _____ **Score** _____

True or False

T   F   1. An insurance policy is a contract that sets forth the rights of the insured and the amount of coverage.

T   F   2. Endowment life insurance has the highest cost per thousand.

T   F   3. The money paid at regular intervals to an insurance company is termed coverage.

T   F   4. The beneficiary will receive the face value of the life insurance policy.

T   F   5. When an insurance company states that the insured must share in the risk of insurance by absorbing part of the losses, the firm is specifying a coinsurance clause.

T   F   6. A term life insurance policy will buy more coverage for a stated amount of dollars than any other type of policy.

T   F   7. The purpose of insurance is to reduce the risk of loss.

T   F   8. If the homeowner has a total loss and the insurance coverage exceeds the value of the property, the insurance company will pay out the value of the policy to the homeowner.

9. The Wilson Manufacturing Company will purchase term life insurance for its employees up to the amount of their individual annual income. The insurance is a fringe benefit for employees after one year of employment. Beverly Seldon, a 40-year-old receptionist, earns $18,000 a year. What is the annual cost of this fringe benefit to the Wilson Manufacturing Company?

10. How much term life insurance can be purchased by a 30-year-old male who has an extra $70 per month? Round off your answer to the nearest $1,000.

11. The Shaw family home burned down at a total loss of $34,000. They had insurance on the property. The insurance company had a 75 percent coinsurance clause in the contract. What will the payout be if the family had only 60 percent coverage?

SIMPLE INTEREST

9

OBJECTIVES

After mastering the material in this chapter, you will be able to:

1. Explain what *interest* means. p. 162

2. Compute the time factor. p. 162

3. Compute both ordinary and exact interest, using the principal, rate, and time factors. p. 166

4. Compute any one unknown factor in the interest formula ($I = P \times R \times T$) when the other three factors are known. p. 168

5. Understand and use the following terms:
   Interest
   Principal
   Rate
   Time
   The Calendar
   Leap Year
   The Interest Formula ($I = P \times R \times T$)
   Ordinary Interest
   Exact Interest

SPORT-A-MERICA

As a result of the conversation with Mark Barink about the discount on insurance, Hal has decided to purchase an additional delivery vehicle for Sport-A-Merica. The cost of the vehicle is $26,000. Hal feels the money can be borrowed from a bank for a short period of time, and he calls Community National Bank for loan information after reading their ad in the business section of the newspaper. Marion McGregor, a loan officer with the bank, responds, "A ninety-day loan, using the vehicle as collateral, could be negotiated with a favorable rate." Hal feels that Sport-A-Merica will have the cash to pay off the loan prior to the end of the ninety-day period. He will need to determine the day the loan must be repaid and the cost of borrowing the money. After checking the rates at First National Bank, where Sport-A-Merica has its checking account, Hal makes an appointment to arrange the loan with Marion McGregor at Community National Bank.

What Interest Means
Objective 1

interest

There are undoubtedly many answers to the question "What is interest?" The lender may say that it is a return to cover the risk on his investment, a borrower may state that it is a penalty for having to borrow, and the economist would probably state that it is related to the supply and demand of risk capital. **Interest** is a rental amount or a charge for the use of money.

The total amount of interest is determined by the amount of money borrowed, the rate of interest or charge rate, and the length of time the money is borrowed.

When borrowing, or renting, money, the borrower is obligated to pay back the amount borrowed at the end of the period, as well as the interest or rent on the amount borrowed. It should be noted that both the amount borrowed and the interest are owed by the borrower.

*principal, rate
time*

In order to be able to compute interest, you must know three factors. The amount of money to be borrowed is termed the **principal.** The charge rate is termed the **rate,** and the length of time the money is to be borrowed is termed the **time.** To solve an interest problem, you must be able to determine the *number of days* that the money has been borrowed. You may in fact be given the date that the money was borrowed, termed the date of loan, and the date the money was paid back, termed the repayment date, or *maturity date.* You will need to compute the number of days that have elapsed.

The Calendar
Objective 2

In order to accomplish this task you must be familiar with the **calendar.** This can be accomplished in at least three ways. You can memorize the number of days per month from the table below.

Days per Month				
January	31	July	31
February	28 or 29	August	31
March	31	September	30
April	30	October	31
May	31	November	30
June	30	December	31

A second way to become familiar with the calendar is to memorize the poem: "Thirty days has September, April, June, and November. All the rest have thirty-one, except for February which has twenty-eight and in leap year has twenty-nine." You probably remember the poem from your childhood. It is a good tool to recall for your work in this chapter.

A third way to become familiar with the calendar is to use the knuckles of your hands. By making a fist with both hands and starting the months on the first knuckle as shown below, you can determine the months with thirty-one days and those which have less. The months on top of a knuckle have 31 days. Those months (other than February) that fall

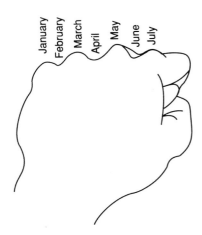
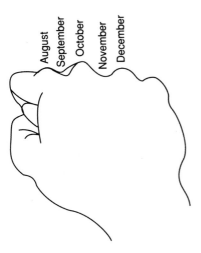

between a knuckle have 30 days. By remembering that February is a special exception depending on leap year, you should be able to determine the correct number of days in a month.

leap year   February has twenty-eight days, with the exception of **leap year,** when it has twenty-nine days. Leap year is any year that is evenly divisible by four. When computing the time factor, you may find that the dates given for the length of time the money was borrowed may not be at the beginning or the end of a month. You must then compute the elapsed time period in order to find the time factor. Below is a model to be used to compute the time.

Time Calculation Model

> days in the month that borrowing took place
> −date money is borrowed
> days that money was borrowed in this month
> +days in each succeeding month
> +date of repayment
> total time factor

Notice that you are actually not counting the day that the money was borrowed, but you are counting, as a full day, the day the money was returned.

Example 1   A loan made on July 16 was repaid on October 22. What is the time of the loan?

July has	31 days
Loan was made on July	−16
Borrowed money for	15 days in July
Borrowed money in August for	31 days
Borrowed money in September for	30 days
Borrowed money in October for	22 days (date of repayment)
Total time of loan =	98 days

(31 − 16 =   +31 + 30 + 22 =)

Example 2   Julie borrows $7,000 at 8 percent on December 4 to be repaid on February 16 of the following year.

December has	31 days
Loan was made on December	−4
Borrowed money for	27 days in December
Borrowed money in January for	31 days
Borrowed money in February for	16 days (date of repayment)
Total time of loan =	74 days

(31 − 4 =   +31 + 16 =)

The time factor can also be determined by using the time table shown on page 164. Each day of the year is assigned a number that begins with the first day of the year. By subtracting the number of the date of the loan from the date to repayment, the number of days in the loan can be determined.

Time Table

Day of Month	Jan.	Feb.	March	April	May	June	July	Aug.	Sept.	Oct.	Nov.	Dec.	Day of Month
1	1	32	60	91	121	152	182	213	244	274	305	335	1
2	2	33	61	92	122	153	183	214	245	275	306	336	2
3	3	34	62	93	123	154	184	215	246	276	307	337	3
4	4	35	63	94	124	155	185	216	247	277	308	338	4
5	5	36	64	95	125	156	186	217	248	278	309	339	5
6	6	37	65	96	126	157	187	218	249	279	310	340	6
7	7	38	66	97	127	158	188	219	250	280	311	341	7
8	8	39	67	98	128	159	189	220	251	281	312	342	8
9	9	40	68	99	129	160	190	221	252	282	313	343	9
10	10	41	69	100	130	161	191	222	253	283	314	344	10
11	11	42	70	101	131	162	192	223	254	284	315	345	11
12	12	43	71	102	132	163	193	224	255	285	316	346	12
13	13	44	72	103	133	164	194	225	256	286	317	347	13
14	14	45	73	104	134	165	195	226	257	287	318	348	14
15	15	46	74	105	135	166	196	227	258	288	319	349	15
16	16	47	75	106	136	167	197	228	259	289	320	350	16
17	17	48	76	107	137	168	198	229	260	290	321	351	17
18	18	49	77	108	138	169	199	230	261	291	322	352	18
19	19	50	78	109	139	170	200	231	262	292	323	353	19
20	20	51	79	110	140	171	201	232	263	293	324	354	20
21	21	52	80	111	141	172	202	233	264	294	325	355	21
22	22	53	81	112	142	173	203	234	265	295	326	356	22
23	23	54	82	113	143	174	204	235	266	296	327	357	23
24	24	55	83	114	144	175	205	236	267	297	328	358	24
25	25	56	84	115	145	176	206	237	268	298	329	359	25
26	26	57	85	116	146	177	207	238	269	299	330	360	26
27	27	58	86	117	147	178	208	239	270	300	331	361	27
28	28	59	87	118	148	179	209	240	271	301	332	362	28
29	29	. . .	88	119	149	180	210	241	272	302	333	363	29
30	30	. . .	89	120	150	181	211	242	273	303	334	364	30
31	31	. . .	90	. . .	151	. . .	212	243	. . .	304	. . .	365	31

Note: One day must be added to dates after February 28 on leap year problems.

Example 3   Using the information in example 1 and the time table, a loan is made on July 16 and repaid on October 22. What is the time of the loan?

October 22 is day   295
July 16 is day   −197
    98 days

(295 − 197 =)

Example 4   Use the time table to determine the time of a loan that was made on January 10, 2004 and repaid on May 23, 2004.

May 23 is day   144 (143 + 1 for leap year)
January 10 is day   −10
    134 days

(144 − 10 =)

Exercise A

In the following problems, find the number of days or the term of the loan.

Date of Loan	Date of Repayment	Days
1. July 17	August 21 (31 − 17 = 14 14 + 21 =)	35
2. May 17	June 30	
3. March 10	June 1	
4. September 16	December 3	
5. November 22	January 3	

When the date of the loan is known and the number of days, or the term, of the loan is known, you have the necessary information to compute the date of repayment. The determination of the repayment date is very similar to the approach used in determining the term of the loan.

Example 5

The $26,000 loan made to Sport-A-Merica for the new vehicle was made on January 16, 2001 for ninety days. Determine the date of repayment or the maturity date.

January has	31 days
Loan was made on January −	16
Borrowed money for	15 days in January
There are	28 days in February
There are	31 days in March
	74 days in total to date
→	16 more days required to make a total of
	90 days

Loan will come due on April 16.

(31 − 16 =   +28 + 31 + 16 =)

Example 6

Mark Watson borrowed $562 on June 25 for sixty days at 9 percent interest. What is the date of repayment, or the maturity date?

June has	30 days
Loan was made on June	−25
Borrowed money for	5 days in June
There are	31 days in July
	36 days in total to date
→	24 more days required to make total
	60 days

Loan will mature on August 24.

(30 − 25 =   +31 + 24 =)

The time table can be used to determine the due date on a loan. Using the date of the loan number from the table, add the term of the loan to the number, and find the new number (sum) in the table. The new number will be the maturity date.

Example 7

Using the information from example 6, find the maturity date of a loan made on June 25 for sixty days.

June 25 is day

176	
+60	term of note
236	number to be found on the time table

Day 236 is August 24.

(176 + 60 =)

Exercise B

In the following problems, find the date of repayment or the maturity date. Watch for leap year!

Date of Loan	Term of Loan	Date of Repayment
1. December 8, 2001	90 days (31 − 8 = 23 23 + 31 + 28 + 8 =)	March 8, 2002
2. October 12, 2001	45	_____
3. June 3, 2004	60	_____
4. August 30, 2002	120	_____
5. January 23, 2004	60	_____

The Formula:
I = P × R × T
Objective 3

time

ordinary interest

exact interest

rate

○ *The old adage of "time is money" is especially true in interest calculations. The more the time, the more the money in interest!*

principal

Time is one of the three factors needed in the formula to compute interest (I = P × R × T). Time is usually expressed as a fraction of a year. The numerator is the length of time that the money is to be borrowed (term of loan). The denominator is either 360 or 365. A denominator of 360 is used to compute what is termed **ordinary interest.** This approach earns more interest for the lender and is used for commercial loans to businesses. When a 365-day denominator is used to compute interest, the answer is termed **exact interest.** Exact interest earns the lender less interest because the denominator for time is larger. By law, exact interest is used for loans to consumers. Time may also be described as the term or term of loan.

The **rate** of interest is always expressed as a percent. The percent or interest rate is an annual rate, that is, for one year. When the loan is for less than one year, the time factor must be used to adjust the annual rate for the shorter time period. The rate factor multiplied by the time factor will result in an adjusted rate factor in the interest formula. (*Note:* If the loan is for a year, the time factor will be 1, or $\frac{1}{1}$; for four years it will be 4 or $\frac{4}{1}$; for six months, or $\frac{1}{2}$ of a year, it will be $\frac{6}{12}$ or $\frac{1}{2}$.)

The amount of money borrowed is termed the **principal**. This is the amount that the borrower is "renting," and therefore, the borrower must pay interest, or "rent," on this amount. The more that is borrowed at a given rate and time factor, the higher the interest.

You should now be able to understand all three elements of the interest formula: principal, rate, and time. To compute interest, you will be multiplying; dividing; using fractions, decimals, and percent; computing time; and lastly, rounding off.

Refer to appendix B for additional information on shortcuts in this area.

Example 8

Principal = $15,000
Rate = 6%
Time = 90 days
Interest = _____? (using ordinary interest)

I =   P   ×   R   ×   T

$$I = \$15{,}000 \times .06 \times \frac{90}{360}$$

You can cancel in the above formula as shown.

$$I = \$\cancel{15{,}000}^{7{,}500} \times \cancel{.06}^{.03} \times \frac{\cancel{90}^{1}}{\cancel{360}_{\cancel{4}_{\cancel{2}_{1}}}}$$

$$I = \frac{\$7{,}500}{1} \times \frac{.03}{1} \times \frac{1}{1}$$

$$I = \$225$$

(7,500 × .03 =)

Example 9

Principal = $1,600
Rate = 7%
Time = June 23 to August 30 (using exact interest)

I =   P   ×   R   ×   T

$$I = \$\cancel{1{,}600}^{320} \times .07 \times \frac{68}{\cancel{365}_{73}}$$

$$I = 320 \times .07 \times \frac{68}{73}$$

$$I = \$20.87$$

Time Calculation

June 30
–June 23
   7 days in June
+31 days in July
+30 days in August
  68 days in total

(320 × .07 × 68 ÷ 73 =)

Exercise C

Compute interest in the following problems (use ordinary interest).

1. Principal = $600
   Rate = 8%
   Time = 42 days

   Interest = __$5.60__ $\left(600 \times .08 \times \frac{42}{360} = \right)$

2. P = $1,000
   R = 14%
   T = 25 days

   I = _____

3. P = $6,000
   $R = 12\frac{1}{4}\%$
   T = 120 days

   I = _____

4. P = $408
   $R = 9\frac{1}{2}\%$
   T = From June 16 to October 20

   I = _____

5. P = $260
   R = 8%
   T = January 17, 2004 to June 3, 2004

   I = _____

6. P = $3,500
   R = 12%
   T = February 29, 2004 to May 19, 2004

   I = _____

Exercise D

In the following problems, compute interest using the exact interest method (365-day year).

1. P = $600
   R = 8%
   T = 45 days
   I = ___$5.92___ $\left(600 \times .08 \times \dfrac{45}{365} = \right)$

2. P = $3,000
   R = $12\frac{1}{4}$%
   T = 180 days
   I = _____

3. P = $365
   R = 15%
   T = 52 days
   I = _____

4. P = $850
   R = $13\frac{1}{2}$%
   T = 36 days
   I = _____

5. P = $1,000
   R = 10%
   T = 36 days
   I = _____

6. P = $450
   R = 14%
   T = 75 days
   I = _____

Other Interest Problems
Objective 4

It may be necessary to determine the amount of principal, rate, or time on a loan when the interest and two other factors of the I = P × R × T formula are known. The formula I = P × R × T can be manipulated to develop the following formulas.

$$P = \frac{I}{R \times T} \qquad R = \frac{I}{P \times T} \qquad T = \frac{I}{P \times R}$$

○ *Algebra can be used with the I = P × R × T formula to find any one of the four unknowns.*

You will need to know these formulas; use the model shown or use your skill at finding the unknown to solve these problems.

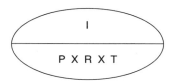

By putting your finger on the factor that you wish to find, the remaining portion of the model reveals the formula.

Example 10

If you know Interest, Principal, and Time, but need to find Rate, place your finger over R for Rate. Rate is found by dividing I by the product of P × T.

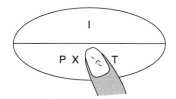

Computing Principal

The principal amount in a loan problem can be found when the other three factors are known.

Example 11 The interest on a loan to Sport-A-Merica was $520. The loan ran for ninety days at an 8 percent interest rate. What was the principal amount?

$$P = \frac{I}{R \times T} \qquad P = \frac{\$520}{.08 \times \frac{90}{360}} = \$26,000$$

(520 ÷ (.08 =   × 90 =   ÷ 360 =))

Computing Rate

When the rate on a loan is to be found, the formula $R = \frac{I}{P \times T}$ must be used.

Example 12 The same data from the earlier example will be used to illustrate this formula.

$$R = \frac{I}{P \times T} \qquad R = \frac{\$520}{\$26,000 \times \frac{90}{360}} = .08 = 8\%$$

(520 ÷ (26,000 =   × 90 =   ÷ 360 =))

Computing Time

The time factor can be determined by using the rate formula with one additional step. The decimal results equal a portion of a year. Multiply the decimal by 360 days to obtain the answer in days.

Example 13 $\quad T = \frac{I}{P \times R} \qquad T = \frac{\$520}{\$26,000 \times .08} = .25 \qquad .25 \times 360 \text{ days} = 90 \text{ days}$

(520 ÷ 26,000 =   ÷ .08 =   × 360 =)

Exercise E

Fill in the missing information in these problems (use ordinary interest). Round off your answer to the nearest percent, day, or cent.

	Interest	Principal	Rate of Interest	Term of Loan
1.	$8.00	$720.00	8% (8 ÷ (.08 × 50 ÷ 360 =))	50 days
2.	_____	5,500.00	10	20
3.	4.50	400.00	___	30
4.	14.63	1,300.00	15	___
5.	17.50	_____	14	60
6.	80.00	2,700.00	___	90
7.	_____	10,000.00	17	45
8.	18.00	_____	10	60
9.	8.98	4,200.00	11	___
10.	40.00	1,200.00	___	90
11.	73.00	_____	12	53
12.	100.00	_____	15	40
13.	3.90	450.00	13	___
14.	70.00	750.00	___	270
15.	83.67	1,000.00	12	___
16.	_____	42,000.00	11	90
17.	12.00	1,000.00	___	45
18.	75.00	_____	13	45
19.	53.37	2,500.00	13	___
20.	41.00	_____	14	60

ASSIGNMENT Chapter 9

Name _____ Date _____ Score _____

Skill Problems

Select the best answer for the statements below.

A. Interest   B. Principal   C. Rate   D. Time
E. Calendar   F. The Interest Formula G. Ordinary Interest   H. Exact Interest

1. _____ The number of days in each month of a year.

2. _____ A financial return to the lender for the risks undertaken during the term of a loan.

3. _____ The expression for finding the annual charge for the use of money.

4. _____ An interest rate that uses a 365-day year; used for personal loans only.

5. _____ An interest rate that uses a 360-day year; used for commercial loans only.

6. _____ The amount borrowed.

In the following problems, find the number of days or the term of the loan.

Date of Loan	Date of Repayment	Days
7. January 3, 2002	August 2, 2002	_____
8. March 27, 2001	December 6, 2003	_____
9. December 5, 1998	February 29, 2000	_____
10. January 2, 2000	March 1, 2000	_____
11. May 23, 2002	October 5, 2002	_____

In the following problems, find the date of repayment or the maturity date.

Date of Loan	Term of Loan	Date of Repayment
12. February 3, 2001	120 days	_____
13. November 24, 2002	60 days	_____
14. April 14, 2003	90 days	_____
15. June 22, 2004	45 days	_____
16. December 18, 2003	120 days	_____

Find the number of days or the term of the loan.

Date of Loan	Date of Repayment	Days
17. July 19	September 10	_____
18. March 23	July 1	_____
19. July 30	October 2	_____
20. May 2	June 1	_____
21. September 1	December 6	_____

Find the date of repayment or maturity date.

Date of Loan	Term of Loan	Date of Repayment
22. January 10, 2000	65 days	_____
23. June 3, 2001	120	_____
24. May 13, 2002	40	_____
25. March 6, 2002	40	_____
26. January 12, 2004	90	_____

Compute interest in the following problems (use ordinary interest).

27. Principal = $4,000
     Rate = 8%
     Time = 45 days

    Interest = _____

28. P = $7,500
    R = 10%
    T = 90 days

    I = _____

29. P = $24,500
    R = 7%
    T = January 5, 2002 to May 23, 2003

    I = _____

30. P = $10,000
    R = 15%
    T = 25 days

    I = _____

31. P = $400
    R = 7 1/4%
    T = From July 1 to November 12

    I = _____

32. P = $15,750
    R = 12.5%
    T = January 4, 2000 to March 29, 2000

    I = _____

Compute interest in the following problems (use exact interest).

33. Principal = $800
     Rate = 8%
     Time = 45 days

    Interest = _____

34. P = $10,740
    R = 10%
    T = 74 days

    I = _____

35. P = $1,000
    R = 11%
    T = 32 days

    I = _____

36. P = $8,760
    R = 11 1/4%
    T = From July 3 to August 29

    I = _____

37. P = $806
    R = 6%
    T = 120 days

    I = _____

38. P = $105,500
    R = 7 1/2%
    T = 90 days

    I = _____

Compute interest in the following problems.

39. Principal = $2,000

   Rate = $12\frac{1}{2}$%

   Time = 30 days

   Interest = $_____ (ordinary)

40. Principal = $1,800

   Rate = 9%

   Time = 13 days

   Interest = $_____ (ordinary)

41. Principal = $30,000

   Rate = 16%

   Time = 62 days

   Interest = $_____ (ordinary)

42. Principal = $600

   Rate = 9%

   Time = 60 days

   Interest = $_____ (exact)

43. Principal = $9,500

   Rate = $15\frac{3}{4}$%

   Time = 1 year

   Interest = $_____ (exact)

Fill in the missing information in these problems (use ordinary interest). Round off your answer to the nearest percent, day, or cent.

	Interest	Principal	Rate of Interest	Term of Loan
44.	$9.14	_____	7%	50 days
45.	_____	4,800.00	11	75
46.	143.00	1,300.00	__	264
47.	323.00	48,450.00	12	__
48.	175.56	_____	10	80
49.	2.33	150.00	__	70
50.	_____	55,000.00	6	40
51.	27.30	_____	14	90
52.	54,078.75	940,500.00	11	__
53.	15.41	460.00	__	170
54.	165.00	_____	9	60
55.	600.00	_____	12	30
56.	87.50	_____	15	30
57.	500.00	20,000.00	__	90
58.	36.00	900.00	16	__
59.	14.70	760.00	__	68
60.	38.33	5,000.00	12	__
61.	46.67	_____	16	30

Business Application Problems

Round off your answer to the nearest unit, percent, or cent.

1. Mari paid $78.90 in interest to her credit union for a ninety-day, 8 percent loan that indicated exact interest. To the nearest dollar, what amount did she borrow?

2. A loan for $50,000 was repaid by Latire Company to their bank through the issuance of a check for $50,489.58. The loan carried an ordinary interest rate of 11.75 percent. What was the term of the loan?

3. A commercial loan in the amount of $14,500 was repaid, including $161.11 in interest at 10 percent, on June 20th. What was the date that the ordinary interest loan was made?

4. Marti Cook lent a friend $500 for a period of 85 days. She charged 15 percent interest. What was the amount due at the end of the 85 days? (Use exact interest.)

5. The National Bank received $15,166.67 on a $15,000 note that was outstanding for 40 days. What was the rate of interest on the note? (Use ordinary interest.)

6. What is the amount of a commercial loan that carried an interest rate of 15.5 percent for a term of sixty days if the total interest on the loan was $4,133.33? (Use ordinary interest.) Round off your answer to the nearest dollar.

7. What is the payoff on the loan in problem 6?

8. What would be the amount of payoff in problem 6 if the loan had been an exact interest loan?

9. Birkhut and Masters Company paid $1,250 on a $25,000, 120-day ordinary interest loan. What was the interest rate on the loan?

10. Alisandra borrowed $400 for a period of six months. She paid off the exact interest loan, including $20 in interest, on time. What rate of interest did she pay?

11. Sight and Sound Video borrowed $15,000 from a supplier for a period of sixty days at 14 percent ordinary interest. What is the amount of interest due at the end of the loan?

12. Chad deposited his bonus check for $2,000 in his savings account on January 15, 2004. He hopes to have enough saved to take a vacation on July 5, 2004. The savings account pays $5\frac{1}{4}$ percent interest. What will his deposit be worth on July 5? (Use ordinary interest.)

13. A ninety-day loan was secured from the First National Bank. The loan was made by Jackie Lee for $12,000. The bank is charging a 14 percent rate. What will be the amount of interest on the loan if it runs the full ninety-day period? (Use exact interest.)

14. James B. Mackie borrowed $9,900 from a bank. The bank must charge 13 percent interest on loans to make a profit. Mr. Mackie's loan was for ninety-three days. How much interest did the bank earn? (Use ordinary interest.)

15. The Severance National Bank of Adrain, Michigan, charged Harold Denlar $9 worth of interest on a ninety-day loan of $1,000 at $10\frac{1}{2}$ percent interest. Mr. Denlar went all the way to the bank president, Mr. Ronald Severance, to complain that he had been overcharged. After careful examination of the facts, Mr. Severance stated that Mr. Denlar was in fact *under*charged. Who was correct—Mr. Denlar or Mr. Severance? (Use exact interest.)

16. A loan was repaid with a check in the amount of $5,758.81. The loan was made on January 31, 2002 and repaid on June 1, 2002. The loan was for $5,500. To the nearest percent, what was the rate of interest on the loan? (Use ordinary interest.)

17. Tony Cox borrowed $1,983 from the Pratt Union. The credit union charges a $12\frac{3}{4}$ percent interest rate on loans. What is the amount due on the loan if it runs for sixty days? (Use exact interest.)

18. A loan for $850 dated July 27 was repaid on December 20. The rate of interest was 16 percent. What was the value of the check? (Use exact interest.)

19. Mike Bruen agreed to repay an education loan of $12,000 to his father plus 9 percent interest. The loan was from a period starting on August 31, 2002 to July 15, 2003. What was the amount due his father on July 15? (Use ordinary interest.)

20. A loan for $22,500 at 10 percent interest was paid back with a check in the amount of $22,950 on July 10, 2001. Using ordinary interest, determine the date the note was made.

21. Orson Ingrahm loaned a business associate $5,000 on December 20, 2000. The loan was for three months at $14\frac{3}{4}$ percent. What is the amount of the interest on the loan? (Use ordinary interest.)

22. What was the principal of a commercial loan that carried a rate of 16 percent for a period of ninety days if the total interest paid was $14,000? (Use ordinary interest.)

23. What is the difference in interest on a loan using ordinary versus exact interest if the loan value is $5,000 for a term of 120 days at $16\frac{1}{2}$ percent?

24. Write a complete sentence in answer to this question. A student loan for $6,000 must be refinanced. The lender had loaned the money to the student two years ago at 14 percent. The student has not made any payments on the principal or interest. How much will the student need to refinance?

25. Write a complete sentence in answer to this question. Sherry borrowed $1,500 from a loan company on June 23. She repaid the loan on August 30. The loan specified a 15 percent rate (ordinary interest). What is the amount of payoff on the loan?

★ **Challenge Problem**

The Malon Corporation borrowed $52,000 from the Manufacturers Commerce Bank. The bank charges $9\frac{3}{4}$ percent interest for loans, and pays $4\frac{1}{2}$ percent compounded semiannually on savings. The bank is forty-seven years old and will soon finance a new $4.3 million Science and Trade Center. How much interest will the Malon Corporation have to pay for their ninety-day loan? (Use ordinary interest.)

✓ MASTERY TEST Chapter 9

Name _____ **Date** _____ **Score** _____

Complete the following problems.

1. Interest can be defined as
   a. The amount of money borrowed.
   b. The amount of money that is to be returned to the lender.
   c. The money that the borrower must pay the lender for the use of the principal.
   d. The results of time multiplied by rate.

2. Date of loan: March 6
   Date of repayment: August 1

   Days: _____

3. Date of loan: February 20, 2002
   Days of term of loan: Sixty days
   Date of repayment: _____

4. P = $50,000
   R = 16%
   T = 30 days

   I = $ _____ (ordinary interest)

5. What is the answer to problem 4 using exact interest?
   I = P × R × T

   I = $50,000 × .16 × $\frac{30}{365}$

   I = _____

6. Viola Austin borrowed $1,000 from a friend at the rate of $10\frac{3}{4}$ percent interest. The loan ran for forty-five days. What is the repayment date for this loan made on May 12?

7. What is the amount of the interest due on the loan in problem 6? (Use ordinary interest.)

8. What is the total amount of money that Viola must repay when the loan comes due in problem 6?

9. Bud borrowed $1,800 from a friend for sixty days at $15\frac{1}{2}$ percent interest. What was the amount of interest on the loan at the repayment date? (Use ordinary interest.)

10. Patty Emerson lent a business associate $5,000 on May 2. The loan was to be repaid on October 12 along with 13 percent interest. What was the amount due on October 12? (Use exact interest.)

11. Julie Snider borrowed $5,000 for sixteen days at 17 percent interest. How much must she pay back? (Use exact interest.)

12. A loan for $900 at 12 percent interest runs for a period of seventy-one days. The borrower asks the bank to indicate the amount of interest owed when the loan comes due. How much interest will be due on the loan at the maturity date? (Use ordinary interest.)

PROMISSORY NOTES 10

OBJECTIVES

After mastering the material in this chapter, you will be able to:

1. Describe the legal implications of a promissory note. p. 180

2. Compute the maturity value of a promissory note using ordinary or exact interest. p. 180

3. Compute the proceeds of a discounted noninterest-bearing or discounted interest-bearing promissory note. p. 181

4. Understand and use the following terms:
   Promissory Note
   Maker
   Payee
   Face Value
   Term of Note
   Maturity Value
   Noninterest-bearing Note
   Interest-bearing Note
   Discount Rate
   Term of Discount
   Discount Amount
   Proceeds

SPORT-A-MERICA

 Although the holiday season is over, the sales volume of Sport-A-Merica continues to increase. Hal has been working hard managing the business in the store, while Kerry has been concentrating her efforts on selling to commercial accounts. After several months of calling on Shape-Up, Inc., Kerry obtains a $7,000 wholesale order for sportswear and equipment. Kerry's persistence has paid off again! Shape-Up had been interested in Sport-A-Merica merchandise for some time, but was unable to purchase in a quantity large enough to qualify for a discount from Sport-A-Merica. This order will be profitable for both firms, but Sport-A-Merica will have to accept a promissory note from Shape-Up in order to make the deal. With the promissory note due soon, Kerry is considering the effect this will have on the cash situation for Sport-A-Merica.

Legal Environment
Objective 1

promissory note

maker

payee

face value

term of note

A **promissory note** is used in business as an IOU. It is a legally enforceable written contract. The **maker** of the note is obligated to repay the amount of the note, along with interest, to the **payee.** The Uniform Commercial Code defines a promissory note as "an unconditional promise in writing made by one person to another, signed by the maker engaging to pay on demand or at a particular future time a certain sum of money to order or to the bearer."

The amount written on the note is known as the **face value** and is equal to the amount borrowed. The note is a negotiable instrument (i.e., it is transferable from one party to another) for cash or merchandise. The time period that the note is to run is known as the **term of note.** This will be the actual number of days from the time the note is made until it is repaid. Some notes run for years and are usually for larger sums of money.

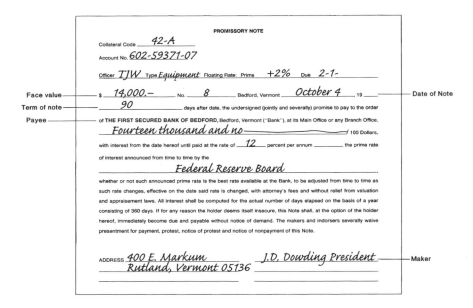

Objective 2

maturity value

A promissory note can be given, as an example, to the seller of merchandise by the buyer or as an extension of credit by the seller to the buyer. In this example, the maker, J. D. Dowding, promises to pay the payee, The First Secured Bank of Bedford, the sum of $14,000 plus 12 percent interest for the 90-day term of the note. This amount to be repaid is due 90 days after October 4, the date the note was made. The value of the note on that date is known as the **maturity value,** and will be $14,420 [$14,000 + ($14,000 \times .12 \times \frac{90}{360} = $420)]. The maker is obligated to pay $14,420, the maturity value, to the payee or bearer on that date.

The numerical values can be shown graphically.

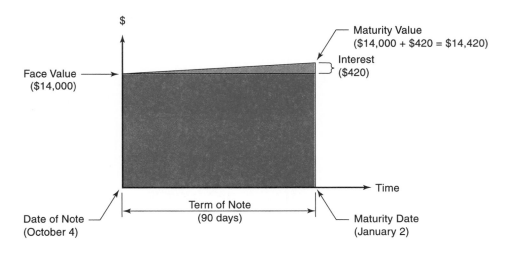

noninterest-bearing note

interest-bearing note

Occasionally, the payee will not charge the drawer interest on the note. This type of note is known as a **noninterest-bearing note.** In the promissory note shown on page 180, the face value of $14,000 would also be the maturity value ($14,000) as a noninterest-bearing note. Otherwise, and most frequently, the note will be an **interest-bearing note,** indicating that the maker must repay both face value ($14,000) and interest ($420) at maturity, a total of $14,420.

Exercise A

Determine the maturity value of the interest-bearing and noninterest-bearing notes described here. Use ordinary interest.

Face Value	Term of Note	Interest Rate	Maturity Value
1. $6,500	60 Days	15% (6,500 × .15 × 60 ÷ 360 =)	$6,662.50
2. 4,000	60	12	
3. 13,600	60	17	
4. 25,000	100	$10\frac{3}{4}$	
5. 9,000	40	None	
6. 7,000	120	13	
7. 8,500	30	$17\frac{1}{2}$	
8. 5,000	30	None	
9. 60,000	73	14	
10. 3,000	1 Year	18	

Discounting
Objective 3

When the payee of a promissory note (noninterest-bearing or interest-bearing) has a shortage of cash, the firm (or person) may decide to sell a promissory note. This will turn a semiliquid asset into spendable cash, allowing the payee to pay invoices, take advantage of cash discounts, and meet payroll obligations. The note is negotiable, and therefore, it can be sold to a bank or another firm that has ready cash. The payee must determine the rate that this other lending institution will charge the payee to gain cash for the note. This rate is called the **discount rate.** The length of time the note will be discounted at this rate is known as the **term of discount.**

discount rate
term of discount

Computing Discount

If a 90-day note dated October 4 has a maturity value of $14,420, and is discounted for cash on December 13 of that year, then the term of discount will be twenty days.

October has	31 days	January	2—Maturity date
−October	4	December	+31
	27	−December	−13—Date of discount
November	30		20 days = Term of discount
December	31		
January	+2—Maturity date		
	90 days		

discount amount

If the discount rate was 10 percent, the **discount amount** would be computed as follows:

○ *The discount is always computed on the maturity value.*

$$\text{discount amount} = \text{maturity value} \times \text{discount rate} \times \text{term of discount}$$
$$\$80.11 \quad = \quad \$14,420 \quad \times \quad .10 \quad \times \quad \frac{20}{360}$$

Computing Proceeds

proceeds

The discount amount is charged by the lending institution for cash and will be subtracted from the maturity value in order to find the amount of cash that the payee will receive, the **proceeds.** The proceeds = $14,420 − $80.11 = $14,339.89. The payee receives $14,339.89 on December 13; the lending institution will receive the $14,420 maturity value of the note from the maker the following year on January 2 as the bearer of the note.

The numerical values to determine proceeds can also be shown graphically.

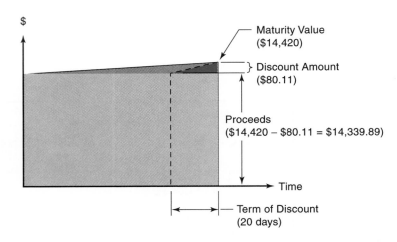

Another example should clarify any questions.

Example 1

Information

Face value of note	$780
Rate of interest	$9\frac{1}{2}\%$
Date of note	June 17, 2001
Term of note	60 days
Interest	$_____
Maturity value	$_____
Discount rate	$10\frac{1}{4}\%$
Date of discount	July 30, 2001
Term of discount	_____
Discount amount	$_____
Proceeds	$_____

Solution

$I = P \times R \times T$

$I = \$780 \times .095 \times \dfrac{60}{360}$

$I = \$12.35$

$\$780.00 + \$12.35 = \$792.35 =$ maturity value

Discount = maturity value × discount rate × term of discount

Discount $= \$792.35 \times .1025 \times \dfrac{17}{360}$

Discount $= \$3.84$

```
  June 30            July has 31
 −June 17           −July 30
     13                  1
 +July 31           +Aug. 16
     44                 17 days = Term of discount
 +Aug. 16—Maturity date
     60
```

Maturity value	=	$792.35
Amount of discount	=	−3.84
Proceeds	=	$788.51

(780 × .095 × 60 =   ÷ 360 =   792.35 × .1025 × 17 =   ÷ 360 =)

In the next example, the calculations are shown in an abbreviated form. This model should be useful for further assignments in this area.

Example 2

Kerry finds that the $7,000 note she accepted from Shape-Up must be discounted for cash in order to pay the current bills on May 17, 2002. The note will be discounted at 9 percent even though it has only earned 8 percent since it was given on March 11, 2002. What are the proceeds on this 90-day note?

Face value of note	$7,000
Rate of interest	8%
Date of note	March 11, 2002
Term of note	90 days
Discount rate	9%
Date of discount	May 17, 2002
Proceeds	$_____

Solution

Interest $= \$7,000 \times .08 \times \dfrac{90}{360} = \$140$

$$\text{Discount} = \underset{\underline{+140 \text{ interest}}}{\$7,140} \times .09 \times \dfrac{23}{360} = \$41.06$$

$$\text{Proceeds} = \underset{\underline{-41.06 \text{ discount}}}{\$7,098.94}$$

Note: The proceeds may be greater or less than the face value.

(7,000 × .08 × 90 =   ÷ 360 =   7,140 × .09 × 23 =   ÷ 360 =)

Exercise B

Determine the term of discount in each of the following exercises.

Date of Note	Term of Note	Date of Discount	Term of Discount
1. June 4, 2002	60 days (20 + 3 =)	July 11, 2002	23 days
2. Aug. 20, 2001	60	Sept. 26, 2001	_____
3. Sept. 14, 2000	90	Oct. 5, 2000	_____
4. May 12, 2002	60	June 20, 2002	_____
5. Jan. 7, 2000	95	Mar. 16, 2000	_____
6. Dec. 14, 2003	150	Feb. 27, 2004	_____
7. April 12, 2002	30	April 20, 2002	_____
8. Jan. 30, 2000	60	Feb. 29, 2000	_____

Exercise C

Compute the answers to these problems. Use ordinary interest (360-day year).

	1.	2.
Face value	$450	$5,000
Rate of interest	10%	16%
Date of note	Oct. 3, 2002	Mar. 6, 2001
Term	45 days	60 days
Interest (450 × .10 × 45 ÷ 360 =)	$5.63	_____
Maturity value	$455.63	_____
Discount rate	10%	11%
Date of discount	Nov. 1, 2002	Mar. 12, 2001
Term of discount	16 days	_____
Discount am't. (455.63 × .10 × 16 ÷ 360 =)	$2.03	_____
Proceeds	$453.60	_____

	3.	4.
Face value	$5,000	$6,000
Rate of interest	14%	12%
Date of note	May 3, 2001	July 16, 2002
Term	90 days	60 days
Interest	_____	_____
Maturity value	_____	_____
Discount rate	13%	13%
Date of discount	July 3, 2001	Aug. 30, 2002
Term of discount	_____	_____
Discount am't.	_____	_____
Proceeds	_____	_____

	5.	6.
Face value	$1,800	$170
Rate of interest	13%	$13\frac{3}{4}$%
Date of note	March 1, 2002	June 16, 2003
Term	120 days	19 days
Interest	_____	_____
Maturity value	_____	_____
Discount rate	12%	14%
Date of discount	May 14, 2002	June 30, 2003
Term of discount	_____	_____
Discount am't.	_____	_____
Proceeds	_____	_____

	7.	8.
Face value	$800	$1,100
Rate of interest	11%	12%
Date of note	January 3, 2004	April 1, 2001
Term	40 days	75 days
Interest	_____	_____
Maturity value	_____	_____
Discount rate	$9\frac{3}{4}$%	9%
Date of discount	Feb. 3, 2004	May 15, 2001
Term of discount	_____	_____
Discount am't.	_____	_____
Proceeds	_____	_____

(Continued on next page)

Exercise C Continued

	9.	10.
Face value	$1,000	$3,500
Rate of interest	13%	$12\frac{1}{2}\%$
Date of note	May 3, 2001	May 31, 2002
Term	60 days	45 days
Interest	_____	_____
Maturity value	_____	_____
Discount rate	$13\frac{1}{2}\%$	15%
Date of discount	June 30, 2001	July 1, 2002
Term of discount	_____	_____
Discount am't.	_____	_____
Proceeds	_____	_____

Name _____ Date _____ Score _____

Skill Problems

Write T for true or F for false in blank for the following statements.

1. _____ A promissory note is not legally enforceable.

2. _____ A promissory note is used in business as an IOU.

3. _____ The actual number of days from the time the note is made until it is repaid is the term of the note.

4. _____ Noninterest-bearing notes are used the most frequently.

5. _____ The face value is the amount a business is able to pay, not necessarily the amount of the note.

Select the best term for the description below.

A. Face Value   B. Maker   C. Maturity Date   D. Payee   E. Term of Note

6. _____ The amount borrowed (written on the note).

7. _____ The length of time that the note is to run.

8. _____ The party lending the face value of the note.

Determine the maturity value of the interest-bearing and noninterest-bearing notes described here. Use ordinary interest. Round off to the nearest cent.

	Face Value	Term of Note	Interest Rate	Maturity Value
9.	$20,000	60 days	14%	_____
10.	6,000	90	11	_____
11.	460,800	120	None	_____
12.	1,250	60	7 1/2	_____
13.	8,500	100	13	_____
14.	60,750	1 Year	15	_____
15.	3,050	30	None	_____

Compute the answers to these problems. Use ordinary interest. Round off to the nearest cent.

Face value of note	$940	**16.** Interest	_____
Rate of interest	10.5%	**17.** Maturity value	_____
Date of note	Jan. 2, 2002	**18.** Term of discount	_____
Term of note	90 days	**19.** Discount amount	_____
Discount rate	11%	**20.** Proceeds	_____
Date of discount	Feb. 25, 2002		

Face value of note	$7,400	**21.** Interest	_____
Rate of interest	7%	**22.** Maturity value	_____
Date of note	May 12, 2001	**23.** Term of discount	_____
Term of note	180 days	**24.** Discount amount	_____
Discount rate	9%	**25.** Proceeds	_____
Date of discount	June 30, 2001		

Face value of note	$10,000	**26.** Interest _____
Rate of interest	12%	**27.** Maturity value _____
Date of note	Oct. 31, 2000	**28.** Term of discount _____
Term of note	30 days	**29.** Discount amount _____
Discount rate	15%	**30.** Proceeds _____
Date of discount	Nov. 10, 2000	

Compute the answers to these problems. Use exact interest.

Face value of note	$5,000	**31.** Interest _____
Rate of interest	16%	**32.** Maturity value _____
Date of note	Nov. 24, 2003	**33.** Term of discount _____
Term of note	120 days	**34.** Discount amount _____
Discount rate	18%	**35.** Proceeds _____
Date of discount	March 13, 2004	

Face value of note	$1,140	**36.** Interest _____
Rate of interest	0%	**37.** Maturity value _____
Date of note	Feb. 3, 2001	**38.** Term of discount _____
Term of note	45 days	**39.** Discount amount _____
Discount rate	10%	**40.** Proceeds _____
Date of discount	March 16, 2001	

	41.	**42.**	**43.**	**44.**	**45.**
Face value	$65,000	$6,400	$4,500	$1,500	$40,000
Rate of interest	15%	11%	10%	15%	$14\frac{1}{2}$%
Date of note	May 3	March 8	Oct. 15	Jan. 13	June 10
Term of note	90 days	60 days	45 days	80 days	30 days
Date of discount	June 10	May 1	Nov. 2	Feb. 1	July 1
Rate of discount	16%	$10\frac{1}{2}$%	11%	12%	12%
Proceeds	_____	_____	_____	_____	_____

Compute the answers to the problems listed. Use exact interest. Round off to the nearest cent.

	46.	**47.**	**48.**	**49.**	**50.**
Face value	$3,000	$7,000	$6,500	$420	$900
Rate of interest	11%	13%	12%	None	10%
Date of note	July 10	Sept. 30	Oct. 5	May 4	Aug. 19
Term of note	90 days	100 days	90 days	120 days	45 days
Date of discount	Aug. 30	Oct. 15	Dec. 16	July 30	Sept. 3
Rate of discount	12%	15%	15%	16%	11%
Proceeds	_____	_____	_____	_____	_____

Business Application Problems

1. On May 12, Jackson Associates agreed to accept a promissory note in exchange for the $16,000 on account owed by Balentine and Company. The note carried a 13 percent interest rate and a term of 90 days. On June 1, Jackson Associates discounted the note at 15 percent for cash at the First Trust Bank. What is the maturity value of the note? (Use ordinary interest.)

2. Determine the amount of discount in problem 1.

3. What are the proceeds in problem 1?

4. Which firm receives the proceeds in problem 1?

5. Which firm is the maker of the note in problem 1?

6. Which firm is the payee of the note in problem 1?

7. What is the due date of the note in problem 1?

8. Jaber Professional Association discounted a promissory note with a maturity value of $4,560 on October 10 at 12 percent. The 120-day note was signed on September 14. What are the proceeds? (Use ordinary interest.)

9. Smithson Corporation agreed to sign a promissory note on May 26 for their $60,000 debt to X-Ron Limited. The 60-day note carried a 15 percent interest rate. On July 21, X-Ron Limited discounted the note at 13.5 percent at Blanchard Federal Bank of Lewiston. What are the proceeds of the note? (Use ordinary interest.)

10. The proceeds of a promissory note with a face value of $23,400 will be used to pay current expenses for the Bay Rite Company. The 9 percent, 45-day note was dated November 10. The Bay Rite Company discounted the note on November 28 at 10.3 percent at the Commercial Bank of Union City. What are the proceeds of the note? (Use ordinary interest.)

11. The Bader Company received a $4,000 note from a customer who was unable to pay for merchandise purchased on account. The note was dated April 7, 2001, and was to run for 90 days without interest. On June 1, 2001, the Bader Company decided to discount the note at 12 percent for cash. What were the proceeds? (Use ordinary interest.)

12. National City Bank received a check for $16,565.31 on November 1, 2002, from a promissory note customer. The customer had two notes outstanding at the time, and bank officials were not sure which one should be cancelled. Both notes had a $9\frac{1}{2}$ percent interest rate; one was dated October 17, 2002, for $16,500; and the other was dated June 17, 2002, for $16,000. Which note was the check to cover? (Use ordinary interest.)

13. Sass and Company accepted a 15 percent 90-day note for $4,000 from a customer on July 2, 2002. On August 20, 2002, Mrs. Jane Sass discounted the note at 18 percent for cash at the Warren State Bank. What were the proceeds from the note? (Use ordinary interest.)

14. A promissory note held by payee Howard Gorman was discounted for cash on December 20, 2000. The face value of the note was $62,000. The note was to run for a term of 60 days at 20 percent. The date on the note was December 10, 2000. Fred Hass agreed to purchase the note by discounting it 22 percent. What were the proceeds? (Use ordinary interest.)

15. The promissory note shown here was discounted for cash on October 12, 2002. What were the proceeds if the discount rate was 15 percent? (Use ordinary interest.)

No. 546776 ___ Date September 1, 2002 , 19. ___ Due March 1, 2002 _____

One hundred eighty-one _____ days after date, I/XX promise to pay to the order of
EMPIRE NATIONAL BANK, BRONX, N.Y. 10460
the sum of Thirteen thousand ------------------------------ Dollars, $ 13,000.00,
plus 12 % simple interest after 9/1/02 .
This note secured by Office equipment
Holder, upon deeming himself insecure, may, without notice declare the unpaid balance to be immediately due and payable. Holder upon default may realize on or dispose of the collateral. Makers severally waive demand, notice and protest, and any defense due to extension of time or other indulgence by holder and to any substitution or release of collateral. This Note will bear interest on the entire unpaid balance at the rate of 12 per centum (12%) per annum during any period of default hereunder.

Addresses: 2309 Taylor Street _____ Signatures: *Bonnie Kurtz* _____
_____   *Pilskton Corp.* _____

10-1 REV. 8/70

16. A firm decided to discount a note at 14 percent on July 16, 2000, that was valued at $21,880 at maturity. The note was made on June 1, 2000, to run for 120 days. What were the proceeds? (Use ordinary interest.)

17. The Works will discount a promissory note with a face value of $20,000 and an interest rate of 10 percent on March 14, 2001. The note was made on February 28, 2001, to run for 120 days. The discount rate agreed to by the lender and The Works is 15 percent. What are the proceeds? (Use ordinary interest.)

18. The Birk Corp. received a note from Alfred Coloni for a service provided on an industrial machine. The note stated that Mr. Coloni promised to pay $1,750 within 90 days after the date of the note, June 14, 2002. The Birk Corporation discounted the note at 14 percent on July 20, 2002, for cash. What was the amount of the check Mr. Coloni used to pay off the note if he paid it off on August 18, 2002?

19. On July 14, 2003, the lawyers in the law office of Peterson, Peterson, & Snell agreed to accept a promissory note for $1,200 for services. The note interest rate was 12 percent and was to be paid off within 90 days. The office manager, Janis Tobian, decided there was a shortage of cash for payroll, and discounted the note on September 11, 2003, at a rate of 14 percent. What were the proceeds? (Use ordinary interest.)

20. A $4,500 face-value promissory note bearing an interest rate of 14 percent for 90 days was discounted on June 3, 2004. The note was dated May 1, 2004. The discount rate was 18 percent. What were the proceeds? (Use ordinary interest.)

21. Billie accepted a 90-day promissory note from a friend on October 22, 2003. The face value of the note was $14,000. The note was interest free. Billie discounted the note on December 1, 2003, at a rate of 14 percent. What is the amount of the discount? (Use exact interest.)

22. Write a complete sentence in answer to this question. In problem 20, what would the interest have been if the note had carried an interest rate of 12 percent?

23. Write a complete sentence in answer to this question. Ted Jameson purchased a promissory note from Leonard Silk Design on May 16, 2002. Mr. Jameson purchased the $40,000 face-value note at a discount of 16 percent. The note carried an interest rate of 10 percent for the 60-day term that began on March 30, 2002. What were the proceeds? (Use exact interest.)

★ Challenge Problem

A consumer promissory note was negotiated on January 2, 2002, for $5,100 at $9\frac{3}{4}$ percent for 120 days. On March 18, 2002, the payee discounted the note at a rate of $13\frac{1}{8}$ percent for cash at the State Bank of Burns. The proceeds were used to meet a 15 percent increase in payroll expenses. The maker of the note won a lottery prize of $10,000 on April 6, 2002, and decided to pay off the note on that date. What was the amount that the payee was obligated to pay on April 6, 2002?

Name _____ Date _____ Score _____

Select the terms and correct answers using the following promissory note.

```
No. 463801          Date  June 23                    , 20 02  Due  One hundred twenty

_____(120)_____ days after date, I/WX promise to pay to the order of

the sum of  One thousand, eight hundred fifty and no/100 _____ Dollars, $ 1,850.00 ,
plus  1 2      % simple interest after  6-23-02   .
This note secured by Pilskton Corporation
Holder, upon deeming himself insecure, may, without notice declare the unpaid balance to be immediately due and payable. Holder
upon default may realize on or dispose of the collateral. Makers severally waive demand, notice and protest, and any defense due to
extension of time or other indulgence by holder and to any substitution or release of collateral. This Note will bear interest on the en-
tire unpaid balance at the rate of  Twelve  per centum ( 1 2 %) per annum during any period of default hereunder.

Addresses:  1401 Main Street                Signatures:  James White, Treasurer

          Your City, New York               _____
10-1 REV. 8/70
```

_____ **1.** Face value of note

_____ **2.** Maturity date

_____ **3.** Maturity value (ordin. int.)

_____ **4.** Date of note

_____ **5.** Maker

_____ **6.** Payee

a. First National Bank
b. $1,850
c. Pilskton Corp.
d. $1,924
e. October 21, 2002
f. 463801
g. 12 percent
h. September 20, 2002
i. $2,072
j. October 20, 2002
k. June 23, 2002

Dalmas Office Equipment Co. accepted a promissory note from Warren Steel for the purchase of $3,580 of office furniture on July 10, 2001. The note had an interest rate of 14 percent and was to run for ninety days. On August 7, Park Dalmas, president, discounted the note at 15 percent for cash. (Use ordinary interest.)

7. What was the term of discount? _____

8. What was the discount amount? _____

9. What were the proceeds? _____

10. How much is the maker required to pay at maturity to the holder of the note? _____

11. The Peters Corp. agreed to sell $900 of industrial supplies to Franton Manufacturing if Franton would sign a promissory note for the amount of the purchase. The interest-free note was to be paid off thirty days after the November 10, 2002 purchase. On November 20, 2002 Peters discounted the note at 14 percent for cash. What were the proceeds? (Use ordinary interest.)

INSTALLMENT LOANS

11

OBJECTIVES

After mastering the material in this chapter, you will be able to:

1. Determine the payment amount on an installment contract using a table or, when given, the amount of a loan, the time period of the loan, and the rate of interest. p. 198

2. Compute a repayment schedule for an installment contract. p. 201

3. Describe the use of an escrow account and its effect on a mortgage. p. 204

4. Describe the effect of an early payoff on an installment contract. p. 205

5. Understand and use the following terms and concepts:
   Down Payment
   Payment Amount
   Principal Balance
   Repayment Schedule
   Fixed Rate
   Variable Rate
   Fixed Payment Amount
   Variable Payment Amount
   Balloon Payment
   Points
   Closing Costs
   Escrow Account
   Rule of 78s

SPORT-A-MERICA

Kerry and Hal recognize that in order to expand the business, Sport-A-Merica needs to borrow money. "We cannot afford to discount a promissory note like we did with Shape-Up recently. The lack of working capital is going to be our next big hurdle to overcome," commented Kerry. They need more money, not only to buy more inventory and to finance the new store, but to establish the firm with a bank so that someday they can make the big push to Eastern Mall, a regional mall two miles away.

Kerry and Hal take the records of Sport-A-Merica to Larry Mataway, a loan officer at First National Bank, to apply for a loan. Now is the time for some serious financial planning. Kerry and Hal have a business plan in which $85,000 may be needed as additional working capital over the next 12 months. Though all of the $85,000 may not be used, they want to have approval for that amount in the event it is needed. Once the merchandise they plan to buy has been sold, a portion of the loan can be repaid. The entire $85,000 will probably not be needed for the full 12 months.

Kerry has also determined that a $120,000 loan is needed for the purchase of the new space. This loan would need to be paid off over a longer period of time, probably several years. She indicates to Larry that the only other loan outstanding for Sport-A-Merica is the $26,000 short-term loan with Community National Bank for the delivery vehicle. The loan will be paid off within two months.

As they discuss the loans with Larry, most of their needs are clear. However, Larry raises some important issues in the initial discussion. Rates, points, down payment, and amortizing are some of the items that must be determined. Hal, in particular, was happy they inquired early about the loans. He will need to know the monthly payment on the big loan to help him with cash planning. There are a lot of details to work out.

Credit Purchases

There are certain circumstances in a consumer's life that result in the purchase of goods and services costing more than the consumer earns. These instances are usually in the early years and again in the later years when, for example, children enter college or are married. Goods and services such as automobiles, furniture, homes, appliances, medical services, petroleum products, and entertainment are often purchased on credit. A credit purchase allows the consumer to own and use the product when it is needed, and to pay for it while it is being used. Without such an opportunity, consumers would have to go without desired and necessary items until they have saved the money for the purchase. Debt through credit purchases is typically incurred in the consumer's early twenties (purchase of an automobile, appliances, and a home), is reduced over a period of the next fifteen to twenty years, and then increases again (weddings, college education, etc.). Most of the debt is paid off by the time the consumer retires.

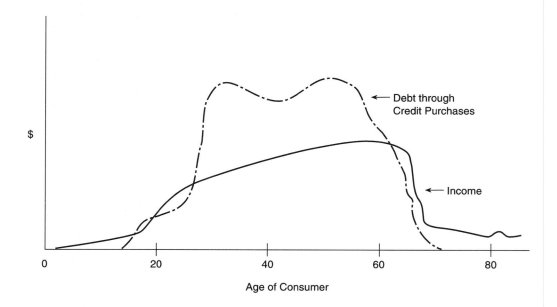

The ability to purchase on credit appeals to most Americans, as the majority use some type of credit financing. As with any freedom, there is also a degree of responsibility. The responsibility of repayment and credit management falls into the hands of the consumer. Currently, more than one million households declare bankruptcy each year, at least partly due to misuse of credit.

Consumers and business firms often make purchases that are in excess of available cash or exceed the limit of the credit card. When this happens, the bank or other lending institution provides a source of funds through an *installment loan*. Installment loans are not only used for larger amounts; they are also used when a longer time period (term) in which to pay back the loan is desired. The longer time period allows for a smaller monthly payment and lower total interest amount than would be the case if the loan were made through a promissory note. The following graph shows a comparison between a promissory note and an installment contract for $500 at 12 percent annual interest over a period of ten months.

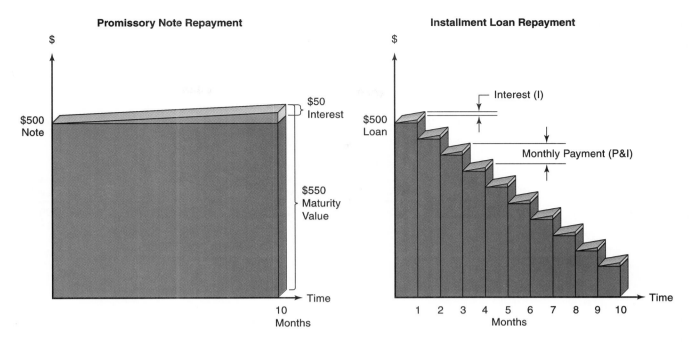

Promissory Note Repayment

$

$500 Note

$50 Interest

$550 Maturity Value

Time

10 Months

Installment Loan Repayment

$

$500 Loan

Interest (I)

Monthly Payment (P&I)

1 2 3 4 5 6 7 8 9 10

Months

Time

The buyer borrowed $500 and decreases the amount of the principal by making monthly payments. Because the amount of the loan is decreased over its life, the amount of interest also is less. The example shows that the buyer would repay a total of $550 on a promissory note but only a total of $527.92 if it were financed on an installment contract. (See page 201 for computations on the interest for the installment contract.) The borrower should also realize there is use of only approximately one-half of the principal over the term of the loan on an installment contract.

down payment

Unlike a promissory note, an installment loan usually requires a **down payment.** The down payment is made to secure the purchase and to prove the buyer's intent to the seller or lender. The borrower indicates the amount of down payment that can be made. The lender subtracts the down payment from the amount of the purchase. The loan is made on the balance (amount to be financed).

○ *The larger the down payment, the lower the monthly payments.*

Example 1

Jenny bought a TV satellite dish for a total purchase price of $386.75. She decided to make a **down payment** of $86.75 and to finance the remainder. Jenny borrowed the $300 at 10 percent interest for four months, to be repaid at $76.60 per month. The repayment schedule is computed as follows:

Month 1 $300.00 $\times .10 \times \dfrac{1}{12} = \$2.50$

$\underline{+2.50}$ = Interest
302.50
$\underline{-76.60}$ = Payment

Month 2 $225.90 $\times .10 \times \dfrac{1}{12} = \$1.88$

$\underline{+1.88}$ = Interest
227.78
$\underline{-76.60}$ = Payment

(Continued on next page)

Example 1
continued

Month 3 $151.18 $\times .10 \times \dfrac{1}{12} = \$1.26$

             $\dfrac{+1.26}{\$152.44}$ = Interest

             -76.60 = Payment

Month 4 $75.84 $\times .10 \times \dfrac{1}{12} = \$0.63$

             $\dfrac{+.63}{\$76.47}$ = Interest

             $\$76.47$ = Last payment

Record of Calculations   (Repayment Schedule)

Payment Number	Payment Amount	Interest Amount	Payment on Principal	Principal Balance
1	$76.60	$2.50	$74.10	$225.90
2	76.60	1.88	74.72	151.18
3	76.60	1.26	75.34	75.84
4	76.47	.63	75.84	0.00

Determining the Payment Amount
Objective 1

payment amount

In example 1, the **payment amount** was determined for you. *You* will need to determine the payment amount as a businessperson. There are two circumstances that will determine which method you will need to use. If the loan is for a few months, you will use the "estimating total interest on an installment contract" method shown below. This will be especially useful if you are selling small-dollar-amount items. If you are involved in the sale of large-dollar-value items such as buildings, equipment, or vehicles, you will want to use the payment table method presented later in this chapter, which will spread the payments out over several years.

Estimating the Total Interest on an Installment Contract

The Truth-in-Lending Act requires that the total amount of finance charges and the effective annual rate be stated on a loan. In order to compute the total amount of interest on a loan and the amount of the monthly payment, it will be necessary to be able to use this formula.

$$\text{estimating total interest on an installment contract} = \frac{\text{Interest for the first month} \times (\text{the months in the contract} + 1)}{2}$$

This estimating formula is used for installment contracts of only a few months. For installment contracts that have a term of years, use the table on page 200.

Note that the formula is different than the $I = P \times R \times T$ formula used for single-payment promissory notes.

Example 2

Betty bought her husband two sport coats for his birthday gift. The merchandise totaled $218. The men's store where the purchase was made requires 20 percent down and charges 11 percent interest. As the store manager, you are asked to determine the monthly payment required to pay the loan off within eight months. The calculations are as follows:

$218.00 = Amount of purchase $218.00
$\times$ .20 -43.60
    $43.60 = Down payment $174.40 = Amount to be financed

$$\text{total interest} = \frac{\$174.40 \times .11 \times \dfrac{1}{12} \times (8 + 1)}{2}$$

$$= \frac{\$1.60 \times 9}{2} = \$7.20$$

$174.40 = Amount to be financed
$\dfrac{+7.20}{\$181.60}$ = Interest on contract over 8 months
$181.60 = Total to be repaid

$181.60 ÷ 8 payments = $22.70 (monthly payment amount)

Example 3   Sandy Fossum purchased kitchen furniture for her condominium from Kirkby Associates, Inc. She decided to take the credit terms that were offered by store manager Carolyn Kirkby. She will finance the balance at 11 percent for twelve months. What will be the monthly payment amount?

Kirkby Associates, Inc.   # INVOICE

P O Box 459
Laingsburg, Mi. 48848-0459   Phone: 517-651-6302
                                     Fax: 517-651-5134

SOLD TO:		
Sandy Fossum		
3216 Emanuel Ave.		
Muskegon, MI 49507		

INVOICE NUMBER	*41623*
INVOICE DATE	*6-16-97*
OUR ORDER NUMBER	*12141*
YOUR ORDER NUMBER	—
TERMS	*installments*
SALES REP	*Carolyn*
SHIPPED VIA	*Nomo Trucking*
F.O.B.	*Muskegon*
PREPAID or COLLECT	*Prepaid*

SHIPPED TO: *Same*

QUANTITY	DESCRIPTION	UNIT PRICE	AMOUNT
1	*Table w/two leaves*	*999.–*	*999.–*
6	*Chairs w/caned backs*	*75.–*	*450.–*
			1,449.–
	Sales Tax		*86.94*
	Total		*$1,535.94*
	less down payment		*– 192.44*
	Charged		*$1,343.50*

THANK YOU FOR YOUR BUSINESS!

Courtesy of Kirkby Associates, Inc. Laingsburg, Michigan.

$1,535.94 = Purchase amount
−192.44 = Down payment
$1,343.50 = Amount to be financed

$$\text{Total interest} = \frac{\$1,343.50 \times .11 \times \frac{1}{12} \times (12 + 1)}{2}$$

$$= \frac{\$12.32 \times 13}{2}$$

$$= \$80.08$$

$1,343.50 = Amount to be financed
+80.08 = Amount of interest
$1,423.58 = Amount to be repaid in 12 payments
        = $118.63 monthly payment amount

Use of Payment Tables

The previous examples were for small loans for such items as office equipment, appliances, recreation equipment, furniture, and engagement rings. Installment loans are also used for larger loans—such items as machinery, buildings, land, and private homes. Because these items cost so much more, the loan amortization is usually extended over a period of years to let the buyer successfully pay off the loan. The process of computing the repayment schedule is exactly the same as in the earlier examples. The amount of the monthly payment can be determined by using the following table.

Monthly Payment Amount per $1,000 to Pay Principal and Interest on a Loan (P & I)

Term of Loan	5%	6%	7%	8%	9%	10%	11%	12%	13%	14%	15%	16%
5 years	$18.90	$19.35	$19.81	$20.28	$20.76	$21.25	$21.75	$22.25	$22.76	$23.27	$23.79	$24.32
10 years	10.61	11.11	11.62	12.14	12.67	13.22	13.78	14.35	14.94	15.53	16.14	16.76
15 years	7.89	8.43	8.99	9.56	10.15	10.75	11.37	12.01	12.66	13.32	14.00	14.69
20 years	6.60	7.17	7.76	8.37	9.00	9.66	10.33	11.02	11.72	12.44	13.17	13.92
25 years	5.82	6.43	7.07	7.72	8.40	9.09	9.81	10.54	11.28	12.04	12.81	13.59
30 years	5.38	6.01	6.66	7.34	8.05	8.78	9.53	10.29	11.07	11.85	12.65	13.45

Example 4

A boat was sold for $23,800. The buyers agreed to make a $2,800 down payment and finance the balance at 10 percent over five years. The amount of monthly payment on the boat would be found as follows:

$23,800 = Purchase price
$\underline{-2,800}$ = Down payment
$21,000 = Amount to be financed

From the above table (Monthly Payment Amount per $1,000), the amount of $21.25 can be found (10 percent, 5 years). This amount should be multiplied by twenty-one (the number of $1,000 in the loan) to find a payment amount of $446.25.

$21.25 × 21 = $446.25

Example 5

Sport-A-Merica will purchase its new space with a 12 percent loan over twenty years. Kerry will make a $30,000 down payment and will finance the $120,000 balance. What will the monthly payment amount be?

$11.02 = Payment per $1,000
$\underline{×120}$ = The number of $1,000s in loan
$1,322.40 = Monthly payment amount

Exercise A

In the following problems, compute the monthly payment amount. Use the table or the estimating formula.

	Amount of Purchase	Down Payment	Rate of Interest	Term of Loan	Monthly Payment Amount
1.	$13,600	$ 3,000	7% (19.81 × 10.6 =)	5 years	$209.99
2.	40,000	6,000	13	15 years	_____
3.	69,900	21,000	10	25 years	_____
4.	3,000	500	14	5 months	_____
5.	21,500	3,100	8	5 years	_____
6.	500	125	12	10 months	_____
7.	800	400	16	15 months	_____
8.	56,500	$10,000	6	30 years	_____

Computing the Principal Balance
Objective 2

Example 6

principal balance
repayment schedule

Let us use an example of a firm that wishes to purchase an office machine. The firm will finance or borrow $500 in order to make the purchase. A lending institution, bank, or finance company has agreed to lend the $500 at an annual rate of 12 percent for a period of ten months. A payment amount of $52.75 is established, which will pay back the $500 principal and the 12 percent interest over the ten months of the loan. A record is kept of payments, interest, and the amount due after each payment, which is known as the **principal balance.** This record is termed a **repayment schedule.** It is developed as follows.

Computing the Interest

$$\text{Interest} = \$500 \text{ (principal)} \times 12\% \text{ (annual rate)} \times \frac{1}{12} \text{ (time} - \text{monthly payments)}$$

$$\$5 \quad = \$500.00 \times .12 \times \frac{1}{12}$$

$\quad\quad\quad \underline{+5.00} = \text{Interest for the first month}$
$\quad\quad\quad \$505.00 = \text{Amount owed at the end of the first month}$
$\quad\quad\quad \underline{-52.75} = \text{Amount of monthly payment}$
$\quad\quad\quad \$452.25 = \text{Amount of loan after the first monthly payment (principal balance)}$

$$\$4.52 = \$452.25 \times .12 \times \frac{1}{12}$$

$\quad\quad\quad \underline{+4.52} = \text{Interest for the second month}$
$\quad\quad\quad \$456.77 = \text{Amount owed at the end of the second month}$
$\quad\quad\quad \underline{-52.75} = \text{Amount of monthly payment}$
$\quad\quad\quad \$404.02 = \text{Amount of loan after the second monthly payment}$

$$\$4.04 = \$404.02 \times .12 \times \frac{1}{12}$$

$\quad\quad\quad \underline{+4.04} = \text{Interest for the third month}$
$\quad\quad\quad \$408.06 = \text{Amount owed at the end of the third month}$
$\quad\quad\quad \underline{-52.75} = \text{Amount of monthly payment}$
$\quad\quad\quad \$355.31 = \text{Amount of loan after the third monthly payment}$

This process is continued until the principal balance is reduced to zero. The completed repayment schedule for the $500 loan follows.

Repayment Schedule

Payment Number	Payment Amount	Interest Amount	Payment on Principal	Principal Balance
1	$52.75	$5.00	$47.75	$452.25
2	52.75	4.52	48.23	404.02
3	52.75	4.04	48.71	355.31
4	52.75	3.55	49.20	306.11
5	52.75	3.06	49.69	256.42
6	52.75	2.56	50.19	206.23
7	52.75	2.06	50.69	155.54
8	52.75	1.56	51.19	104.35
9	52.75	1.04	51.71	52.64
10	53.17	.53	52.64	0.00

Note that the last payment may be a few cents more or less than the established payment amount.

Home Mortgages

Developed property is purchased by both business (stores, office buildings, research centers, manufacturing plants, and warehouses) and individuals (single-family homes, multi-unit complexes, and condominiums). The information and problems that follow, although applicable to both business and individuals, deal with purchasing a home—a decision relevant to most Americans.

Maximum Monthly Cost

As a guide, the maximum amount that should be spent on shelter by a family is about one-quarter of the gross income of the household. A family with a combined income of $60,000 could spend a maximum of $15,000 per year on shelter, or $1,250 per month. That amount must include the payment, taxes (if applicable), insurance, utilities, and repairs. This guide applies to those renting or buying.

Advantages

There are some advantages to both buying and renting. Examples follow.

Buying:

1. Interest and tax expenses are deductible for federal income tax purposes.
2. There may be some appreciation in the value of the property over the period that it is owned.
3. The property is being purchased. At some point in time, there will be no more mortgage payments!

Renting:

1. Little or no effort is needed to maintain the property.
2. Renter is free to move when the lease expires. There is no need to sell. This is an important factor for those who have jobs that require frequent relocation.

Many of the homes sold in the United States in recent years have been financed by the seller. This is an important factor to consider, because as a potential buyer through this method, you will need to know exactly what you are facing with respect to interest rates, terms of the loan, and how the balance is affected by each payment. The homes that are not financed by the seller are financed by savings and loan associations and commercial banks, which usually have more conservative loan terms and payment schedules.

The average single-family home costs about $150,000 with new homes somewhat higher. Most buyers will put a 10 or 20 percent down payment on the property in order to reduce the loan amount and, thereby, the monthly payment amount.

Effect of Interest Rate Difference

Interest rates are competitive, and therefore, a prospective buyer should shop around or try to negotiate a lower rate. The difference of one percentage point can result in several thousand dollars in additional payments over the term of the agreement.

Example 7

A family decides to purchase a small used home that is priced at $75,000. They put a $25,000 down payment on the home. The balance of $50,000 will be financed by a commercial bank. One bank will provide 7 percent financing on a twenty-year term. Another bank will provide 8 percent financing for the same term. What is the difference in the amount of the total payments if the loan goes to term?

$7.76 (7%, 20 years)	$8.37 (8%, 20 years)
$\underline{\times 50}$ = Thousands	$\underline{\times 50}$ = Thousands
$388 = Payment per month	$418.50 = Payment per month
$\underline{\times 240}$ = (12 months × 20 years)	$\underline{\times 240}$ = (12 months × 20 years)
$93,120 = Total repayment	$100,440 = Total repayment

$100,440 = (8%, 20 years)
$\underline{-93,120}$ = (7%, 20 years)
$7,320   difference

The family in example 7 has a combined household income of $30,000. The mortgage payment (along with taxes and insurance) total $650 per month. This family may not be able to afford this home, because the total shelter expense exceeds the one-quarter maximum guideline (*$30,000 × $\frac{1}{4}$ = $7,500, and $7,500 ÷ 12 = $625 per month*).

Effect of Term Difference

Example 8   Another lender was found that would provide financing at 7 percent for a thirty-year term. Would the extra ten years of term allow the buyers to fit the guideline of one-quarter of gross income?

$6.66 (7%, 30 years)
×50 = Thousands
$333 = Payment per month

  $388 = (7%, 20 years)
−$333 = (7%, 30 years)
  $55 = Less per month

$650 = Monthly household expenses with 20-year mortgage
 −55 = Reduction in monthly mortgage payment
$595 = This would allow the buyer to meet the guidelines!

Example 9   How much additional cost would be incurred over the life of the loan in example 8 for the expanded amortization period?

    $333 = Payment per month
    ×360 = (12 months × 30 years)
$119,880 = Total repayment over 30 years
 −93,120 = Total repayment over 20 years   ($388 payment × 12 months × 20 years)
 $26,760 = Additional cost over the term of the loan

Example 9 points out very clearly that time does cost money—in this case, $26,760 more! To put the information into proper perspective, another factor must be considered: The average family moves every six years; therefore, the family in examples 7, 8, and 9 probably won't make all the payments on the property to term. The family will be more than likely to sell the home after making only 72 payments.

Because lenders are aware that consumers are looking for a variety of financing plans to fit their individual family needs, they have formulated several plans that will fit both their needs and the needs of the borrower. Some of these variations are outlined here.

Fixed Rate

The rate on a **fixed-rate** loan remains the same over the term of the loan. This is a traditional type of loan and the type that will be used in the assigned problems unless otherwise indicated.

Variable Rate

The rate on a **variable-rate** loan (sometimes termed an adjustable rate loan) will usually change annually. The change in the rate will usually be based on some interest indicator, such as treasury bills. There may be limits on the amount of the adjustment.

Fixed Payment Amount

The traditional method of amortizing a loan involves using a **fixed payment amount,** and this is the method used in the assignments unless otherwise indicated. The payment amount remains the same over the term of the loan.

Variable Payment Amount

A loan with a **variable payment amount** begins with low payments that increase in amount as the loan begins to mature. This enables the new (and usually younger) buyer to buy a home early in life. The loan is designed to anticipate the increase in household income in future years. In this type of loan, the balance may actually increase as payments are made because they are so small. This is termed *negative amortization.*

Balloon Payment

A **balloon payment** is often used with the other loan variations. When there is a balloon payment, the total amount of the balance of the loan must be paid on the balloon date. The balloon date is usually five years after the original loan begins. The payment amount is for the loan balance at that point and is, therefore, a large amount. The loan will therefore have to be refinanced by the balloon date or paid in full.

Home Equity Loan

The Tax Reform Act of 1986 made a change in the way the federal government looked at the interest expenses of property buyers. The government disallowed the interest on what was termed *personal loans*. The law then made the *home equity loan* a way to finance family expenditures. The home buyer could use the amount of the home that was paid for, the equity, as a basis for a loan. This allowed the interest to be deductible for tax purposes. This type of loan allows some tax advantages, but also places the home in jeopardy if the buyer defaults on payments.

Points

When the buyer applies for a loan from a financial institution (commercial bank or savings and loan), there may be an application fee. The lender may also be allowed to charge a fee for a legal review of the papers, an appraisal fee, filing fees, prorated taxes, and points. A **point** is one percent of the amount of the loan. The points assessed on the loan may be paid by the buyer, by the seller, or by both. Points can be negotiable and should be reviewed by legal counsel.

Closing Costs

When the buyer makes the final financial commitment to purchase the property, the buyer and seller go through what is known as the *closing*. At the closing meeting, the appropriate papers are signed, and fees and monies are transferred to the lender, seller, and other parties. The total of these fees is termed the **closing costs,** because they are paid at closing. These costs are in excess of and not included in the amount of the loan. Often, the buyer is not aware of the exact amount of the closing costs until the day or week before the closing date.

Escrow Account
Objective 3

When a piece of property is mortgaged, the lending institution may, at the request of the buyer, collect money from the buyer to cover the yearly property taxes and fire insurance. A lump sum is collected at closing for the portion of taxes prorated to that point in the tax year. Each monthly payment thereafter will include a predetermined amount for taxes and fire insurance. This predetermined amount, which is over and above the regular monthly payment, is put into an **escrow account** for the buyer. The escrow account money, which does not earn or pay interest, is a reserve for the buyer to ensure that there will always be money to make the tax and insurance payments.

When the tax and insurance payments are due, the lender draws money from the escrow account and makes the payments for the buyer. In this way, too, the escrow account provides some protection to the lender by ensuring that the payments will be made.

Example 10

Ken and Linda purchased a home for $143,000. They made a down payment of $30,000, and financed the balance on a mortgage at 7 percent for thirty years. The lending institution stated that the home must be insured at a cost of $1,860 over three years. The property taxes were $1,470 per year. What would the monthly payment amount be in order to pay the principal (P), interest (I), taxes (T), and insurance (I) (i.e., PITI)?

$143,000 = Purchase price
−30,000 = Down payment
$113,000 = Amount of loan

$6.66 = Payment per $1,000 (from table)
113 = The number of $1,000s in loan
$752.58 = Monthly payment amount for principal and interest (P and I)

$752.58 = Monthly payment amount (P & I)
$1,470 (taxes for 1 year) ÷ 12 =   122.50 = Taxes per month (T)
$1,860 (insurance for 3 years) ÷ 36 = +51.67 = Insurance per month (I)
$926.75 = Monthly payment amount including
escrow amount (PITI)

To develop a repayment schedule or to determine the balance on the loan, the payment amount without escrow is used.

Exercise B

In the following problems, compute the monthly payment amount including the amount for escrow. Use the table.

	Amount of Purchase	Down Payment	Rate of Interest	Term of Loan	Insurance Cost per Year	Taxes per Year	Monthly Payment Amount Including Escrow (i.e., PITI)
1.	$ 31,500	$ 0	9%	15 years	$147	$ 600	$381.98 (10.15 × 31.5 =)
2.	149,000	30,000	5	25	458	1,942	_____
3.	18,000	3,000	7	5	92	480	_____
4.	41,600	5,600	13	10	186	1,520	_____
5.	194,000	40,000	6	25	—	1,800	_____
6.	21,000	13,000	10	15	108	165	_____
7.	233,500	45,000	5	20	160	478	_____

Early Payoff Penalties
Objective 4

The development of a repayment schedule earlier in the chapter (examples 1 and 6) provided a complete analysis of the amount of interest, payment on the principal, and payment amount, as well as the balance after each monthly payment. This information is useful and necessary on an installment contract. Sometimes a loan is paid off before the end of the term. In this case, usually the borrower must pay some type of penalty to receive credit for complete payment. For example, on a long-term mortgage, the penalty may be a simple 1 percent of the original amount of the loan. This early payoff penalty must be spelled out in the loan agreement.

Example 11

A 25-year $42,000 loan at 8 percent has a principal balance of $7,860. The borrower has found enough money to pay off the loan before the end of the twenty-five years. After reviewing the contract, the borrower finds a penalty clause, which states that the borrower must pay a $\frac{3}{4}$ percent penalty on the current principal balance if an early payoff is made. The payoff amount would be computed as follows:

$$\$7,860.00 = \text{Principal balance}$$

$$\$7,860 \times .08 \times \frac{1}{12} = \quad +52.40 = \text{Interest for month}$$

$$\underline{+58.95} = \frac{3}{4}\% \text{ on principal balance for early payoff penalty}$$

$$\$7,971.35 = \text{Payoff amount}$$

Exercise C

Determine the payoff amount in the following exercises using the penalty information indicated.

	Amount Borrowed	Current Principal Balance	Interest Rate		Penalty	Payoff Amount
1.	$40,000	$28,960	10%	(28,960 × .005 =)	$\frac{1}{2}$% of principal balance	$29,346.13
2.	372,000	13,643	12		1% of principal balance over $10,000	_____
3.	125,000	103,970	13		$250 if principal balance is over $80,000 plus $\frac{1}{2}$% of principal balance	_____
4.	52,500	8,950	10		1% of amount borrowed	_____
5.	185,000	13,805	7		$\frac{1}{4}$% of principal balance and $300	_____

Rule of 78s

On a shorter-term loan a bank may use the Rule of 78s. This rule goes along with the concept that more of the interest is earned in the earlier months of a loan's life. To compute the amount of payoff on a loan using the **Rule of 78s,** the number of the months in the term of the loan must be totaled. In a twelve-month loan, the total is 78; that is,

$$12 + 11 + 10 + 9 + 8 + 7 + 6 + 5 + 4 + 3 + 2 + 1 = 78$$

The bank uses $\frac{12}{78}$ of the principal to compute the first month's interest for payoff purposes; the second month uses $\frac{11}{78}$, etc. This results in a higher payoff amount than most consumers expect.

Note: The formula $\frac{n(n+1)}{2}$ can be used to find the sum of the months, where n is equal to the number of months of the loan. In a twelve-month loan, the formula gives $\frac{12(12+1)}{2} = 78$.

To find the payoff amount, the lender will

1. Determine the amount of unearned interest.

2. Determine the amount still owed (payments remaining $\times$ payment amount).

3. Subtract unearned interest from amount still owed.

Example 12

Fred borrowed $780, which was to be repaid in eighteen monthly payments of $47 each. The total to be repaid was $846. After making the fifteenth payment, Fred inquired about the payoff amount. By using the Rule of 78s, the bank indicated the amount to be $138.68. The amount was found as follows.

```
18                  $846 = Amount to be repaid (18 × $47)
17                 −780 = Amount of loan
16                  $66 = finance charge
15
14
13
12     first 15
11     months of         6
10     loan (earned    ─── × $66 = $2.32 Unearned interest
 9     interest)       171
 8
 7
 6
 5
 4
 3     remaining 3 months
 2     of unearned interest
+1     total = 6
───
171
```

```
  $47 = Payment amount
×  3 = Number of months remaining in loan
$141.00 = Balance of loan after the fifteenth payment (still owed)
 −2.32 = Unearned interest
$138.68 = Payoff amount
```

Exercise D

Determine the payoff amount in the following exercises using the Rule of 78s.

Amount Borrowed	Term (Months)	After Payment No.	Payment Amount	Payoff Amount
1. $400 (420 − 400 = 20)	10 (6 ÷ 55 × 20 =)	7	$ 42	$123.82
2. 1,600	7	3	240	_____
3. 750	15	10	53	_____
4. 1,000	12	9	90	_____
5. 625	6	4	111	_____

Computer Applications

It should be realized that even though the calculations in this chapter seem lengthy, they are very easily programmed into a computer. The repayment schedule, for example, can be developed for the full term of a 25-year loan in a matter of seconds. The computer needs only the principal, the interest rate, and the payment amount, as well as a simple set of instructions, in order to develop the complete repayment schedule.

This task is typical of many business problems that are repetitious, and therefore, it is a task that is easily automated. Computer applications also include determinations of depreciation, payroll, compound interest, and monthly billing.

Firms that have their own computers and computer service firms need trained employees to help program the computers. Individual consumers need training in the topic of installment contracts in case they purchase or sell property or merchandise on a land contract.

National bank credit cards and retail merchants that sell on credit all use a similar installment contract repayment analysis. The variations, though small, are numerous and should be understood by the consumer before authorizing a purchase on credit.

Name _____ Date _____ Score _____

Skill Problems

Using the chart within the chapter, compute the payment amount (P&1) of the following loan amounts.

1. Amount of purchase = $74,000
   Rate of interest = 10%
   Term of loan = 25 years

   Monthly payment = _____

2. Amount of purchase = $1,500
   Rate of interest = 7%
   Term of loan = 5 years

   Monthly payment = _____

3. Amount of purchase = $16,000
   Rate of interest = 16%
   Term of loan = 15 years

   Monthly payment = _____

4. Amount of purchase = $47,540
   Rate of interest = 11%
   Term of loan = 10 years

   Monthly payment = _____

In the following problems, compute the monthly payment amount (P&I). Use the table for long-term loans and the estimating formula for short-term loans.

	Amount of Purchase	Down Payment	Rate of Interest	Term of Loan	Monthly Payment Amount
5.	$15,800	$2,000	8%	5 years	_____
6.	60,000	6,000	10	10 years	_____
7.	750	0	18	5 months	_____
8.	48,500	$10,000	13	10 months	_____
9.	64,400	5,400	7	25 years	_____

10. Amount of purchase = $64,000
    Rate of interest = 7%
    Term of loan = 10 years

    Monthly payment = _____ (P&I)

11. Amount of purchase = $50,000
    Down payment = $8,000
    Rate of interest = 5%
    Term of loan = 25 years

    Monthly payment = _____ (P&I)

12. Amount of purchase = $29,500
    Down payment = $4,000
    Rate of interest = 13%
    Term of loan = 30 years

    Monthly payment = _____ (P&I)

13. Amount of purchase = $160,000
    Down payment = $12,000
    Rate of interest = 12%
    Term of loan = 15 years
    Taxes per year = $540
    Insurance for three years = $1,008

    Monthly payment = _____ (PITI)

14. Amount of purchase = $25,000
    Down payment = $8,000
    Rate of interest = 14%
    Term of loan = 15 years
    Taxes per year = $768
    Insurance per year = $261

    Monthly payment = _____ (PITI)

15. Amount of purchase = $286

    Rate of interest = $16\frac{1}{2}$%

    Term of loan = 8 months
    Down payment = $36

    Monthly payment = _____
    (Use the Estimating Total Interest on an Installment Contract method.)

In the following problems, compute the monthly payment amount including the amount for escrow. Use the table in the chapter.

	Amount of Purchase	Down Payment	Rate of Interest	Term of Loan	Insurance Cost per Year	Taxes per Year	Monthly Payment Amount Including Escrow (PITI)
16.	$ 45,500	$ 0	13%	15 yr.	$225	$ 457	_____
17.	150,000	10,000	10	25	483	1,594	_____
18.	46,750	3,250	11	10	140	0	_____
19.	50,050	5,050	12	5	645	2,870	_____
20.	250,000	15,000	7	30	915	3,450	_____

Cost of item = $548.25
Down payment = 48.25
Loan period = 6 months
Interest rate = 10%
Complete the repayment schedule below.

Payment Amount	Interest Amount	Payment on Principal	Principal Balance
$85.78	**21.** _____	**22.** _____	**23.** _____
85.78	**24.** _____	**25.** _____	**26.** _____
85.78	**27.** _____	**28.** _____	**29.** _____
85.78	**30.** _____	**31.** _____	**32.** _____
85.78	**33.** _____	**34.** _____	**35.** _____
85.79	**36.** _____	**37.** _____	**38.** _____

Cost of item = $247
Down payment = 15%
Loan period = 6 months
Interest rate = 18%

39. Down payment amount = _____

40. Amount to be financed = _____

41. Total interest = _____

42. Monthly payment = _____

Cost of item = $658
Down payment = 0
Loan period = 12 months
Interest rate = $14\frac{1}{4}$%

43. Total interest for 12 months = _____

44. Monthly payment for 12 months = _____

45. Amount of purchase = $8,000
Rate of interest = 12%
Monthly payment = $180 (P&I)
Principal balance after the second monthly payment =

_____

46. Amount of purchase = $8,640
Rate of interest = 9%
Monthly payment = $120 (P&I)

Interest amount for fourth month = _____

47. Amount of purchase = $26,000
Rate of interest = 8%
Monthly payment = $305 (P&I)
Principal balance after the third monthly payment =

_____

48. Amount of purchase = $13,100
Rate of interest = 16%
Monthly payment = $260 (P&I)

Interest for first two months = _____

49. Amount of purchase = $26,000
Rate of interest = 9%
Monthly payment = $390 (including $106 for escrow account)
Principal balance after the third monthly payment =

_____

50. Amount of purchase = $26,000
Down payment = $0
Rate of interest = 10%
Monthly payment = $236.34

Term of loan = ___ years

51. Amount of purchase = $380
Monthly payments = $48
Amount to be repaid = $384
8 monthly payments
Amount of payoff after the sixth payment = _____
(Use Rule of 78s.)

In the following problems, compute the payoff amount.

	Amount Borrowed	Current Principal Balance	Interest Rate	Early Payoff Penalty	Payoff Amount
52.	$ 45,500	$20,737	13%	$1/2$% of principal balance	_____
53.	150,000	7,635.60	10	1% of amount borrowed	_____
54.	46,750	7,193.16	11	$300 and $1/4$% of principal balance	_____
55.	50,050	48,060.00	12	$1/2$% of principal balance over $10,000	_____
56.	$250,000	31,302.00	7	$1/4$% of balance	_____

Determine the payoff amount in exercises 57–62 using the Rule of 78s.

	Amount Borrowed	Term (Months)	After Payment No.	Payment Amount	Payoff Amount
57.	$600	7	4	$104.00	_____
58.	4,600	14	10	375.80	_____
59.	750	6	3	130.00	_____
60.	845	11	5	89.20	_____
61.	1,500	18	8	102.00	_____
62.	678	24	4	33.16	_____

Business Application Problems

1. A company bought a building for $82,500. The building was financed through a mortgage at 11 percent for 20 years. What is the amount of monthly payment on the mortgage loan?

2. In problem 1, what is the total amount of interest charges over the life of the loan?

3. In problem 1, what is the balance of the mortgage after the second monthly payment?

4. In problem 1, how much interest has the company paid in the first two months of the mortgage?

5. A business has a balance of $14,540 on a $15,000, five-year installment contract. The business has made three monthly payments of $318. What is the penalty on the contract if there is an early payoff penalty clause of 1 percent of the current balance?

6. In problem 5, how much interest has the company paid in the first three months?

7. What is the difference in the total interest paid on a 25-year, $30,000 mortgage at 6 percent versus at 5 percent?

8. What is the difference in the mortgage balance using 5 percent versus 6 percent, after the first monthly payment in problem 7?

9. Kelly and Carl have a combined household annual income of $72,000. They have accumulated a down payment of $14,000. What is the maximum amount they can spend on shelter per month?

10. In problem 9, the current interest rate is 10 percent and escrow payments are estimated to be $150 per month. To the nearest thousand dollars, what is the maximum value of shelter they can purchase through a 30-year mortgage?

11. Candy's Confectionary will add on to their existing store with a loan from the American State Bank. The loan will be for $125,000. The owner negotiated a 10 percent rate and an amortization over twenty years. What is the payment amount?

12. As cash payments clerk of the Accounting Department, you are asked to develop a repayment schedule on a loan of $22,800. The rate of interest is 12 percent and the loan will run for fifteen years. Taxes on the property are $640 a year. The insurance coverage for a year will be $230. Complete the repayment schedule through the third month.

		Repayment Schedule		
Payment Number	(PITI) Payment Amount	Interest Amount	Payment on Principal	Principal Balance
1	_____	_____	_____	_____
2	_____	_____	_____	_____
3	_____	_____	_____	_____

13. The teller of the Mid-State Bank referred a customer to the loan manager to explain the payoff amount on an installment loan for $600. The amount to be repaid was $684 over eight months. The customer has paid four payments and feels that the loan balance should be $258 because four payments have been made on the $600 loan. Using the Rule of 78s, determine the correct payoff amount.

14. Alice Baid paid off the mortgage on her home early. The amount of the mortgage was initially $93,000. The balance at the date of payoff was $56,328.40. The mortgage contract stipulated a 1 percent penalty charge on the current balance for early payoff or for payments in excess of 10 percent of the initial mortgage amount. What is the payoff amount?

15. Four months ago, Predmore Associates purchased a piece of land for future development at a price of $46,000. They paid $10,000 down and financed the balance at 15 percent interest. The payments are $600 per month. No formal record or repayment schedule has been established to date. As assistant to the Budget Director, you are asked to prepare the figures for this month's interest expense on the loan. What is the interest for the fourth month?

16. Blanche Olsen selected a condominium to purchase. She decided to pay $12,500 as a down payment on the purchase price of $79,500. She will amortize the 11 percent loan over a period of fifteen years. What will be the principal balance after the second monthly payment?

17. If Blanche, in problem 16, had elected to take a 10 percent, 30-year loan with Marlin Mortgage, how much more would it have cost her over the term of the loan?

18. Write a complete sentence in answer to this question. Lloyd must obtain a mortgage on a home in the amount of $80,000. He has the choice of a 6-percent 30-year mortgage, or a 5-percent 25-year mortgage. What is the difference on the first month's interest?

19. Write a complete sentence in answer to this question. What is the difference in the interest over the terms of the mortgages in problem 18?

20. Write a complete sentence in answer to this question. Marin and Associates will build a new home on a piece of land. The finished home will sell for $184,900. A loan company will finance 90 percent of the value of the home. If a buyer finances the 90 percent at 7 percent interest over a twenty-year period, what will be the amount of interest for the first month?

★ **Challenge Problem**

The accountant for the Vilper Corporation must determine the principal balance of a loan negotiated six months ago on a piece of property. The property purchase price was $24,000. The rate of interest on the loan was 9 percent. The payment amount of $263 per month will pay off the loan and the $300 annual tax bill. The vacant land required a down payment of $6,000 before the bank would finance the loan. A penalty charge of 1 percent of the balance of the loan was agreed upon by Vilper should they decide to pay off the loan early. What is the principal balance?

Complete the following problems.

1. In computing the principal balance on an installment loan, the payment amount is first applied to the interest. Any additional amount is applied to the reduction of the principal.   True   False

2. When computing a principal balance on an installment loan, the escrow amount of the payment is added to the beginning principal before computing the interest. True   False

3. Amount of purchase = $3,000
Down payment = $500
Interest rate = 7%
Term of loan = 15 years
Monthly payment amount = $ _____

4. Amount of purchase = $91,000
Down payment = $5,000
Interest rate = 6%
Term of loan = 25 years
Insurance per year = $1,200
Taxes per year = $1,890
Monthly payment = $ _____

5. Amount of loan = $28,000
Payment amount = $300
Interest rate = 9%
Principal balance at the end of the second
month = $ _____

6. Mr Boldt borrowed $900 for six months. The monthly payments were $158.87. At the end of the fifth month he paid off the loan. Using the Rule of 78s, what was the payoff amount?

7. Determine the amount of the final payment on a property loan that was: (1) originally for $64,000, (2) has a current balance of $4,907.35, and (3) had a clause in the contract that any payment in excess of 5 percent of the original loan amount be penalized at the rate of 1 percent of the current balance. The mortgage carries a 10 percent interest rate.

CONSUMER CREDIT

12

OBJECTIVES

After mastering the material in this chapter, you will be able to:

1. Describe the obligations of buying with a credit card. p. 218
2. Determine the balance for a credit card statement. p. 219
3. Determine the annual percentage rate on a loan. p. 221
4. Describe how a business owner turns credit sales into cash. p. 224
5. Understand and use the following terms and concepts:
   Credit Card
   Store Card
   Gasoline Card
   Bank Card
   Travel or Entertainment Card
   Monthly Statement
   Minimum Payment Amount
   Amortizing
   Regulation Z
   Annual Percentage Rate

SPORT-A-MERICA

 Sales last weekend set a new record for the store. Kerry saw firsthand the huge role that credit plays at Sport-A-Merica's checkout counter over the weekend. Nearly 70 percent of all sales made were on credit. She was convinced of the importance of the ability to accept credit cards. She does wish that the cost of credit to Sport-A-Merica were lower. The credit card company takes 4 percent off the top. That hurts! She asks herself how much Sport-A-Merica gives up on each deposit for the opportunity to sell on credit using credit cards. Kerry occasionally offers store credit to a good customer. Though this takes some figuring, it saves the store the 4 percent that the credit card company would have charged on that sale. She knows that a store credit card could be offered to the best customers. However, that would require a billing-and-collections operation. She recognizes that she is the strong sales arm in the company, and that there will be opportunity in the future to expand into offering more store credit. She thinks a lot about the cost of credit.

Credit Cards

Since World War II, the acceptance of **credit cards** has grown to the point that today more than one-half of the families in the United States use them. This widespread acceptance is based at least in part on the functions that credit cards can provide. Today, a credit card acts to identify a person, to allow for credit payment of an item or service, and to provide a means to obtain additional credit from other sources. Having one credit card with a good payment record opens up the door to other credit arrangements. There are four types of credit cards:

Store Card

The **store card** can be used in one store or one chain of stores.

Gasoline Card

The **gasoline card** was initially issued by gasoline companies to attempt to gain a greater market share and customer loyalty. Today, the oil and gas companies have a select group of customers, often business people, who use their card exclusively.

Bank Card

The **bank card** is the most frequently used type of credit card. It is being sought because of its versatility. It can be used to purchase merchandise; most gasoline stations will accept it; and entertainment establishments often accept it as a form of payment. This type of card is often offered through organizations other than banks.

Travel or Entertainment Card

The **travel** or **entertainment card** is issued on a more selective basis to business people or professionals who are apt to use it more often. The charges on the card per month will exceed $100 on most occasions.

Objective 1

The obligation for payment begins when a purchase is made with a credit card. The finance charge or service charge agreement is usually printed on the card or on the credit card application. The charge for the use of credit with a credit card ranges from 1 to $1\frac{3}{4}$ percent per month.

monthly statement

The credit card customer receives a **monthly statement** of purchases from the card-issuing institution. On this statement the date of the purchase and amount are itemized. Charges for credit will also be shown at the bottom of the statement. The

minimum payment amount

amount owed, or balance, must be paid within a specified time period, or a **minimum payment amount** will be specified on the statement. The minimum payment amount will be a percent of the balance or a flat amount such as $10.

MINIMUM AMOUNT DUE	PAST DUE AMOUNT	PAYMENT DUE DATE	NEW BALANCE	ACCOUNT NUMBER	PLEASE WRITE AMOUNT ENCLOSED
10.00	.00	10/15/01	206.90	0000–000000–0000	

☐ IF ADDRESS AS SHOWN IS INCORRECT PLEASE INDICATE CHANGE ON BACK

Brian D. Fitzpatrick
38 Bromfield Street
Boston, Massachusetts
02144

BANKCARD CENTER
P.O. BOX 2040
BOSTON, MA 02143–9917

PLEASE DETACH AND ENCLOSE THIS PORTION WITH REMITTANCE   MAKE CHECK PAYABLE TO: MASTERCARD

IMPORTANT: PLEASE REPORT ANY DISCREPANCIES ON THIS BILL TO THE BANK IMMEDIATELY.

DATE OF TRANS.	POSTING	REFERENCE NUMBER	DESCRIPTION OF TRANSACTION		AMOUNT
08/31	09/02	784853OU7JK13GBA498KLHI642	Restaurant St. Michel	Coral Gables, FL	122.10
09/08	09/09	753604BL8BD34IOS444DMKP738	Dolphins Plus, Inc.	Key Largo, FL	84.80

ACCOUNT NUMBER	TO REPORT LOST OR STOLEN CREDIT CARDS OR INQUIRIES CALL:	SEND INQUIRIES TO: BANKCARD CENTER P.O. BOX 2040 BOSTON, MA 02143–9917	PAYMENTS MUST BE RECEIVED BY 2:00 PM. OR THEY WILL BE CREDITED AS OF THE NEXT BUSINESS DAY. PAYMENTS RECEIVED AT ANY OTHER BANK LOCATION OR ADDRESS WILL BE CREDITED PROMPTLY BUT NO LATER THAN 5 DAYS FROM RECEIPT.
0000–000000–0000	800–345–8571		

PREVIOUS BALANCE	CASH ADVANCES	PURCHASES	FINANCE CHARGE	ANNUAL FEE	INSURANCE PREMIUM	CREDITS	PAYMENTS	NEW BALANCE
.00	.00	206.90	.00	.00	.00	.00	.00	206.90

CREDIT LIMIT	MINIMUM PAYMENT	BILLING DATE	PAYMENT DUE DATE	TO AVOID ADDITIONAL FINANCE CHARGES, PAYMENT OF NEW BALANCE MUST BE RECEIVED BY DUE DATE.	THE MINIMUM PAYMENT	AS A PERCENT OF THE NEW BALANCE IS:	BUT NOT LESS THAN:
2000		09/20/01	10/15/01			%	$ 10.00

PERIODIC RATES	ON BALANCES UP TO:	THE MONTHLY PERIODIC RATE IS	WHICH IS AN ANNUAL PERCENTAGE RATE OF:	ON BALANCES OVER:	THE MONTHLY PERIODIC RATE IS:	WHICH IS AN ANNUAL PERCENTAGE RATE OF:	AVERAGE DAILY BALANCE ON WHICH FINANCE CHARGE IS COMPUTED:
	$ ALL	1.650 %	19.80%	$	%	%	

Objective 2

The finance charge is usually based on the unpaid balance at the end of the month. An example of the effects of the minimum payment amount and finance charge are shown in example 1.

Example 1

Martin Stahl has a credit card. The past four monthly statements show the following information.

Month	Previous Balance	Purchases this Month	Payments	Finance Charges	New Balance
1	$42.10	$37.20	$20	_____	_____
2	_____	85.00	20	_____	_____
3	_____	. . .	20	_____	_____
4	_____	7.50	50	_____	_____

The minimum payment amount is $20 and the finance charge is 1 percent. What is the amount of total finance charge for the four months on Martin's credit card and what is the new balance at the end of the fourth month?

Solution

The new balance for any month is found by (1) computing the finance charge on the previous balance, (2) adding the finance charge to the previous balance, (3) adding the amount of the purchases for the month, and (4) subtracting the payments.

Month

1
$42.10 \times .01 = $.42
$42.10 + $.42 = $42.52
$42.52 + $37.20 = $79.72
$79.72 - $20 = 59.72—New balance

2
$59.72 \times .01 = $.60
$59.72 + $.60 = $60.32
$60.32 + $85 = $145.32
$145.32 - $20 = 125.32—New balance

3
$125.32 \times .01 = $1.25
$125.32 + $1.25 = $126.57
$126.57 + 0 = $126.57
$126.57 - $20 = 106.57—New balance

4
$106.57 \times .01 = $1.07
$106.57 + $1.07 = $107.64
$107.64 + $7.50 = $115.14
$115.14 - $50 = 65.14—New balance

The total finance charge is $3.34 ($.42 + $.60 + $1.25 + $1.07). The new balance at the end of the fourth month is $65.14.

Some card-issuing banks and institutions are adding an annual service charge if the card is unused for a year or so. Other firms add interest charges that begin at the date of the purchase. These variations would increase the cost of a credit purchase to the consumer. Other card-issuing banks and institutions offer incentives to use their card, such as a card with no annual fee to the cardholder during the first year.

Example 2

A record of Malinda's use of her credit card is shown below. The monthly purchases are for gasoline, clothes, and jewelry. There is a 1 percent finance charge on the unpaid monthly balance, a minimum payment amount of $10, and a flat service charge of $1.00 per month. What is the amount of the new balance at the end of the third month?

Month	Previous Balance	Purchases this Month	Payments	Flat Service Charge and Finance Charges	New Balance
1	$86.40	—	$10	$1.86 ($86.40 × .01 + $1)	$78.26
2	78.26	$17.58	40	1.78 ($78.26 × .01 + $1)	57.62
3	57.62	—	10	1.58 ($57.62 × .01 + $1)	49.20

The new balance at the end of the third month is $49.20.

Exercise A

Compute the new balance in these exercises.

Month	Previous Balance	Purchases this Month	Payments	Finance Charges (1%)	New Balance
1. 1	$48.20	—	$10	$.48 ($48.20 × .01)	$38.68
2	38.68	$35.00	10	.39 ($38.68 × .01)	64.07
3	64.07	22.60	10	.64 ($64.07 × .01)	77.31
2. 1	187.00	13.00	30	_____	_____
2	_____	—	20	_____	_____
3	_____	70.00	50	_____	_____
3. 1	36.50	10.00	30	_____	_____
2	_____	46.10	10	_____	_____
3	_____	51.80	50	_____	_____
4	_____	—	10	_____	_____
4. 1	0.00	53.00	_____	_____	_____
2	_____	65.50	10	_____	_____
3	_____	—	10	_____	_____
5. 1	56.10	—	20	_____	_____
2	_____	186.50	15	_____	_____
3	_____	21.50	20	_____	_____
4	_____	—	30	_____	_____

Annual Percentage Rate (APR)

amortizing

Credit buying requires the buyer, to whom credit is extended, to first pay the amount of interest due on the loan at the end of each month, as well as some amount on the principal. The amount of payment on the principal will reduce the amount of the loan to zero over a period of time. This payment process is termed **amortizing** the loan. This differs from a promissory note that requires only one payment, which includes the face value of the note as well as the interest, at the end of the term of the loan.

The Truth-in-Lending Act states that (1) the buyer must be made aware of all of the finance charges on a loan, and that (2) the annual rate of interest on the loan must be clearly stated. A monthly rate of interest of $1\frac{1}{2}$ percent is equal to 18 percent annually, 1 percent is equal to 12 percent annually, etc. This rate of interest is determined differently on an installment contract which involves a declining amount of principal borrowed over the life of the loan.

regulation Z   **Regulation Z,** the implementation phase of the Truth-in-Lending Act, states that total financing charges on credit purchases must be disclosed to the borrower. The amount of financing charges is found by subtracting the amount that is borrowed from the amount that is to be paid back. The difference is the amount of financing charges.

Objective 3

The **annual percentage rate (APR)** on an installment contract must be stated, according to Regulation Z. The Federal Reserve Board has prepared a table that is used to determine the APR. For example, an 8-month loan for $700 has interest charges of $28, an application fee of $6, and a filing fee of $8. The borrower will pay back

$$
\begin{aligned}
\$700 &= \text{amount of loan} \\
28 &= \text{interest} \\
6 &= \text{application fee} \\
+8 &= \text{filing fee} \\
\hline
\$742 &= \text{amount to be paid back} \\
-700 &= \text{amount borrowed} \\
\hline
\$42 &= \text{finance charges}
\end{aligned}
$$

$$\frac{\text{Finance charges}}{\text{Number of hundreds in loan}} = \text{Finance charge per \$100 of loan}$$

$$\frac{\$42}{7} = \$6 \text{ per hundred}$$

By referring to the table on page 222, the annual percentage rate can be determined. Using eight payments along the left-hand edge, look across until you find the $6 amount. The annual percentage rate is 15.75 percent.

Example 3

Max charged a purchase of $500 for 10 months on store credit from Sport-A-Merica. Kerry indicated that the interest charges would be $32.50. There would also be a service charge of $3 and a filing fee of $2.50. What is the annual percentage rate on the loan?

$$
\begin{aligned}
\$500.00 &= \text{Loan} \\
32.50 &= \text{Interest} \\
3.00 &= \text{Service charge} \\
+2.50 &= \text{Filing fee} \\
\hline
\$538.00 &= \text{Amount to be paid back} \\
-500.00 &= \text{Amount borrowed} \\
\hline
\$38.00 &= \text{Total finance charges}
\end{aligned}
$$

$$\frac{\text{Total finance charges}}{\text{Number of \$100 in loan}} = \text{Finance charge per \$100 of loan}$$

$$\frac{\$38}{5} = \$7.60 \text{ finance charge per \$100 of loan} = 16.25\% \text{ from the APR table listed}$$

To find the annual percentage rate, look down the number of payments column on the table until you find the number of months for the loan. Proceed across this line until you locate the amount that was charged on the loan for each $100. When you find the amount, look at the percent column heading. This will be the annual percentage rate. In this example, the rate is 16.25%. Kerry knows that this is much less than the rates that some credit card companies charge.

The annual percentage rate table can also be used to determine the amount of finance charge, if the rate is known. The amount per hundred ($7.60 in example 3) is found in the 16.25 percent column across from the ten payments level. By multiplying $7.60 times the number of hundreds (five), the finance charge of $38.00 is found.

Annual Percentage Rate

Number of Payments	14.00%	14.25%	14.50%	14.75%	15.00%	15.25%	15.50%	15.75%	16.00%	16.25%	16.50%	16.75%
1	1.17	1.19	1.21	1.23	1.25	1.27	1.29	1.31	1.33	1.35	1.38	1.40
2	1.75	1.78	1.82	1.85	1.88	1.91	1.94	1.97	2.00	2.04	2.07	2.10
3	2.34	2.38	2.43	2.47	2.51	2.55	2.59	2.64	2.68	2.72	2.76	2.80
4	2.93	2.99	3.04	3.09	3.14	3.20	3.25	3.30	3.36	3.41	3.46	3.51
5	3.53	3.59	3.65	3.72	3.78	3.84	3.91	3.97	4.04	4.10	4.16	4.23
6	4.12	4.20	4.27	4.35	4.42	4.49	4.57	4.64	4.72	4.79	4.87	4.94
7	4.72	4.81	4.89	4.98	5.06	5.15	5.23	5.32	5.40	5.49	5.58	5.66
8	5.32	5.42	5.51	5.61	5.71	5.80	5.90	6.00	6.09	6.19	6.29	6.38
9	5.92	6.03	6.14	6.25	6.35	6.46	6.57	6.68	6.78	6.89	7.00	7.11
10	6.53	6.65	6.77	6.88	7.00	7.12	7.24	7.36	7.48	7.60	7.72	7.84
11	7.14	7.27	7.40	7.53	7.66	7.79	7.92	8.05	8.18	8.31	8.44	8.57
12	7.74	7.89	8.03	8.17	8.31	8.45	8.59	8.74	8.88	9.02	9.16	9.30
13	8.36	8.51	8.66	8.81	8.97	9.12	9.27	9.43	9.58	9.73	9.89	10.04
14	8.97	9.13	9.30	9.46	9.63	9.79	9.96	10.12	10.29	10.45	10.62	10.78
15	9.59	9.76	9.94	10.11	10.29	10.47	10.64	10.82	11.00	11.17	11.35	11.53
16	10.20	10.39	10.58	10.77	10.95	11.14	11.33	11.52	11.71	11.90	12.09	12.28
17	10.82	11.02	11.22	11.42	11.62	11.82	12.02	12.22	12.42	12.62	12.83	13.03
18	11.45	11.66	11.87	12.08	12.29	12.50	12.72	12.93	13.14	13.35	13.57	13.78
19	12.07	12.30	12.52	12.74	12.97	13.19	13.41	13.64	13.86	14.09	14.31	14.54
20	12.70	12.93	13.17	13.41	13.64	13.88	14.11	14.35	14.59	14.82	15.06	15.30
21	13.33	13.58	13.82	14.07	14.32	14.57	14.82	15.06	15.31	15.56	15.81	16.06
22	13.96	14.22	14.48	14.74	15.00	15.26	15.52	15.78	16.04	16.30	16.57	16.83
23	14.59	14.87	15.14	15.41	15.68	15.96	16.23	16.50	16.78	17.05	17.32	17.60
24	15.23	15.51	15.80	16.08	16.37	16.65	16.94	17.22	17.51	17.80	18.09	18.37
25	15.87	16.17	16.46	16.76	17.06	17.35	17.65	17.95	18.25	18.55	18.85	19.15
26	16.51	16.82	17.13	17.44	17.75	18.06	18.37	18.68	18.99	19.30	19.62	19.93
27	17.15	17.47	17.80	18.12	18.44	18.76	19.09	19.41	19.74	20.06	20.39	20.71
28	17.80	18.13	18.47	18.80	19.14	19.47	19.81	20.15	20.48	20.82	21.16	21.50
29	18.45	18.79	19.14	19.49	19.83	20.18	20.53	20.88	21.23	21.58	21.94	22.29
30	19.10	19.45	19.81	20.17	20.54	20.90	21.26	21.62	21.99	22.35	22.72	23.08
31	19.75	20.12	20.49	20.87	21.24	21.61	21.99	22.37	22.74	23.12	23.50	23.88
32	20.40	20.79	21.17	21.56	21.95	22.33	22.72	23.11	23.50	23.89	24.28	24.68
33	21.06	21.46	21.85	22.25	22.65	23.06	23.46	23.86	24.26	24.67	25.07	25.48
34	21.72	22.13	22.54	22.95	23.37	23.78	24.19	24.61	25.03	25.44	25.86	26.28
35	22.38	22.80	23.23	23.65	24.08	24.51	24.94	25.36	25.79	26.23	26.66	27.09
36	23.04	23.48	23.92	24.35	24.80	25.24	25.68	26.12	26.57	27.01	27.46	27.90
48	31.17	31.77	32.37	32.98	33.59	34.20	34.81	35.42	36.03	36.65	37.27	37.88
60	39.61	40.39	41.17	41.95	42.74	43.53	44.32	45.11	45.91	46.71	47.51	48.31

Annual Percentage Rate

Number of Payments	17.00%	17.25%	17.50%	17.75%	18.00%	18.25%	18.50%	18.75%	19.00%	19.25%	19.50%	19.75%
1	1.42	1.44	1.46	1.48	1.50	1.52	1.54	1.56	1.58	1.60	1.63	1.65
2	2.13	2.16	2.19	2.22	2.26	2.29	2.32	2.35	2.38	2.41	2.44	2.48
3	2.85	2.89	2.93	2.97	3.01	3.06	3.10	3.14	3.18	3.23	3.27	3.31
4	3.57	3.62	3.67	3.73	3.78	3.83	3.88	3.94	3.99	4.04	4.10	4.15
5	4.29	4.35	4.42	4.48	4.54	4.61	4.67	4.74	4.80	4.86	4.93	4.99
6	5.02	5.09	5.17	5.24	5.32	5.39	5.46	5.54	5.61	5.69	5.76	5.84
7	5.75	5.83	5.92	6.00	6.09	6.18	6.26	6.35	6.43	6.52	6.60	6.69
8	6.48	6.58	6.67	6.77	6.87	6.96	7.06	7.16	7.26	7.35	7.45	7.55
9	7.22	7.32	7.43	7.54	7.65	7.76	7.87	7.97	8.08	8.19	8.30	8.41
10	7.96	8.08	8.19	8.31	8.43	8.55	8.67	8.79	8.91	9.03	9.15	9.27
11	8.70	8.83	8.96	9.09	9.22	9.35	9.49	9.62	9.75	9.88	10.01	10.14
12	9.45	9.59	9.73	9.87	10.02	10.16	10.30	10.44	10.59	10.73	10.87	11.02
13	10.20	10.35	10.50	10.66	10.81	10.97	11.12	11.28	11.43	11.59	11.74	11.90
14	10.95	11.11	11.28	11.45	11.61	11.78	11.95	12.11	12.28	12.45	12.61	12.78
15	11.71	11.88	12.06	12.24	12.42	12.59	12.77	12.95	13.13	13.31	13.49	13.67
16	12.46	12.65	12.84	13.03	13.22	13.41	13.60	13.80	13.99	14.18	14.37	14.56
17	13.23	13.43	13.63	13.83	14.04	14.24	14.44	14.64	14.85	15.05	15.25	15.46
18	13.99	14.21	14.42	14.64	14.85	15.07	15.28	15.49	15.71	15.93	16.14	16.36
19	14.76	14.99	15.22	15.44	15.67	15.90	16.12	16.35	16.58	16.81	17.03	17.26
20	15.54	15.78	16.01	16.25	16.49	16.73	16.97	17.21	17.45	17.69	17.93	18.17
21	16.31	16.56	16.81	17.07	17.32	17.57	17.82	18.07	18.33	18.58	18.83	19.09
22	17.09	17.36	17.62	17.88	18.15	18.41	18.68	18.94	19.21	19.47	19.74	20.01
23	17.88	18.15	18.43	18.70	18.98	19.26	19.54	19.81	20.09	20.37	20.65	20.93
24	18.66	18.95	19.24	19.53	19.82	20.11	20.40	20.69	20.98	21.27	21.56	21.86
25	19.45	19.75	20.05	20.36	20.66	20.96	21.27	21.57	21.87	22.18	22.48	22.79
26	20.24	20.56	20.87	21.19	21.50	21.82	22.14	22.45	22.77	23.09	23.41	23.73
27	21.04	21.37	21.69	22.02	22.35	22.68	23.01	23.34	23.67	24.00	24.33	24.67
28	21.84	22.18	22.52	22.86	23.20	23.55	23.89	24.23	24.58	24.92	25.27	25.61
29	22.64	22.99	23.35	23.70	24.06	24.41	24.77	25.13	25.49	25.84	26.20	26.56
30	23.45	23.81	24.18	24.55	24.92	25.29	25.66	26.03	26.40	26.77	27.14	27.52
31	24.26	24.64	25.02	25.40	25.78	26.16	26.55	26.93	27.32	27.70	28.09	28.47
32	25.07	25.46	25.86	26.25	26.65	27.04	27.44	27.84	28.24	28.64	29.04	29.44
33	25.88	26.29	26.70	27.11	27.52	27.93	28.34	28.75	29.16	29.57	29.99	30.40
34	26.70	27.12	27.54	27.97	28.39	28.81	29.24	29.66	30.09	30.52	30.95	31.37
35	27.52	27.96	28.39	28.83	29.27	29.71	30.14	30.58	31.02	31.47	31.91	32.35
36	28.35	28.80	29.25	29.70	30.15	30.60	31.05	31.51	31.96	32.42	32.87	33.33
48	38.50	39.13	39.75	40.37	41.00	41.63	42.26	42.89	43.52	44.15	44.79	45.43
60	49.12	49.92	50.73	51.55	52.36	53.18	54.00	54.82	55.64	56.47	57.30	58.13

Annual Percentage Rate

Number of Payments	20.00%	20.25%	20.50%	20.75%	21.00%	21.25%	21.50%	21.75%	22.00%	22.25%	22.50%	22.75%
1	1.67	1.69	1.71	1.73	1.75	1.77	1.79	1.81	1.83	1.85	1.88	1.90
2	2.51	2.54	2.57	2.60	2.63	2.66	2.70	2.73	2.76	2.79	2.82	2.85
3	3.35	3.39	3.44	3.48	3.52	3.56	3.60	3.65	3.69	3.73	3.77	3.82
4	4.20	4.25	4.31	4.36	4.41	4.47	4.52	4.57	4.62	4.68	4.73	4.78
5	5.06	5.12	5.18	5.25	5.31	5.37	5.44	5.50	5.57	5.63	5.69	5.76
6	5.91	5.99	6.06	6.14	6.21	6.29	6.36	6.44	6.51	6.59	6.66	6.74
7	6.78	6.86	6.95	7.04	7.12	7.21	7.29	7.38	7.47	7.55	7.64	7.73
8	7.64	7.74	7.84	7.94	8.03	8.13	8.23	8.33	8.42	8.52	2	8.72
9	8.52	8.63	8.73	8.84	8.95	9.06	9.17	9.28	9.39	9.50	9.61	9.72
10	9.39	9.51	9.63	9.75	9.88	10.00	10.12	10.24	10.36	10.48	10.60	10.72
11	10.28	10.41	10.54	10.67	10.80	10.94	11.07	11.20	11.33	11.47	11.60	11.73
12	11.16	11.31	11.45	11.59	11.74	11.88	12.02	12.17	12.31	12.46	12.60	12.75
13	12.05	12.21	12.36	12.52	12.67	12.83	12.99	13.14	13.30	13.46	13.61	13.77
14	12.95	13.11	13.28	13.45	13.62	13.79	13.95	14.12	14.29	14.46	14.63	14.80
15	13.85	14.03	14.21	14.39	14.57	14.75	14.93	15.11	15.29	15.47	15.65	15.83
16	14.75	14.94	15.14	15.33	15.52	15.71	15.90	16.10	16.29	16.48	16.68	16.87
17	15.66	15.86	16.07	16.27	16.48	16.68	16.89	17.09	17.30	17.50	17.71	17.92
18	16.57	16.79	17.01	17.22	17.44	17.66	17.88	18.09	18.31	18.53	18.75	18.97
19	17.49	17.72	17.95	18.18	18.41	18.64	18.87	19.10	19.33	19.56	19.79	20.02
20	18.41	18.66	18.90	19.14	19.38	19.63	19.87	20.11	20.36	20.60	20.84	21.09
21	19.34	19.60	19.85	20.11	20.36	20.62	20.87	21.13	21.38	21.64	21.90	22.16
22	20.27	20.54	20.81	21.08	21.34	21.61	21.88	22.15	22.42	22.69	22.96	23.23
23	21.21	21.49	21.77	22.05	22.33	22.61	22.90	23.18	23.46	23.74	24.03	24.31
24	22.15	22.44	22.74	23.03	23.33	23.62	23.92	24.21	24.51	24.80	25.10	25.40
25	23.10	23.40	23.71	24.02	24.32	24.63	24.94	25.25	25.56	25.87	26.18	26.49
26	24.04	24.36	24.68	25.01	25.33	25.65	25.97	26.29	26.62	26.94	27.26	27.59
27	25.00	25.33	25.67	26.00	26.34	26.67	27.01	27.34	27.68	28.02	28.35	28.69
28	25.96	26.30	26.65	27.00	27.35	27.70	28.05	28.40	28.75	29.10	29.45	29.80
29	26.92	27.28	27.64	28.00	28.37	28.73	29.09	29.46	29.82	30.19	30.55	30.92
30	27.89	28.26	28.64	29.01	29.39	29.77	30.14	30.52	30.90	31.28	31.66	32.04
31	28.86	29.25	29.64	30.03	30.42	30.81	31.20	31.59	31.98	32.38	32.77	33.17
32	29.84	30.24	30.64	31.05	31.45	31.85	32.26	32.67	33.07	33.48	33.89	34.30
33	30.82	31.23	31.65	32.07	32.49	32.91	33.33	33.75	34.17	34.59	35.01	35.44
34	31.80	32.23	32.67	33.10	33.53	33.96	34.40	34.83	35.27	35.71	36.14	36.58
35	32.79	33.24	33.68	34.13	34.58	35.03	35.47	35.92	36.37	36.83	37.28	37.73
36	33.79	34.25	34.71	35.17	35.63	36.09	36.56	37.02	37.49	37.95	38.42	38.89
48	46.07	46.71	47.35	47.99	48.64	49.28	49.93	50.58	51.23	51.88	52.54	53.19
60	58.96	59.80	60.64	61.48	62.32	63.17	64.01	64.86	65.71	66.57	67.42	68.28

Annual Percentage Rate

Number of Payments	23.00%	23.25%	23.50%	23.75%	24.00%	24.25%	24.50%	24.75%	25.00%	25.25%	25.50%	25.75%
1	1.92	1.94	1.96	1.98	2.00	2.02	2.04	2.06	2.08	2.10	2.12	2.15
2	2.88	2.92	2.95	2.98	3.01	3.04	3.07	3.10	3.14	3.17	3.20	3.23
3	3.86	3.90	3.94	3.98	4.03	4.07	4.11	4.15	4.20	4.24	4.28	4.32
4	4.84	4.89	4.94	5.00	5.05	5.10	5.16	5.21	5.26	5.32	5.37	5.42
5	5.82	5.89	5.95	6.02	6.08	6.14	6.21	6.27	6.34	6.40	6.46	6.53
6	6.81	6.89	6.96	7.04	7.12	7.19	7.27	7.34	7.42	7.49	7.57	7.64
7	7.81	7.90	7.99	8.07	8.16	8.25	8.33	8.42	8.51	8.59	8.68	8.77
8	8.82	8.91	9.01	9.11	9.21	9.31	9.40	9.50	9.60	9.70	9.80	9.90
9	9.83	9.94	10.04	10.15	10.26	10.37	10.48	10.59	10.70	10.81	10.92	11.03
10	10.84	10.96	11.08	11.21	11.33	11.45	11.57	11.69	11.81	11.93	12.06	12.18
11	11.86	12.00	12.13	12.26	12.40	12.53	12.66	12.80	12.93	13.06	13.20	13.33
12	12.89	13.04	13.18	13.33	13.47	13.62	13.76	13.91	14.05	14.20	14.34	14.49
13	13.93	14.08	14.24	14.40	14.55	14.71	14.87	15.03	15.18	15.34	15.50	15.66
14	14.97	15.13	15.30	15.47	15.64	15.81	15.98	16.15	16.32	16.49	16.66	16.83
15	16.01	16.19	16.37	16.56	16.74	16.92	17.10	17.28	17.47	17.65	17.83	18.02
16	17.06	17.26	17.45	17.65	17.84	18.03	18.23	18.42	18.62	18.81	19.01	19.21
17	18.12	18.33	18.53	18.74	18.95	19.16	19.36	19.57	19.78	19.99	20.20	20.40
18	19.19	19.41	19.62	19.84	20.06	20.28	20.50	20.72	20.95	21.17	21.39	21.61
19	20.26	20.49	20.72	20.95	21.19	21.42	21.65	21.89	22.12	22.35	22.59	22.82
20	21.33	21.58	21.82	22.07	22.31	22.56	22.81	23.05	23.30	23.55	23.79	24.04
21	22.41	22.67	22.93	23.19	23.45	23.71	23.97	24.23	24.49	24.75	25.01	25.27
22	23.50	23.77	24.04	24.32	24.59	24.86	25.13	25.41	25.68	25.96	26.23	26.50
23	24.60	24.88	25.17	25.45	25.74	26.02	26.31	26.60	26.88	27.17	27.46	27.75
24	25.70	25.99	26.29	26.59	26.89	27.19	27.49	27.79	28.09	28.39	28.69	29.00
25	26.80	27.11	27.43	27.74	28.05	28.36	28.68	28.99	29.31	29.62	29.94	30.25
26	27.91	28.24	28.56	28.89	29.22	29.55	29.87	30.20	30.53	30.86	31.19	31.52
27	29.03	29.37	29.71	30.05	30.39	30.73	31.07	31.42	31.76	32.10	32.45	32.79
28	30.15	30.51	30.86	31.22	31.57	31.93	32.28	32.64	33.00	33.35	33.71	34.07
29	31.28	31.65	32.02	32.39	32.76	33.13	33.50	33.87	34.24	34.61	34.98	35.36
30	32.42	32.80	33.18	33.57	33.95	34.33	34.72	35.10	35.49	35.88	36.26	36.65
31	33.56	33.96	34.35	34.75	35.15	35.55	35.95	36.35	36.75	37.15	37.55	37.95
32	34.71	35.12	35.53	35.94	36.35	36.77	37.18	37.60	38.01	38.43	38.84	39.26
33	35.86	36.29	36.71	37.14	37.57	37.99	38.42	38.85	39.28	39.71	40.14	40.58
34	37.02	37.46	37.90	38.34	38.78	39.23	39.67	40.11	40.56	41.01	41.45	41.90
35	38.18	38.64	39.09	39.55	40.01	40.47	40.92	41.38	41.84	42.31	42.77	43.23
36	39.35	39.82	40.29	40.77	41.24	41.71	42.19	42.66	43.14	43.61	44.09	44.57
48	53.85	54.51	55.16	55.83	56.49	57.15	57.82	58.49	59.15	59.82	60.50	61.17
60	69.14	70.01	70.87	71.74	72.61	73.48	74.35	75.23	76.11	76.99	77.87	78.76

Annual Percentage Rate

Number of Payments	26.00%	26.25%	26.50%	26.75%	27.00%	27.25%	27.50%	27.75%	28.00%	28.25%	28.50%	28.75%
1	2.17	2.19	2.21	2.23	2.25	2.27	2.29	2.31	2.33	2.35	2.37	2.40
2	3.26	3.29	3.32	3.36	3.39	3.42	3.45	3.48	3.51	3.54	3.58	3.61
3	4.36	4.41	4.45	4.49	4.53	4.58	4.62	4.66	4.70	4.74	4.79	4.83
4	5.47	5.53	5.58	5.63	5.69	5.74	5.79	5.85	5.90	5.95	6.01	6.06
5	6.59	6.66	6.72	6.79	6.85	6.91	6.98	7.04	7.11	7.17	7.24	7.30
6	7.72	7.79	7.87	7.95	8.02	8.10	8.17	8.25	8.32	8.40	8.48	8.55
7	8.85	8.94	9.03	9.11	9.20	9.29	9.37	9.46	9.55	9.64	9.72	9.81
8	9.99	10.09	10.19	10.29	10.39	10.49	10.58	10.68	10.78	10.88	10.98	11.08
9	11.14	11.25	11.36	11.47	11.58	11.69	11.80	11.91	12.03	12.14	12.25	12.36
10	12.30	12.42	12.54	12.67	12.79	12.91	13.03	13.15	13.28	13.40	13.52	13.64
11	13.46	13.60	13.73	13.87	14.00	14.13	14.27	14.40	14.54	14.67	14.81	14.94
12	14.64	14.78	14.93	15.07	15.22	15.37	15.51	15.66	15.81	15.95	16.10	16.25
13	15.82	15.97	16.13	16.29	16.45	16.61	16.77	16.93	17.09	17.24	17.40	17.56
14	17.00	17.17	17.35	17.52	17.69	17.86	18.03	18.20	18.37	18.54	18.72	18.89
15	18.20	18.38	18.57	18.75	18.93	19.12	19.30	19.48	19.67	19.85	20.04	20.22
16	19.40	19.60	19.79	19.99	20.19	20.38	20.58	20.78	20.97	21.17	21.37	21.57
17	20.61	20.82	21.03	21.24	21.45	21.66	21.87	22.08	22.29	22.50	22.71	22.92
18	21.83	22.05	22.27	22.50	22.72	22.94	23.16	23.39	23.61	23.83	24.06	24.28
19	23.06	23.29	23.53	23.76	24.00	24.23	24.47	24.71	24.94	25.18	25.42	25.65
20	24.29	24.54	24.79	25.04	25.28	25.53	25.78	26.03	26.28	26.53	26.78	27.04
21	25.53	25.79	26.05	26.32	26.58	26.84	27.11	27.37	27.63	27.90	28.16	28.43
22	26.78	27.05	27.33	27.61	27.88	28.16	28.44	28.71	28.99	29.27	29.55	29.82
23	28.04	28.32	28.61	28.90	29.19	29.48	29.77	30.07	30.36	30.65	30.94	31.23
24	29.30	29.60	29.90	30.21	30.51	30.82	31.12	31.43	31.73	32.04	32.34	32.65
25	30.57	30.89	31.20	31.52	31.84	32.16	32.48	32.80	33.12	33.44	33.76	34.08
26	31.85	32.18	32.51	32.84	33.18	33.51	33.84	34.18	34.51	34.84	35.18	35.51
27	33.14	33.48	33.83	34.17	34.52	34.87	35.21	35.56	35.91	36.26	36.61	36.96
28	34.43	34.79	35.15	35.51	35.87	36.23	36.59	36.96	37.32	37.68	38.05	38.41
29	35.73	36.10	36.48	36.85	37.23	37.61	37.98	38.36	38.74	39.12	39.50	39.88
30	37.04	37.43	37.82	38.21	38.60	38.99	39.38	39.77	40.17	40.56	40.95	41.35
31	38.36	38.76	39.16	39.57	39.97	40.38	40.79	41.19	41.60	42.01	42.42	42.83
32	39.68	40.10	40.52	40.94	41.36	41.78	42.20	42.62	43.05	43.47	43.90	44.32
33	41.01	41.44	41.88	42.31	42.75	43.19	43.62	44.06	44.50	44.94	45.38	45.82
34	42.35	42.80	43.25	43.70	44.15	44.60	45.05	45.51	45.96	46.42	46.87	47.33
35	43.69	44.16	44.62	45.09	45.56	46.02	46.49	46.96	47.43	47.90	48.37	48.85
36	45.05	45.53	46.01	46.49	46.97	47.45	47.94	48.42	48.91	49.40	49.88	50.37
48	61.84	62.52	63.20	63.87	64.56	65.24	65.92	66.60	67.29	67.98	68.67	69.36
60	79.64	80.53	81.42	82.32	83.21	84.11	85.01	85.91	86.81	87.72	88.63	89.54

Exercise B

Determine the annual percentage rate or the interest from the table in the following problems.

Amount Borrowed	Interest	Other Charges		Term of Loan (in Months)	Annual Percentage Rate
1. $600	$22.42	$5.00	(22.42 + 5 = ÷ 6 =)	6	15.5%
2. 1,000	232.00	25.00		24	_____
3. 450	_____	0		9	18.75
4. 800	27.00	6.60		6	_____
5. 2,500	859.75	31.00		36	_____

Credit Card Deposits
Objective 4

When a store accepts a credit card form of payment from a customer, the sales clerk follows a standard procedure: The customer's credit card is run through an electronic reader or an imprinting machine (which has a plate that indicates the name of the store) with a blank standard charge form; the clerk completes the form by filling in the correct amounts; the clerk usually calls the credit card company to verify the card and the chargeable amount; the customer then signs the charge form, the clerk gives a copy to the customer and keeps two copies.

When the store is ready to make a deposit at the bank, they fill out a deposit slip which is presented at the bank along with one copy of each charge form. The bank computes the amount of discount charged by the credit card company, subtracts this from the deposit amount, and credits the store's account with the difference. The store automatically receives payment for the difference to their account. This may also be accomplished electronically. The credit card company has the responsibility for billing and collection.

Some credit card companies require the store to send the charge forms to them. They determine the discount and return the difference to the store. This may take longer, but the results are the same.

Though nearly all stores are charged some discount by the credit card company, the rate of discount is negotiable. Some credit card companies charge smaller stores as much as 5 percent of the credit sale.

Example 4

Kerry made out a deposit slip for Tuesday's credit sales. She took the deposit slip to the bank with copies of the charge forms and asked the bank teller to credit the account of Sport-A-Merica with the amount remaining after the discount charge was subtracted. The credit forms totaled $2,180. Sport-A-Merica pays a discount charge of 4 percent. What is the amount of the deposit to the account after the discount charge?

$2,180
×.04
$87.20 Discount charge

$2,180.00
−87.20
$2,092.80 Deposit

Exercise C

Determine the discount charge amount and the deposit amount for the following stores.

Store	Total of Charge Forms	Discount Charge Rate	Discount Charge Amount	Deposit Amount
1. Pick-up	$480.00	5% (480 × .05 = 480 − 24 =)	$24.00	$456.00
2. Sarah's T's	1,900.00	$4\frac{1}{2}$		
3. Pringle & Tamold	600.65	2		
4. The Fast Hut	1,200.00	3		
5. Ann's Shop for Men	89.20	$3\frac{3}{4}$		

ASSIGNMENT Chapter 12

Name _____ **Date** _____ **Score** _____

Skill Problems

1. When does the obligation for payment begin when one owns a credit card?
   (a) The time of the application.
   (b) The time of credit approval.
   (c) The time of a purchase.
   (d) The receipt of the monthly statement.

2. When a store accepts a credit card form of payment, what will the sales clerk do?
   (a) Fill out form by hand or cash register.
   (b) Verify the card and the amount.
   (c) Run card through imprinting machine or slide magnetic strip through machine.
   (d) All of the above.

3. When the store is ready to make a deposit of credit sales at the bank, what amount does the store put in its check register?
   (a) The store records the total credit sales and a sales commission from credit card companies.
   (b) The store records the total credit sales less a fee charged by the credit card company.
   (c) The store records just the total credit sales, no fees are ever charged.
   (d) None of the above.

Select the best term for the description below.

A. Credit Card   B. Annual Percentage Rate   C. Monthly Statement
D. Regulation Z   E. Amortizing   F. Minimum Payment Amount

4. _____ A purchase device issued to those with a good credit history.

5. _____ The smallest amount that can be paid on a credit card balance to maintain a good credit rating.

6. _____ This specifies that interest rates must be calculated using a set method.

7. _____ A record of the account balance from the viewpoint of the issuer.

8. _____ Paying off a loan by making monthly payments.

Compute the last balance in the following problems.

Month	Previous Balance	Purchases this Month	Payments	Finance Charges (1%)	Balance
1	$36.50	$21.50	$10	**9.** _____	**10.** _____
2	_____	—	10	**11.** _____	**12.** _____
3	_____	12.00	20	**13.** _____	**14.** _____
1	$24	$24.95	$40	**15.** _____	**16.** _____
2	_____	65.62	20	**17.** _____	**18.** _____
3	_____	—	40	**19.** _____	**20.** _____
1	$1,467	$82.00	$150	**21.** _____	**22.** _____
2	_____	$135.00	150	**23.** _____	**24.** _____
3	_____	50.83	300	**25.** _____	**26.** _____

Complete the following chart for the finance charge and new balance. There is a 1.8 percent finance charge on unpaid monthly balance and a minimum payment amount of $15.00.

Month	Previous Balance	Purchases this Month	Payments	Finance Charges	New Balance
Jan.	$ 0.00	$ 79.95	$ 0.00	27. _____	28. _____
Feb.		50.50	15.00	29. _____	30. _____
Mar.		0.00	30.00	31. _____	32. _____
Apr.		46.15	15.00	33. _____	34. _____
May		495.21	60.00	35. _____	36. _____
June		0.00	300.00	37. _____	38. _____
July		25.31	60.00	39. _____	40. _____
Aug.		0.00	200.00	41. _____	42. _____
Sep.		15.59	15.00	43. _____	44. _____

Determine the annual percentage rate or the interest from the table in the following problems.

45.
$1,000 = Loan
62.00 = Interest
25.00 = Filing fee
11 = Number of payments
_____ = Amount to be paid back
_____ = Total finance charge
_____ = Finance charge per $100 of loan
_____ = APR

47.
$450 = Loan
13.57 = Interest
2.50 = Filing fee
4 = Number of payments
_____ = Amount to be paid back
_____ = Total finance charge
_____ = Finance charge per $100 of loan
_____ = APR

46.
$500.00 = Loan
21.25 = Interest
20.00 = Filing fee
7 = Number of payments
_____ = Amount to be paid back
_____ = Total finance charge
_____ = Finance charge per $100 of loan
_____ = APR

48.
$10,000 = Loan
4,199 = Interest
75.00 = Filing fee
60 = Number of payments
_____ = Amount to be paid back
_____ = Total finance charge
_____ = Finance charge per $100 of loan
_____ = APR

49.
$500 = Loan
38.90 = Interest
10.50 = Filing fee
10 = Number of payments
_____ = Amount to be paid back
_____ = Total finance charge
_____ = Finance charge per $100 of loan
_____ = APR

50.
$100.00 = Loan
25.84 = Interest
0.00 = Filing fee
29 = Number of payments
_____ = Amount to be paid back
_____ = Total finance charge
_____ = Finance charge per $100 of loan
_____ = APR

51.
$105,200 = Loan
42,365.28 = Interest
125.00 = Filing fee
60 = Number of payments
_____ = Amount to be paid back
_____ = Total finance charge
_____ = Finance charge per $100 of loan
_____ = APR

52.
$7,840 = Loan
1,877.07 = Interest
50.00 = Filing fee
28 = Number of payments
_____ = Amount to be paid back
_____ = Total finance charge
_____ = Finance charge per $100 of loan
_____ = APR

53. Amount borrowed = $1,500
Interest charges = $99.40
Service charges = $20
Term of loan = 10 months
Annual percentage rate = _____ %

54. Amount borrowed = $600
Interest charges = $44.86
Filing fee = $5
Term of loan = 12 months
Annual percentage rate = _____ %

55. Amount borrowed = $4,000
Interest charges = $2,104
Service charges = $40
Filing fee = $10
Term of loan = 48 months
Annual percentage rate = _____ %

56. Amount borrowed = $1,200
Interest charges = $44.92
Other fees = $26
Term of loan = 6 months
Annual percentage rate = _____ %

Determine the discount charge amount and the deposit amount for the following problems.

57.
$1,280 = Charge sales
3% = Discount charge rate
_____ = Discount charge amount
= Deposit amount

58.
$2,040.49 = Charge sales
3 1/2% = Discount charge rate
_____ = Discount charge amount
_____ = Deposit amount

59. $87.23 = Charge sales
     5% = Discount charge rate
_____ = Discount charge amount
_____ = Deposit amount

61. $84,020 = Charge sales
     3% = Discount charge rate
_____ = Discount charge amount
_____ = Deposit amount

60. $842.11 = Charge sales
     2.6% = Discount charge rate
_____ = Discount charge amount
_____ = Deposit amount

62. $674.61 = Charge sales
     4% = Discount charge rate
_____ = Discount charge amount
_____ = Deposit amount

Determine the deposit amount on credit card sales in the following problems.

Store	Total of Charge Forms	Discount Charge Rate	Deposit Amount
63. Confectionaires	$1,853	3%	_____
64. Kelly's Pub	2,940	$2\frac{1}{2}$	_____
65. Benzie's	1,560	$1\frac{3}{4}$	_____

Business Application Problems

1. Ed and Lou heard an ad on the radio indicating a $1,995 living room suite could be bought for just $59 down plus 12 monthly payments of $178.94. What is the difference in cost between buying on credit and paying cash?

5. A credit card balance of $400 was paid off in payments of $150 each month. The credit card company charged 2 percent per month. What is the amount of the finance charges on the loan?

2. In problem 1, what is the APR on the loan?

6. What is the amount of the last monthly payment in problem 5?

3. Bilison Supply Company sold an office desk and chair for a total cash price of $570. The buyer made a down payment of $40 and three monthly payments of $183.49. What is the amount of interest and finance charges earned by Bilison Supply Company on the installment sale?

7. Benny deposited the daily receipts from his restaurant at the bank. The total was $1,689.40. Cash and currency were $1,434.52 of the total. The bank charges 2 percent on credit card deposits. What is the value of the deposit to Benny?

4. What is the APR in problem 3?

8. Bob Marsh paid service plus finance charges of $2,703.20 on an $8,000 monthly installment loan over a three-year term. What is the annual percentage rate on the loan?

9. What is the amount of interest on the loan in the first month in problem 8?

13. What will be the last monthly payment amount in problem 12?

10. Waterslide Resort, Inc. receipts are an average of $642 a day for the 98 days a year that the firm operates. One fourth of the receipts are credit card charges. The credit card company charges 3.5 percent for deposits at the bank. What is the cost of credit to the Waterslide Resort, Inc. over the length of the season?

14. Marilyn bought a television for $580. She paid $80 down and charged the balance on her credit card. She paid $150 a month until the balance was paid off. The service charge was $1\frac{1}{2}$ percent per month. What was the amount of the last payment?

11. Julie Smith purchased a video camera on a credit card. The video camera was priced at $302. She paid $50 down and financed the balance. The credit card stipulated a finance charge of 1 percent plus a flat charge of $2 per month. Her previous balance was $63. What is the balance that she will owe after a $30 payment is made on the balance that includes the video camera purchase?

15. What were the finance charges in problem 14?

12. Ricardo Perez purchased patio furniture for $860. He paid $150 down and financed the balance through the furniture store credit card at 1 percent interest per month. He paid $200 a month on the loan. What will be the finance charges on the loan?

16. Sally found a store willing to provide credit for six months on a purchase of $450. The store manager indicated that the store would charge $22.95 for interest and a $5 service fee. Sally knows that her credit card's APR is 18.5 percent. Should she borrow that money through the store credit or charge it on her credit card?

17. Gerry had a previous balance of $653 on his credit card. He decided to not use the card and to pay off the balance as quickly as possible by making payments of $200 each month. The credit card company charges his card at the rate of 1 percent per month. What is the last payment amount?

18. Ken has a balance on his credit card of $1,426.50 from last month. The credit card company charges $1\frac{3}{4}$ percent per month on the balance from the previous month. His charges for the month were $239.95. He will make a payment of $200 this month. What will his balance be after the payment this month?

19. Write a complete sentence in answer to this question. Amy took $457 in currency, $5.23 in coin, and $389 in credit card charge forms to the bank for deposit. The credit card company uses a discount charge rate of 3 percent for her firm. What is the total amount of the deposit for the firm?

20. Write a complete sentence in answer to this question. Another credit card company has offered the firm in problem 19 a discount charge rate of $2\frac{1}{2}$ percent. If each day the firm makes the same amount of credit card deposits as they did in problem 19 for the 312 days that they are open each year, how much would the firm save annually?

★ **Challenge Problem**

Lee had a credit card balance of $369 on her statement in September. She made monthly payments of $200. The credit card company charged $1\frac{3}{4}$ percent interest per month. On October 7, Anna made an additional purchase of $328.60. What is the balance on the statement that she received at the end of November?

Name _____ Date _____ Score _____

1. The annual percentage rate includes all charges that are used for the use of the borrowed funds. Amount paid back less amount borrowed equals finance charges.
   True   False

2. A credit card can be a source of identification.
   True   False

3. The APR is a banker's formula to identify interest rates.
   True   False

4. Amortization means to increase the credit card loan limit.   True   False

5. Bill took $327 in currency, $4.10 in coins, and $824.80 in credit card slips to the bank for deposit. The charge on credit card deposits is 4%. What is the value of the deposit made by Bill?

6. Kern Fredrick purchased a coat for $160. She paid $20 down and repaid the balance at $1\frac{1}{2}$ percent interest on her credit card. She made payments of $50 a month. What was the last payment amount?

7. Determine the APR on a loan for $750 that is paid back in eight monthly installments of $100 each.

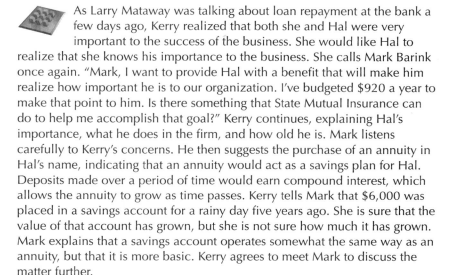

OBJECTIVES

After mastering the material in this chapter, you will be able to:

1. Compute the compounded amount in an investment problem and use compound interest tables. p. 236

2. Compute the balance on a savings account. p. 237

3. Compute the present value in an investment problem using present value tables. p. 240

4. Compute the cash value of two types of annuities in an investment problem using an annuity table. p. 244

5. Determine the investment amount in a sinking fund problem. p. 250

6. Understand and use the following terms and concepts:
Compound Interest
Compounded Amount
Present Value
Annuity
Ordinary Annuity
Annuity Due
Sinking Fund

SPORT-A-MERICA

As Larry Mataway was talking about loan repayment at the bank a few days ago, Kerry realized that both she and Hal were very important to the success of the business. She would like Hal to realize that she knows his importance to the business. She calls Mark Barink once again. "Mark, I want to provide Hal with a benefit that will make him realize how important he is to our organization. I've budgeted $920 a year to make that point to him. Is there something that State Mutual Insurance can do to help me accomplish that goal?" Kerry continues, explaining Hal's importance, what he does in the firm, and how old he is. Mark listens carefully to Kerry's concerns. He then suggests the purchase of an annuity in Hal's name, indicating that an annuity would act as a savings plan for Hal. Deposits made over a period of time would earn compound interest, which allows the annuity to grow as time passes. Kerry tells Mark that $6,000 was placed in a savings account for a rainy day five years ago. She is sure that the value of that account has grown, but she is not sure how much it has grown. Mark explains that a savings account operates somewhat the same way as an annuity, but that it is more basic. Kerry agrees to meet Mark to discuss the matter further.

Compound Interest
Objective 1

Compound interest is another way the chapter on simple interest can be applied to solve business problems. To calculate **compound interest,** the interest from the previous period becomes part of the new principal amount; that is, the interest is added to the principal before the interest for the next period is calculated. See the Compound Interest and Present Value Model on page 241 for a visual presentation of compound interest.

Example 1

compounded amount

An example will help illustrate the dramatic effect compounding has on the principal. Under three sets of circumstances, $1,000 invested at 4 percent for a year will show a difference in the amount returned to the investor, termed the **compounded amount.**

$1,000 at 4% simple interest

Interest = $40 = $1,000 × .04 × 1

$$\begin{aligned}\$40 &= \text{Interest} \\ +1,000 &= \text{Principal} \\ \hline \$1,040 &= \text{Return to investor}\end{aligned}$$

$1,000 at 4% interest compounded semiannually

Int = $20.00 = $1,000 × .04 × $\frac{1}{2}$ (first six months)
    +20

Int = $20.40 = $1,020 × .04 × $\frac{1}{2}$ (second six months)
    +20.40
    $1,040.40 = Return to investor (compounded amount)

$1,000 at 4% interest compounded quarterly

Int = $10.00 = $1,000 × .04 × $\frac{1}{4}$ (first three months)
    +10

Int = $10.10 = $1,010 × .04 × $\frac{1}{4}$ (second three months)
    +10.10

Int = $10.20 = $1,020.10 × .04 × $\frac{1}{4}$ (third three months)
    +10.20

Int = $10.30 = $1,030.30 × .04 × $\frac{1}{4}$ (fourth three months)
    +10.30
    $1,040.60 = Return to investor (compounded amount)

In this first-year period, the investor has earned an additional 40 cents by selecting an opportunity to invest at 4 percent compounded semiannually, as opposed to 4 percent at simple interest. The investor would receive 60 cents more by compounding quarterly rather than using simple interest, or 20 cents more than by compounding semiannually. Notice that the more frequently the principal is compounded in a year, the greater the return or compounded amount. Notice also that the *increase* in earnings gained from compounding interest twice a year to compounding interest four times a year is not as great as the increase in earnings gained by going from simple interest to compounding interest twice a year. The amount of increase in interest earnings declines as the principal is compounded more frequently.

Objective 2
Savings Accounts

○ *It pays to shop for the best overall rate.*

Often one finds banks, savings and loan associations, and credit unions that compound monthly or even daily interest. Daily compounding provides a small increase in the amount of return over quarterly compounding. A consumer should not be persuaded on this fact alone. The computer has made such frequent compoundings fast and inexpensive for these firms. The resulting additional return to the investor is small compared to the additional return gained from compounding semiannually or quarterly.

Compound interest applies very well to saving accounts, money market funds, and certificates of deposit (Cds).

Use of Tables

This type of problem is repetitious and, therefore, open to a shortcut. Tables have been developed to reduce the number of these calculations.

Example 2

Kerry had placed $6,000 in a savings bank five years ago. It could be used for a rainy day for Sport-A-Merica. The bank paid 4 percent interest compounded semiannually. What is the compounded amount?
An investment of $6,000 at 4 percent compounded semiannually for a period of five years can be calculated as follows using the table.

$$I = P \qquad\qquad \times R \qquad\qquad \times T$$
Interest = $6,000 (principal) × 4 percent interest (annual rate) × $\frac{1}{2}$ (compounded semiannually)

○ *What would the compounded amount have been if the interest rate was 6 percent compounded quarterly?*

Consider for a moment only the rate and time factors in this formula.

R = annual rate ÷ compoundings per year
R = 4% ÷ 2 = 2%
T = compoundings per year × number of years
T = 2 × 5 = 10

The reduced factors indicate that the investment is actually compounded at a 2 percent rate each six-month period. Therefore, the *effective rate* of interest is 2 percent *per period* for the ten 6-month periods of the five-year investment. In using the compound interest table on page 238, the effective rate of 2 percent (find across the top), and the ten periods (along the edge), the investment to be compounded is a factor of 1.21899442. This factor is the amount returned to the investor for each dollar invested. In the example, $6,000 was invested.

$6,000 = Investment   *(6,000 × 1.21899442 =)*
×1.21899442 = Factor/$1.00
$7,313.97 = Returned to investor (compounded amount)

Note: If you are using a calculator, enter the maximum number of digits possible.

Example 3

Let's check our answer in example 1 of an investment of $1,000 at 4 percent compounded quarterly.

R = 4% ÷ 4 = 1% effective rate
T = 4 × 1 = 4 periods

One percent is the effective rate for the four quarter periods in one year. Using the table, 1 percent (across top) and four periods (along the edge) show a factor equal to 1.04060401 for each dollar invested.

$1,000 = Investment
×1.04060401 = Factor/$1.00
$1,040.60 = Returned to investor (compounded amount)

(1,000 × 1.04060401 =)

Compound Interest

Periods	¾%	1%	1½%	2%	2½%	3%
1	1.00750000	1.01000000	1.01500000	1.02000000	1.02500000	1.03000000
2	1.01505625	1.02010000	1.03022500	1.04040000	1.05062500	1.06090000
3	1.02266917	1.03030100	1.04567837	1.06120800	1.07689062	1.09272700
4	1.03033919	1.04060401	1.06136355	1.08243216	1.10381289	1.12550881
5	1.03806673	1.05101005	1.07728400	1.10408080	1.13140821	1.15927407
6	1.04585224	1.06152015	1.09344326	1.12616242	1.15969342	1.19405230
7	1.05369613	1.07213535	1.10984491	1.14868567	1.18868575	1.22987387
8	1.06159885	1.08285671	1.12649259	1.17165938	1.21840290	1.26677008
9	1.06956084	1.09368527	1.14338998	1.19509257	1.24886297	1.30477318
10	1.07758255	1.10462213	1.16054083	1.21899442	1.28008454	1.34391638
11	1.08566441	1.11566835	1.17794894	1.24337431	1.31208666	1.38423387
12	1.09380690	1.12682503	1.19561817	1.26824179	1.34488882	1.42576089
13	1.10201045	1.13809328	1.21355244	1.29360663	1.37851104	1.46853371
14	1.11027553	1.14947421	1.23175573	1.31947876	1.41297382	1.51258972
15	1.11860259	1.16096896	1.25023207	1.34586834	1.44829817	1.55796742
16	1.12699211	1.17257864	1.26898555	1.37278571	1.48450562	1.60470644
17	1.13544455	1.18430443	1.28802033	1.40024142	1.52161826	1.65284763
18	1.14396039	1.19614748	1.30734064	1.42824625	1.55965872	1.70243306
19	1.15254009	1.20810895	1.32695075	1.45681117	1.59865019	1.75350605
20	1.16118414	1.22019004	1.34685501	1.48594740	1.63861644	1.80611123
21	1.16989302	1.23239194	1.36705783	1.51566634	1.67958185	1.86029457
22	1.17866722	1.24471586	1.38756370	1.54597967	1.72157140	1.91610341
23	1.18750723	1.25716302	1.40837715	1.57689926	1.76461068	1.97358651
24	1.19641353	1.26973465	1.42950281	1.60843725	1.80872595	2.03279411
30	1.25127176	1.34784892	1.56308022	1.81136158	2.09756758	2.42726247

Compound Interest

Periods	8%	9%	10%	11%	12%	13%
1	1.08000000	1.09000000	1.10000000	1.11000000	1.12000000	1.13000000
2	1.16640000	1.18810000	1.21000000	1.23210000	1.25440000	1.27690000
3	1.25971200	1.29502900	1.33100000	1.36763100	1.40492800	1.44289700
4	1.36048896	1.41158161	1.46410000	1.51807041	1.57351936	1.63047361
5	1.46932808	1.53862395	1.61051000	1.68505816	1.76234168	1.84243518
6	1.58687432	1.67710011	1.77156100	1.87041455	1.97382269	2.08195175
7	1.71382427	1.82803912	1.94871710	2.07616015	2.21068141	2.35260548
8	1.85093021	1.99256264	2.14358881	2.30453777	2.47596318	2.65844419
9	1.99900463	2.17189328	2.35794769	2.55803692	2.77307876	3.00404194
10	2.15892500	2.36736367	2.59374246	2.83942099	3.10584821	3.39456739
11	2.33163900	2.58042641	2.85311671	3.15175729	3.47854999	3.83586115
12	2.51817012	2.81266478	3.13842838	3.49845060	3.89597599	4.33452310
13	2.71962373	3.06580461	3.45227121	3.88328016	4.36349311	4.89801110
14	2.93719362	3.34172703	3.79749834	4.31044098	4.88711229	5 53475255
15	3.17216911	3.64248246	4.17724817	4.78458949	5.47356576	6.25427038
16	3.42594264	3.97030588	4.59497299	5.31089433	6.13039365	7.06732553
17	3.70001805	4.32763341	5.05447028	5.89509271	6.86604089	7.98607785
18	3.99601950	4.71712042	5.55991731	6.54355291	7.68996580	9.02426797
19	4.31570106	5.14166125	6.11590904	7.26334373	8.61276169	10.19742280
20	4.66095714	5.60441077	6.72749995	8.06231154	9.64629309	11.52308776
21	5.03383372	6.10880774	7.40024994	8.94916581	10.80384826	13.02108917
22	5.43654041	6.65860043	8.14027494	9.93357404	12.10031006	14.71383077
23	5.87146365	7.25787447	8.95430243	11.02626719	13.55234726	16.62662877
24	6.34118074	7.91108317	9.84973268	12.23915658	15.17862893	18.78809051
30	10.06265689	13.26767847	17.44940227	22.89229657	29.95992212	39.11589796

Periods	3½%	4%	4½%	5%	5½%	6%	7%
1	1.03500000	1.04000000	1.04500000	1.05000000	1.05500000	1.06000000	1.07000000
2	1.07122500	1.08160000	1.09202500	1.10250000	1.11302500	1.12360000	1.14490000
3	1.10871788	1.12486400	1.14116612	1.15762500	1.17424137	1.19101600	1.22504300
4	1.14752300	1.16985856	1.19251860	1.21550625	1.23882465	1.26247696	1.31079601
5	1.18768631	1.21665290	1.24618194	1.27628156	1.30696001	1.33822558	1.40255173
6	1.22925533	1.26531902	1.30226012	1.34009564	1.37884281	1.41851911	1.50073035
7	1.27227926	1.31593178	1.36086183	1.40710042	1.45467916	1.50363026	1.60578148
8	1.31680904	1.36856905	1.42210061	1.47745544	1.53468651	1.59384807	1.71818618
9	1.36289735	1.42331181	1.48609514	1.55132822	1.61909427	1.68947896	1.83845921
10	1.41059876	1.48024428	1.55296942	1.62889463	1.70814446	1.79084770	1.96715136
11	1.45996972	1.53945406	1.62285305	1.71033936	1.80209240	1.89829856	2.10485195
12	1.51106866	1.60103222	1.69588143	1.79585633	1.90120749	2.01219647	2.25219159
13	1.56395606	1.66507351	1.77219610	1.88564914	2.00577390	2.13292826	2.40984500
14	1.61869452	1.73167645	1.85194492	1.97993160	2.11609146	2.26090396	2.57853415
15	1.67534883	1.80094351	1.93528244	2.07892818	2.23247649	2.39655819	2.75903154
16	1.73398604	1.87298125	2.02237015	2.18287459	2.35526270	2.54035168	2.95216375
17	1.79467555	1.94790050	2.11337681	2.29201832	2.48480215	2.69277279	3.15881521
18	1.85748920	2.02581652	2.20847877	2.40661923	2.62146627	2.85433915	3.37993228
19	1.92250132	2.10684918	2.30786031	2.52695020	2.76564691	3.02559950	3.61652754
20	1.98978886	2.19112314	2.41171402	2.65329771	2.91775749	3.20713547	3.86968446
21	2.05943147	2.27876807	2.52024116	2.78596259	3.07823415	3.39956360	4.14056237
22	2.13151158	2.36991879	2.63365201	2.92526072	3.24753703	3.60353742	4.43040174
23	2.20611448	2.46471554	2.75216635	3.07152376	3.42615157	3.81974966	4.74052986
24	2.28332849	2.56330416	2.87601383	3.22509994	3.61458990	4.04893464	5.07236695
30	2.80679370	3.24339751	3.74531813	4.32194238	4.98395129	5.74349117	7.61225504

Periods	14%	15%	16%	17%	18%	19%	20%
1	1.14000000	1.15000000	1.16000000	1.17000000	1.18000000	1.19000000	1.20000000
2	1.29960000	1.32250000	1.34560000	1.36890000	1.39240000	1.41610000	1.44000000
3	1.48154400	1.52087500	1.56089600	1.60161300	1.64303200	1.68515900	1.72800000
4	1.68896016	1.74900625	1.81063936	1.87388721	1.93877776	2.00533921	2.07360000
5	1.92541458	2.01135719	2.10034166	2.19244804	2.28775776	2.38635366	2.48832000
6	2.19497262	2.31306077	2.43639632	2.56516420	2.69955415	2.83976086	2.98598400
7	2.50226879	2.66001988	2.82621973	3.00124212	3.18547390	3.37931542	3.58318080
8	2.85258642	3.05902286	3.27841489	3.51145328	3.75885920	4.02138535	4.29981696
9	3.25194852	3.51787629	3.80296127	4.10840033	4.43545386	4.78544856	5.15978035
10	3.70722131	4.04555774	4.41143508	4.80682839	5.23383555	5.69468379	6.19173642
11	4.22623230	4.65239140	5.11726469	5.62398922	6.17592595	6.77667371	7.43008371
12	4.81790482	5.35025011	5.93602704	6.58006738	7.28759263	8.06424172	8.91610045
13	5.49241149	6.15278762	6.88579137	7.69867884	8.59935930	9.59644764	10.69932054
14	6.26134910	7.05570576	7.98751799	9.00745424	10.14724397	11.41977269	12.83918465
15	7.13793798	8.13706163	9.26552087	10.53872146	11.97374789	13.58952950	15.40702157
16	8.13724930	9.35762087	10.74800420	12.33030411	14.12902251	16.17154011	18.48842589
17	9.27646420	10.76126400	12.46768488	14.42645581	16.67224656	19.24413273	22.18611107
18	10.57516918	12.37545361	14.46251446	16.87895329	19.67325094	22.90051795	26.62333328
19	12.05569287	14.23177165	16.77651677	19.74837535	23.21443611	27.25161636	31.94799994
20	13.74348987	16.36653739	19.46075945	23.10559916	27.39303460	32.42942347	38.33759992
21	15.66757845	18.82151800	22.57448097	27.03355102	32.32378083	38.59101393	46.00511991
22	17.86103944	21.64474570	26.18639792	31.62925470	38.14206138	45.92330658	55.20614389
23	20.36158496	24.89145756	30.37622159	37.00622799	45.00763243	54.64873482	66.24737267
24	23.21220685	28.62517619	35.23641704	43.29728675	53.10900627	65.03199444	79.49684720
30	50.95015858	66.21177196	85.84987691	111.06465001	143.37063844	184.67531215	237.37631380

Exercise A

Complete the following problems to find the compounded amount.

	Amount Invested	Annual Interest Rate	Term of Investment (Years)	Compound Frequency	Compounded Amount
1.	$ 1,400	5%	5	Semi. (1,400 × 1.28008454 =)	$1,792.12
2.	4,000	5	12	Semi.	_____
3.	5,500	4	8	Semi.	_____
4.	8,950	4	4	Qtr.	_____
5.	20,000	4	20	Annual	_____
6.	7,500	3	5	Semi.	_____
7.	19,500	6	$5\frac{1}{2}$	Qtr.	_____
8.	5,000	4	6	Qtr.	_____
9.	5,000	5	18	Annual	_____
10.	9,000	4	9	Semi.	_____

Present Value
Objective 3

present value

○ *A back to the future viewpoint of the value of your investment!*

Money has a "time value." That is, a given amount of money can be invested today with the expectation that the compounded amount or future value will be worth more in the future. That was just proven in exercise A. There is also a converse situation. We could ask, "What is the current value of an amount of money to be received at some date in the future?" The current, now or today, value of the future amount is termed its **present value.** Present value is the exact opposite of future value, or compound interest. We are asking, "What amount of money today is equal to the amount that will be received in the future, if today's amount (whatever it is) could be invested now and allowed to grow to that future (known) amount?" See the Compound Interest and Present Value Model on page 241 for a visual presentation of this concept.

Example 4

Acrodin, Incorporated, will receive the sum of $15,600 three years from this date. The treasurer realizes that money can safely be invested at an 8 percent annual interest rate compounded semiannually. What is the equivalent amount (present value) of $15,600 to be received three years from today?

The present value table will aid in the solution to the example problem in very much the same way as did the compound interest table.

R = 8% ÷ 2 = 4% effective rate
T = 3 × 2 = 6 periods

The effective rate for the six 6-month periods is 4 percent. Using the present value table, 4 percent (across top) and six periods (along edge) gives a factor of .79031453.

$15,600 = Future value
×.79031453 = Factor/$1.00
$12,328.91 = Present value   *(15,600 × .79031453 =)*

This means that $12,328.91 received today is equal to $15,600 received three years from now if the $12,328.91 can be invested at 8 percent compounded semiannually for three years. This can be checked by using the compound interest table.

$12,328.91 × 1.26531902 (4 percent, six periods) = $15,600

Example 5   What is the present value of $2,680 to be received five years from today if an investment can be made at 6 percent compounded quarterly?

R = 6 % ÷ 4 = $1\frac{1}{2}$% effective rate
T = 5 × 4 = 20 periods

The effective rate for twenty periods is $1\frac{1}{2}$ percent. Using the table, the factor equals .74247042.

$2,680 = Expected sum
×.74247042 = Factor/$1.00
$1,989.82 = Present value

(2,680 × .74247042 =)

Compound Interest and Present Value Model

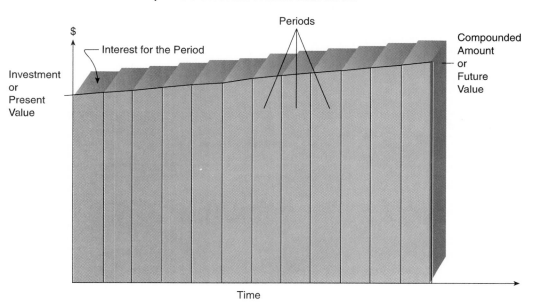

Present Value of 1

Periods	¾%	1%	1½%	2%	2½%	3%
1	0.99255583	0.99009901	0.98522167	0.98039216	0.97560976	0.97087379
2	0.98516708	0.98029605	0.97066175	0.96116878	0.95181440	0.94259591
3	0.97783333	0.97059015	0.95631699	0.94232233	0.92859941	0.91514166
4	0.97055417	0.96098034	0.94218423	0.92384543	0.90595064	0.88848705
5	0.96332920	0.95146569	0.92826033	0.90573081	0.88385429	0.86260878
6	0.95615802	0.94204524	0.91454219	0.88797138	0.86229687	0.83748426
7	0.94904022	0.93271805	0.90102679	0.87056018	0.84126524	0.81309151
8	0.94197540	0.92348322	0.88771112	0.85349037	0.82074657	0.78940923
9	0.93496318	0.91433982	0.87459224	0.83675527	0.80072836	0.76641673
10	0.92800315	0.90528695	0.86166723	0.82034830	0.78119840	0.74409391
11	0.92109494	0.89632372	0.84893323	0.80426304	0.76214478	0.72242128
12	0.91423815	0.88744923	0.83638742	0.78849318	0.74355589	0.70137988
13	0.90743241	0.87866260	0.82402702	0.77303253	0.72542038	0.68095134
14	0.90067733	0.86996297	0.81184928	0.75787502	0.70772720	0.66111781
15	0.89397254	0.86134947	0.79985150	0.74301473	0.69046556	0.64186195
16	0.88731766	0.85282126	0.78803104	0.72844581	0.67362493	0.62316694
17	0.88071231	0.84437749	0.77638526	0.71416256	0.65719506	0.60501645
18	0.87415614	0.83601731	0.76491159	0.70015937	0.64116591	0.58739461
19	0.86764878	0.82773992	0.75360747	0.68643076	0.62552772	0.57028603
20	0.86118985	0.81954447	0.74247042	0.67297133	0.61027094	0.55367575
21	0.85477901	0.81143017	0.73149795	0.65977582	0.59538629	0.53754928
22	0.84841589	0.80339621	0.72068763	0.64683904	0.58086467	0.52189250
23	0.84210014	0.79544179	0.71003708	0.63415592	0.56669724	0.50669175
24	0.83583140	0.78756613	0.69954392	0.62172149	0.55287535	0.49193374
30	0.79918690	0.74192292	0.63976243	0.55207089	0.47674269	0.41198676

Present Value of 1

Periods	8%	9%	10%	11%	12%	13%
1	0.92592593	0.91743119	0.90909091	0.90090090	0.89285714	0.88495575
2	0.85733882	0.84167999	0.82644628	0.81162243	0.79719388	0.78314668
3	0.79383224	0.77218348	0.75131480	0.73119138	0.71178025	0.69305016
4	0.73502985	0.70842521	0.68301346	0.65873097	0.63551808	0.61331873
5	0.68058320	0.64993139	0.62092132	0.59345133	0.56742686	0.54275994
6	0.63016963	0.59626733	0.56447393	0.53464084	0.50663112	0.48031853
7	0.58349040	0.54703424	0.51315812	0.48165841	0.45234922	0.42506064
8	0.54026888	0.50186628	0.46650738	0.43392650	0.40388323	0.37615986
9	0.50024897	0.46042778	0.42409762	0.39092477	0.36061002	0.33288483
10	0.46319349	0.42241081	0.38554329	0.35218448	0.32197324	0.29458835
11	0.42888286	0.38753285	0.35049390	0.31728331	0.28747610	0.26069765
12	0.39711376	0.35553473	0.31863082	0.28584082	0.25667509	0.23070589
13	0.36769792	0.32617865	0.28966438	0.25751426	0.22917419	0.20416450
14	0.34046104	0.29924647	0.26333125	0.23199482	0.20461981	0.18067655
15	0.31524170	0.27453804	0.23939205	0.20900435	0.18269626	0.15989075
16	0.29189047	0.25186976	0.21762914	0.18829220	0.16312166	0.14149624
17	0.27026895	0.23107318	0.19784467	0.16963262	0.14564434	0.12521791
18	0.25024903	0.21199374	0.17985879	0.15282218	0.13003959	0.11081231
19	0.23171206	0.19448967	0.16350799	0.13767764	0.11610678	0.09806399
20	0.21454821	0.17843089	0.14864363	0.12403391	0.10366677	0.08678229
21	0.19865575	0.16369806	0.13513057	0.11174226	0.09255961	0.07679849
22	0.18394051	0.15018171	0.12284597	0.10066870	0.08264251	0.06796327
23	0.17031528	0.13778139	0.11167816	0.09069252	0.07378796	0.06014448
24	0.15769934	0.12640494	0.10152560	0.08170498	0.06588210	0.05322521
30	0.09937733	0.07537114	0.05730855	0.04368282	0.03337792	0.02556505

Periods	3½%	4%	4½%	5%	5½%	6%	7%
1	0.96618357	0.96153846	0.95693780	0.95238095	0.94786730	0.94339623	0.93457944
2	0.93351070	0.92455621	0.91572995	0.90702948	0.89845242	0.88999644	0.87343873
3	0.90194271	0.88899636	0.87629660	0.86383760	0.85161366	0.83961928	0.81629788
4	0.87144223	0.85480419	0.83856134	0.82270247	0.80721674	0.79209366	0.76289521
5	0.84197317	0.82192711	0.80245105	0.78352617	0.76513435	0.74725817	0.71298618
6	0.81350064	0.79031453	0.76789574	0.74621540	0.72524583	0.70496054	0.66634222
7	0.78599096	0.75991781	0.73482846	0.71068133	0.68743681	0.66505711	0.62274974
8	0.75941156	0.73069021	0.70318513	0.67683936	0.65159887	0.62741237	0.58200910
9	0.73373097	0.70258674	0.67290443	0.64460892	0.61762926	0.59189846	0.54393374
10	0.70891881	0.67556417	0.64392768	0.61391325	0.58543058	0.55839478	0.50834929
11	0.68494571	0.64958093	0.61619874	0.58467929	0.55491050	0.52678753	0.47509280
12	0.66178330	0.62459705	0.58966386	0.55683742	0.52598152	0.49696936	0.44401196
13	0.63940415	0.60057409	0.56427164	0.53032135	0.49856068	0.46883902	0.41496445
14	0.61778179	0.57747508	0.53997286	0.50506795	0.47256937	0.44230096	0.38781724
15	0.59689062	0.55526450	0.51672044	0.48101710	0.44793305	0.41726506	0.36244602
16	0.57670591	0.53390818	0.49446932	0.45811152	0.42458109	0.39364628	0.33873460
17	0.55720378	0.51337325	0.47317639	0.43629669	0.40244653	0.37136442	0.31657439
18	0.53836114	0.49362812	0.45280037	0.41552065	0.38146590	0.35034379	0.29586392
19	0.52015569	0.47464242	0.43330179	0.39573396	0.36157906	0.33051301	0.27650833
20	0.50256588	0.45638695	0.41464286	0.37688948	0.34272896	0.31180473	0.25841900
21	0.48557090	0.43883360	0.39678743	0.35894236	0.32486158	0.29415540	0.24151309
22	0.46915063	0.42195539	0.37970089	0.34184987	0.30792567	0.27750510	0.22571317
23	0.45328563	0.40572633	0.36335013	0.32557131	0.29187267	0.26179726	0.21094688
24	0.43795713	0.39012147	0.34770347	0.31006791	0.27665656	0.24697855	0.19714662
30	0.35627841	0.30831867	0.26700002	0.23137745	0.20064402	0.17411013	0.13136712

Periods	14%	15%	16%	17%	18%	19%	20%
1	0.87719298	0.86956522	0.86206897	0.85470085	0.84745763	0.84033613	0.83333333
2	0.76946753	0.75614367	0.74316290	0.73051355	0.71818443	0.70616482	0.69444444
3	0.67497152	0.65751623	0.64065767	0.62437056	0.60863087	0.59341581	0.57870370
4	0.59208028	0.57175325	0.55229110	0.53365005	0.51578888	0.49866875	0.48225309
5	0.51936866	0.49717674	0.47611302	0.45611115	0.43710922	0.41904937	0.40187757
6	0.45558655	0.43232760	0.41044225	0.38983859	0.37043154	0.35214233	0.33489798
7	0.39963732	0.37593704	0.35382953	0.33319538	0.31392503	0.29591792	0.27908165
8	0.35055905	0.32690177	0.30502546	0.28478237	0.26603816	0.24867052	0.23256804
9	0.30750794	0.28426241	0.26295298	0.24340374	0.22545607	0.20896683	0.19380670
10	0.26974381	0.24718471	0.22668360	0.20803738	0.19106447	0.17560238	0.16150558
11	0.23661738	0.21494322	0.19541690	0.17780973	0.16191904	0.14756502	0.13458799
12	0.20755910	0.18690715	0.16846284	0.15197413	0.13721953	0.12400422	0.11215665
13	0.18206939	0.16252796	0.14522659	0.12989242	0.11628773	0.10420523	0.09346388
14	0.15970999	0.14132866	0.12519534	0.11101916	0.09854893	0.08756742	0.07788657
15	0.14009648	0.12289449	0.10792701	0.09488817	0.08351604	0.07358606	0.06490547
16	0.12289165	0.10686477	0.09304053	0.08110100	0.07077630	0.06183703	0.05408789
17	0.10779969	0.09292589	0.08020735	0.06931709	0.05997992	0.05196389	0.04507324
18	0.09456113	0.08080512	0.06914427	0.05924538	0.05083044	0.04366713	0.03756104
19	0.08294836	0.07026532	0.05960713	0.05063708	0.04307664	0.03669507	0.03130086
20	0.07276172	0.06110028	0.05138546	0.04327955	0.03650563	0.03083619	0.02608405
21	0.06382607	0.05313068	0.04429781	0.03699107	0.03093698	0.02591277	0.02173671
22	0.05598778	0.04620059	0.03818776	0.03161630	0.02621778	0.02177544	0.01811393
23	0.04911209	0.04017443	0.03292049	0.02702248	0.02221845	0.01829869	0.01509494
24	0.04308078	0.03493428	0.02837973	0.02309614	0.01882920	0.01537705	0.01257912
30	0.01962702	0.01510305	0.01164824	0.00900376	0.00697493	0.00541491	0.00421272

Exercise B

Complete the following problems to find the present value.

Investment Opportunity

	Future Value or Expected Sum	Annual Interest Rate	Compound Frequency	Term before Receipt	Present Value
1.	$1,600	4%	Semi.	4 years (1,600 × .85349037 =)	$1,365.58
2.	2,000	6	Qtr.	3	_____
3.	5,000	5	Semi.	5	_____
4.	2,800	9	Annually	7	_____
5.	1,000	8	Semi.	6	_____
6.	16,000	6	Semi.	8	_____
7.	6,000	5	Semi.	2	_____
8.	8,500	10	Qtr.	$3\frac{1}{2}$	_____
9.	10,000	5	Semi.	8	_____
10.	12,500	4	Semi.	$5\frac{1}{2}$	_____

Annuities
Objective 4

The concepts that concern an annuity are very similar to the concept of compound interest. Money invested over a period of time accumulates interest at a compound rate on the investment. The important difference is that the **annuity** involves a *series* of investments, usually of an equal amount and at regular intervals. The value of an annuity will grow at the stated compound rate and the amount of regular investment.

⭘ *Compound interest involves one investment. An annuity involves several equal investments.*

Annuities are used by business, religious, and financial firms to save money to pay off a bond issue, a large note, or a piece of property. Individuals use annuities to save money for their children's education, for a planned trip, or for retirement. An annuity can be purchased from an insurance agent or a securities broker.

There are two types of annuities. An *annuity certain* is the type in which the number of payments is stated in the annuity agreement. A *contingent annuity* involves an open-ended agreement, and the investor can make as many payments as he or she chooses to make.

ordinary annuity
annuity due

The point at which the payment is made in the time period further classifies an annuity. In an **ordinary annuity,** the payment is made at the end of the period. In an **annuity due,** the payment is made at the beginning of the period. This can make a difference of one period of interest when using the annuity table.

Ordinary Annuity

An annuity can be calculated over a 4-year period at 4 percent, as is shown in example 6.

Example 6   An ordinary annuity for $200 at 4 percent for four years will result in what amount or cash value?

Investment	End of First Year Value	End of Second Year Value	End of Third Year Value	End of Fourth Year Value
Year 1 $200	(no interest) $200			
Year 2 $200		$200 × 1.04 = $208 (no interest) +200 $408		
Year 3 $200			$408 × 1.04 = $424.32 (no interest) +200.00 $624.32	
Year 4 $200				$624.32 × 1.04 = $649.29 (no interest) +200.00 $849.29 (cash value)

Growth of an Annuity

The compound interest table can be used to compute the same cash value.

Investment
Year 1 $200 invested for 3 years
    at 4% compounded annually = 1.12486400 (factor) × $200 = $224.97
Year 2 $200 invested for 2 years
    at 4% compounded annually = 1.08160000 (factor) × $200 =   216.32
Year 3 $200 invested for 1 year
    at 4% compounded annually = 1.04000000 (factor) × $200 =   208.00
Year 4 $200 invested for 0 years
    at 4% compounded annually = 1.00000000 (factor) × $200 =   200.00
                            4.24646400 = factor or return/$1.00 $849.29

The ordinary annuity table simplifies the problem, as does the table in the compound interest and present value problems.

The effective interest rate (across top) of 4 percent and four periods (along the edge) give you a factor of 4.24646400.

              $200 = Payment
      ×4.24646400 = Factor or return/$1.00
          $849.29 = Cash value

Example 7   Kerry is willing to set $460 aside every six months in an annuity for Hal. The annuity will earn 4 percent over the next six years, according to Mark Barink of State Mutual. This should encourage Hal to stay with Sport-A-Merica for at least six years.

What will be the cash value of an ordinary annuity for $460 at 4 percent compounded for six years if payments are made semiannually?

The effective interest rate (across top) is 2 percent, and using 12 periods (along the edge) will show a factor of 13.41208973.

            $460 = Payment
      ×13.41208973 = Factor
       $6,169.56 = Cash value

(460 × 13.41208973 =)

Annuity Due

The cash value of an annuity due can also be found by using an ordinary annuity table. Remember that an annuity due requires that the payment be made at the beginning of the period. This will make a difference in the cash value compared to that of an ordinary annuity.

Example 8

Using example 6, an ordinary annuity for $200 at 4 percent interest with four annual payments, it is possible to show the comparison to an annuity due.

Investment
Year 1 $200 invested for 4 years
        at 4% compounded annually = 1.16985856 (factor)
Year 2 $200 invested for 3 years
        at 4% compounded annually = 1.12486400 (factor)
Year 3 $200 invested for 2 years
        at 4% compounded annually = 1.08160000 (factor)
Year 4 $200 invested for 1 year
        at 4% compounded annually = 1.04000000 (factor)
                                     4.41632256 = factor or return/$1.00

$$
\begin{aligned}
\$200 &= \text{Payment} \\
\times 4.41632256 &= \text{Factor or return/\$1.00} \\
\hline
\$883.26 &= \text{Cash value}
\end{aligned}
$$

(200 × 4.41632256 =)

○ *Add one period, subtract one for an annuity due.*

By referring to the ordinary annuity table, you will find the effective interest rate of 4 percent (across top) and five periods (along the edge). Five periods is used because the payment is being made at the beginning of the period. The effect is the same as starting an ordinary annuity one period earlier. The factor is 5.41632256. Subtract 1.00000000 from this factor, which will compensate for *not* making the fifth payment with no interest, and the factor 4.41632256 will be found. This compares to the sum of the factors found by using the compound interest table (4%, 1, 2, 3, & 4 periods, i.e., 1.04000000 + 1.08160000 + 1.12486400 + 1.16985856 = 4.41632256).

Example 9

Find the cash value of an annuity due for $380 at 3 percent with annual payments for eleven years. The effective interest rate (across top) is 3 percent, and there are 12 periods (along the edge), so the factor is 14.19202956.

○ *An annuity **due** means you **do** subtract 1.00000000!*

$$
\begin{aligned}
14.19202956 & \\
-1.00000000 & \\
\hline
13.19202956 &= \text{Factor or return/\$1.00} \\
\times \$380 &= \text{Payment} \\
\hline
\$5,012.97 &= \text{Cash value}
\end{aligned}
$$

(13.19202956 × 380 =)

Annuity or Sinking Funds Model

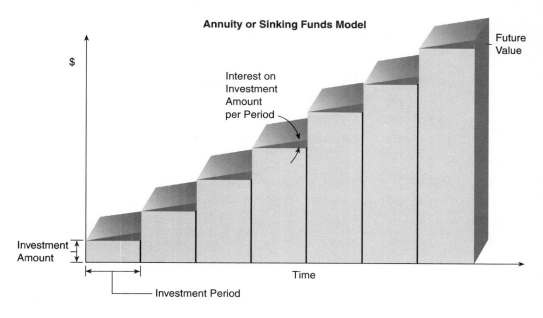

Example 10

○ *In an annuity due, you add one period and subtract 1.00000000!*

To make the distinction between an ordinary annuity and annuity due even clearer, this example will solve the answer to both using the same information.

Find the cash value of an annuity for $200 at 7 percent with annual payments for five years.

Annuity Due		Ordinary Annuity	
7.15329074	Factor for 7%,	5.75073901	Factor for 7%,
−1.00000000	6 periods	×$200	5 periods
6.15329074		$1,150.15 cash value	
×$200			
$1,230.66 cash value			

                                        (6.15329074 × 200 =)   *(5.75073901 × 200 =)*

Example 11

Find the cash value of an annuity for $350 at 16 percent with semiannual payments for six years.

Annuity Due		Ordinary Annuity	
21.49529658	Factor for 8%,	18.97712646	Factor for 8%,
−1.00000000	13 periods	×$350	12 periods
20.49529658		$6,641.99 cash value	
×$350			
$7,173.35 cash value			

                                        (20.49529658 × 350 =)   *(18.97712646 × 350 =)*

Ordinary Annuity Table Rate for Each Interest Period

Periods	¾%	1%	1½%	2%	2½%	3%
1	1.00000000	1.00000000	1.00000000	1.00000000	1.00000000	1.00000000
2	2.00750000	2.01000000	2.01500000	2.02000000	2.02500000	2.03000000
3	3.02255625	3.03010000	3.04522500	3.06040000	3.07562500	3.09090000
4	4.04522542	4.06040100	4.09090338	4.12160800	4.15251563	4.18362700
5	5.07556461	5.10100501	5.15226693	5.20404016	5.25632852	5.30913581
6	6.11363135	6.15201506	6.22955093	6.30812096	6.38773673	6.46840988
7	7.15948358	7.21353521	7.32299419	7.43428338	7.54743015	7.66246218
8	8.21317971	8.28567056	8.43283911	8.58296905	8.73611590	8.89233605
9	9.27477856	9.36852727	9.55933169	9.75462843	9.95451880	10.15910613
10	10.34433940	10.46221254	10.70272167	10.94972100	11.20338177	11.46387931
11	11.42192194	11.56683467	11.86326249	12.16871542	12.48346631	12.80779569
12	12.50758636	12.68250301	13.04121143	13.41208973	13.79555297	14.19202956
13	13.60139325	13.80932804	14.23682960	14.68033152	15.14044179	15.61779045
14	14.70340370	14.94742132	15.45038205	15.97393815	16.51895284	17.08632416
15	15.81367923	16.09689554	16.68213778	17.29341692	17.93192666	18.59891389
16	16.93228183	17.25786449	17.93236984	18.63928525	19.38022483	20.15688130
17	18.05927394	18.43044314	19.20135539	20.01207096	20.86473045	21.76158774
18	19.19471849	19.61474757	20.48937572	21.41231238	22.38634871	23.41443537
19	20.33867888	20.81089504	21.79671636	22.84055863	23.94600743	25.11686844
20	21.49121897	22.01900399	23.12366710	24.29736980	25.54465761	26.87037449
21	22.65240312	23.23919403	24.47052211	25.78331719	27.18327405	28.67648572
22	23.82229614	24.47158598	25.83757994	27.29898354	28.86285590	30.53678030
23	25.00096336	25.71630183	27.22514564	28.84496321	30.58442730	32.45288370
24	26.18847059	26.97346485	28.63352080	30.42186247	32.34903798	34.42647022
25	27.38488412	28.24319950	30.06302361	32.03029972	34.15776393	36.45926432
26	28.59027075	29.52563150	31.51396896	33.67090572	36.01170803	38.55304225
27	29.80469778	30.82088781	32.98667850	35.34432383	37.91200073	40.70963352
28	31.02823301	32.12909669	34.48147867	37.05121031	39.85980075	42.93092252
29	32.26094476	33.45038766	35.99870085	38.79223451	41.85629577	45.21885020
30	33.50290184	34.78489153	37.53868137	40.56807921	43.90270316	47.57541571

Ordinary Annuity Table Rate for Each Interest Period

Periods	8%	9%	10%	11%	12%	13%
1	1.00000000	1.00000000	1.00000000	1.00000000	1.00000000	1.00000000
2	2.08000000	2.09000000	2.10000000	2.11000000	2.12000000	2.13000000
3	3.24640000	3.27810000	3.31000000	3.34210000	3.37440000	3.40690000
4	4.50611200	4.57312900	4.64100000	4.70973100	4.77932800	4.84979700
5	5.86660096	5.98471061	6.10510000	6.22780141	6.35284736	6.48027061
6	7.33592904	7.52333456	7.71561000	7.91285957	8.11518904	8.32270579
7	8.92280336	9.20043468	9.48717100	9.78327412	10.08901173	10.40465754
8	10.63662763	11.02847380	11.43588810	11.85943427	12.29969314	12.75726302
9	12.48755784	13.02103644	13.57947691	14.16397204	14.77565631	15.41570722
10	14.48656247	15.19292972	15.93742460	16.72200896	17.54873507	18.41974915
11	16.64548746	17.56029339	18.53116706	19.56142995	20.65458328	21.81431654
12	18.97712646	20.14071980	21.38428377	22.71318724	24.13313327	25.65017769
13	21.49529658	22.95338458	24.52271214	26.21163784	28.02910926	29.98470079
14	24.21492030	26.01918919	27.97498336	30.09491800	32.39260238	34.88271190
15	27.15211393	29.36091622	31.77248169	34.40535898	37.27971466	40.41746444
16	30.32428304	33.00339868	35.94972986	39.18994847	42.75328042	46.67173482
17	33.75022569	36.97370456	40.54470285	44.50084281	48.88367407	53.73906035
18	37.45024374	41.30133797	45.59917313	50.39593551	55.74971496	61.72513819
19	41.44626324	46.01845839	51.15909045	56.93948842	63.43968075	70.74940616
20	45.76196430	51.16011964	57.27499949	64.20283215	72.05244244	80.94682896
21	50.42292144	56.76453041	64.00249944	72.26514368	81.69873554	92.46991672
22	55.45675516	62.87333815	71.40274939	81.21430949	92.50258380	105.49100590
23	60.89329557	69.53193858	79.54302433	91.14788353	104.60289386	120.20483667
24	66.76475922	76.78981305	88.49732676	102.17415072	118.15524112	136.83146543
25	73.10593995	84.70089623	98.34705943	114.41330730	133.33387006	155.61955594
26	79.95441515	93.32397689	109.18176538	127.99877110	150.33393446	176.85009821
27	87.35076836	102.72313481	121.09994191	143.07863592	169.37400660	200.84061098
28	95.33882983	112.96821694	134.20993611	159.81728587	190.69888739	227.94989040
29	103.96593622	124.13535646	148.63092972	178.39718732	214.58275388	258.58337616
30	113.28321111	136.30753855	164.49402269	199.02087793	241.33268434	293.19921506

Periods	3½%	4%	4½%	5%	5½%	6%	7%
1	1.00000000	1.00000000	1.00000000	1.00000000	1.00000000	1.00000000	1.00000000
2	2.03500000	2.04000000	2.04500000	2.05000000	2.05500000	2.06000000	2.07000000
3	3.10622500	3.12160000	3.13702500	3.15250000	3.16802500	3.18360000	3.21490000
4	4.21494287	4.24646400	4.27819112	4.31012500	4.34226638	4.37461600	4.43994300
5	5.36246588	5.41632256	5.47070973	5.52563125	5.58109103	5.63709296	5.75073901
6	6.55015218	6.63297546	6.71689166	6.80191281	6.88805103	6.97531854	7.15329074
7	7.77940751	7.89829448	8.01915179	8.14200845	8.26689384	8.39383765	8.65402109
8	9.05168677	9.21422626	9.38001362	9.54910888	9.72157300	9.89746791	10.25980257
9	10.36849581	10.58279531	10.80211423	11.02656432	11.25625951	11.49131598	11.97798875
10	11.73139316	12.00610712	12.28820937	12.57789254	12.87535379	13.18079494	13.81644796
11	13.14199192	13.48635141	13.84117879	14.20678716	14.58349825	14.97164264	15.78359932
12	14.60196164	15.02580546	15.46403184	15.91712652	16.38559065	16.86994120	17.88845127
13	16.11303030	16.62683768	17.15991327	17.71298285	18.28679814	18.88213767	20.14064286
14	17.67698636	18.29191119	18.93210937	19.59863199	20.29257203	21.01506593	22.55048786
15	19.29568088	20.02358764	20.78405429	21.57856359	22.40866350	23.27596988	25.12902201
16	20.97102971	21.82453114	22.71933673	23.65749177	24.64113999	25.67252808	27.88805355
17	22.70501575	23.69751239	24.74170689	25.84036636	26.99640269	28.21287976	30.84021730
18	24.49969130	25.64541288	26.85508370	28.13238467	29.48120483	30.90565255	33.99903251
19	26.35718050	27.67122940	29.06356246	30.53900391	32.10267110	33.75999170	37.37896479
20	28.27968181	29.77807858	31.37142277	33.06595410	34.86831801	36.78559120	40.99549232
21	30.26947068	31.96920172	33.78313680	35.71925181	37.78607550	39.99272668	44.86517678
22	32.32890215	34.24796979	36.30337795	38.50521440	40.86430965	43.39229028	49.00573916
23	34.46041373	36.61788858	38.93702996	41.43047512	44.11184669	46.99582769	53.43614090
24	36.66652821	39.08260412	41.68919631	44.50199887	47.53799825	50.81557735	58.17667076
25	38.94985669	41.64590829	44.56521015	47.72709882	51.15258816	54.86451200	63.24903772
26	41.31310168	44.31174462	47.57064460	51.11345376	54.96598051	59.15638272	68.67647036
27	43.75906024	47.08421440	50.71132361	54.66912645	58.98910943	63.70576568	74.48382328
28	46.29062734	49.96758298	53.99333317	58.40258277	63.23351045	68.52811162	80.69769091
29	48.91079930	52.96628630	57.42303316	62.32271191	67.71135353	73.63979832	87.34652927
30	51.62267728	56.08493775	61.00706966	66.43884750	72.43547797	79.05818622	94.46078632

Periods	14%	15%	16%	17%	18%	19%	20%
1	1.00000000	1.00000000	1.00000000	1.00000000	1.00000000	1.00000000	1.00000000
2	2.14000000	2.15000000	2.16000000	2.17000000	2.18000000	2.19000000	2.20000000
3	3.43960000	3.47250000	3.50560000	3.53890000	3.57240000	3.60610000	3.64000000
4	4.92114400	4.99337500	5.06649600	5.14051300	5.21543200	5.29125900	5.36800000
5	6.61010416	6.74238125	6.87713536	7.01440021	7.15420976	7.29659821	7.44160000
6	8.53551874	8.75373844	8.97747702	9.20684825	9.44196752	9.68295187	9.92992000
7	10.73049137	11.06679920	11.41387334	11.77201245	12.14152167	12.52271273	12.91590400
8	13.23276016	13.72681908	14.24009307	14.77325456	15.32699557	15.90202814	16.49908480
9	16.08534658	16.78584195	17.51850797	18.28470784	19.08585477	19.92341349	20.79890176
10	19.33729510	20.30371824	21.32146924	22.39310817	23.52130863	24.70886205	25.95868211
11	23.04451641	24.34927597	25.73290432	27.19993656	28.75514419	30.40354584	32.15041853
12	27.27074871	29.00166737	30.85016901	32.82392578	34.93107014	37.18021955	39.58050224
13	32.08865353	34.35191748	36.78619605	39.40399316	42.21866276	45.24446127	48.49660269
14	37.58106503	40.50470510	43.67198742	47.10267200	50.81802206	54.84090891	59.19592323
15	43.84241413	47.58041086	51.65950541	56.11012623	60.96526603	66.26068160	72.03510787
16	50.98035211	55.71747249	60.92502627	66.64884769	72.93901392	79.85021111	87.44212945
17	59.11760141	65.07509336	71.67303048	78.97915180	87.06803642	96.02175122	105.93055534
18	68.39406560	75.83635737	84.14071536	93.40560761	103.74028298	115.26588395	128.11666640
19	78.96923479	88.21181097	98.60322981	110.28456290	123.41353392	138.16640190	154.73999969
20	91.02492766	102.44358262	115.37974658	130.03293626	146.62797002	165.41801826	186.68799962
21	104.76841753	118.81012001	134.84050604	153.13853542	174.02100463	197.84744173	225.02559955
22	120.43599598	137.63163801	157.41498700	180.17208644	206.34478546	236.43845566	271.03071946
23	138.29703542	159.27638372	183.60138492	211.80134114	244.48684684	282.36176223	326.23686335
24	158.65862038	184.16784127	213.97760651	248.80756913	289.49447928	337.01049706	392.48423602
25	181.87082723	212.79301747	249.21402355	292.10485588	342.60348554	402.04249150	471.98108322
26	208.33274304	245.71197009	290.08826732	342.76268138	405.27211294	479.43056488	567.37729986
27	238.49932707	283.56876560	337.50239009	402.03233722	479.22109327	571.52237221	681.85275984
28	272.88923286	327.10408044	392.50277200	471.37783454	566.48089006	681.11162293	819.22331180
29	312.09372546	377.16969250	456.30321610	552.51206642	669.44745027	811.52283129	984.06797417
30	356.78684702	434.74514638	530.31173068	647.43911771	790.94799132	966.71216923	1181.88156900

Exercise C

Using the ordinary annuity table, determine the cash value in the following problems. Be careful to distinguish the type of annuity that is specified!

	Payment Amount	Payment Frequency	Annuity Term	Type of Annuity	Interest Rate	Cash Value
1.	$3,000	Semi.	6 yr	Ordin.	6%	$42,576.09 (3,000 × 14.19202956 =)
2.	500	Annual	10	Due	5	_____
3.	1,800	Semi.	4	Ordin.	4	_____
4.	2,000	Semi.	5	Due	4	_____
5.	750	Annual	8	Ordin.	6	_____
6.	1,500	Semi.	3	Ordin.	6	_____
7.	3,600	Annual	3	Due	4	_____
8.	5,000	Annual	12	Due	3	_____
9.	875	Semi.	5	Ordin.	4	_____
10.	3,750	Semi.	3	Due	6	_____

Sinking Funds
Objective 5

sinking fund

When a company has plans to replace worn equipment, expand their facilities, or retire a bond issue, they often establish a sinking fund. An investment in the form of a series of payments is made at the end of each period into a fund that pays interest. These payments, along with interest, accumulate over several periods in order to provide the desired amount. The money set aside in the form of payments and its accumulation is termed a **sinking fund.**

A sinking fund is very much like an annuity, except:

1. Sinking fund problems require a solution to the payment question, not the maturity or future value question.

2. Sinking fund problems usually require payments at the end of each period and are therefore similar to ordinary annuities.

To compute the payment amount in a sinking fund problem the following information is needed:

1. the future amount desired

2. the interest rate

3. the frequency that the investment is to be compounded

4. the length of time before the desired amount is required

Example 12

What payment amount will be required if $58,000 is needed ten years from now, and the interest rate is 6 percent compounded annually?

$58,000	= Amount desired
×.07586796	= (6%, 10 periods)—sinking funds table
$4,400.34	= Payment at the end of each year for ten years

(58,000 × .07586796 =)

Example 13

What payment amount will be required every six months if a $150,000 office addition will be needed in three years, and the interest rate is 8 percent compounded semiannually?

$150,000 = Amount desired
×.15076190 = (4%, 6 periods)
$22,614.29 = Payment at the end of each six months for three years

(150,000 × .15076190 =)

Exercise D

Using the sinking funds table, determine the payment amount in the following problems.

	Required Future Value	Payment Frequency	Sinking Fund Term	Interest Rate	Payment Amount
1.	$43,000	Annual	14 yr	6%	$2,046.15 (43,000 × .04758491 =)
2.	16,500	Semi.	5	6	
3.	122,000	Annual	12	5	
4.	42,500	Semi.	6	4	
5.	65,000	Semi.	10	6	
6.	25,000	Annual	3	3	
7.	140,000	Annual	12	5	
8.	75,000	Semi.	7	4	
9.	62,000	Annual	4	5	
10.	57,500	Semi.	12	6	

Sinking Funds Table

Periods	¾%	1%	1½%	2%	2½%	3%
1	1.00000000	1.00000000	1.00000000	1.00000000	1.00000000	1.00000000
2	0.49813200	0.49751244	0.49627792	0.49504950	0.49382716	0.49261084
3	0.33084579	0.33002211	0.32838296	0.32675467	0.32513717	0.32353036
4	0.24720501	0.24628109	0.24444479	0.24262375	0.24081788	0.23902705
5	0.19702242	0.19603980	0.19408932	0.19215839	0.19024686	0.18835457
6	0.16356891	0.16254837	0.16052521	0.15852581	0.15654997	0.15459750
7	0.13967488	0.13862828	0.13655616	0.13451196	0.13249543	0.13050635
8	0.12175552	0.12069029	0.11858402	0.11650980	0.11446735	0.11245639
9	0.10781929	0.10674036	0.10460982	0.10251544	0.10045689	0.09843386
10	0.09667123	0.09558208	0.09343418	0.09132653	0.08925876	0.08723051
11	0.08755094	0.08645408	0.08429384	0.08217794	0.08010596	0.07807745
12	0.07995148	0.07884879	0.07667999	0.07455960	0.07248713	0.07046209
13	0.07352188	0.07241482	0.07024036	0.06811835	0.06604827	0.06402954
14	0.06801146	0.06690117	0.06472332	0.06260197	0.06053652	0.05852634
15	0.06323639	0.06212378	0.05994436	0.05782547	0.05576646	0.05376658
16	0.05905879	0.05794460	0.05576508	0.05365013	0.05159899	0.04961085
17	0.05537321	0.05425806	0.05207966	0.04996984	0.04792777	0.04595253
18	0.05209766	0.05098205	0.04880578	0.04670210	0.04467008	0.04270870
19	0.04916740	0.04805175	0.04587847	0.04378177	0.04176062	0.03981388
20	0.04653063	0.04541531	0.04324574	0.04115672	0.03914713	0.03721571
21	0.04414543	0.04303075	0.04086550	0.03878477	0.03678733	0.03487178
22	0.04197748	0.04086372	0.03870332	0.03663140	0.03464661	0.03274739
23	0.03999846	0.03888584	0.03673075	0.03466810	0.03269638	0.03081390
24	0.03818474	0.03707347	0.03492410	0.03287110	0.03091282	0.02904742
30	0.02984816	0.02874811	0.02663919	0.02464992	0.02277764	0.02101926

Sinking Funds Table

Periods	8%	9%	10%	11%	12%	13%
1	1.00000000	1.00000000	1.00000000	1.00000000	1.00000000	1.00000000
2	0.48076923	0.47846890	0.47619048	0.47393365	0.47169811	0.46948357
3	0.30803351	0.30505476	0.30211480	0.29921307	0.29634898	0.29352197
4	0.22192080	0.21866866	0.21547080	0.21232635	0.20923444	0.20619420
5	0.17045645	0.16709246	0.16379748	0.16057031	0.15740973	0.15431454
6	0.13631539	0.13291978	0.12960738	0.12637656	0.12322572	0.12015323
7	0.11207240	0.10869052	0.10540550	0.10221527	0.09911774	0.09611080
8	0.09401476	0.09067438	0.08744402	0.08432105	0.08130284	0.07838672
9	0.08007971	0.07679880	0.07364054	0.07060166	0.06767889	0.06486890
10	0.06902949	0.06582009	0.06274539	0.05980143	0.05698416	0.05428956
11	0.06007634	0.05694666	0.05396314	0.05112101	0.04841540	0.04584145
12	0.05269502	0.04965066	0.04676332	0.04402729	0.04143681	0.03898608
13	0.04652181	0.04356656	0.04077852	0.03815099	0.03567720	0.03335034
14	0.04129685	0.03843317	0.03574622	0.03322820	0.03087125	0.02866750
15	0.03682954	0.03405888	0.03147378	0.02906524	0.02682424	0.02474178
16	0.03297687	0.03029991	0.02781662	0.02551675	0.02339002	0.02142624
17	0.02962943	0.02704625	0.02466413	0.02247148	0.02045673	0.01860844
18	0.02670210	0.02421229	0.02193022	0.01984287	0.01793731	0.01620085
19	0.02412763	0.02173041	0.01954687	0.01756250	0.01576300	0.01413439
20	0.02185221	0.01954648	0.01745962	0.01557564	0.01387878	0.01235379
21	0.01983225	0.01761663	0.01562439	0.01383793	0.01224009	0.01081433
22	0.01803207	0.01590499	0.01400506	0.01231310	0.01081051	0.00947948
23	0.01644217	0.01438188	0.01257181	0.01097118	0.00955996	0.00831913
24	0.01497796	0.01302256	0.01129978	0.00978721	0.00846344	0.00730826
30	0.00882743	0.00733635	0.00607925	0.00502460	0.00414366	0.00341065

Periods	3½%	4%	4½%	5%	5½%	6%	7%
1	1.00000000	1.00000000	1.00000000	1.00000000	1.00000000	1.00000000	1.00000000
2	0.49140049	0.49019608	0.48899756	0.48780488	0.48661800	0.48543689	0.48309179
3	0.32193418	0.32034854	0.31877336	0.31720856	0.31565407	0.31410981	0.31105167
4	0.23725114	0.23549005	0.23374365	0.23201183	0.23029449	0.22859149	0.22522812
5	0.18648137	0.18462711	0.18279164	0.18097480	0.17917644	0.17739640	0.17389069
6	0.15266821	0.15076190	0.14887839	0.14701747	0.14517895	0.14336263	0.13979580
7	0.12854449	0.12660961	0.12470147	0.12281982	0.12096442	0.11913502	0.11555322
8	0.11047665	0.10852783	0.10660965	0.10472181	0.10286401	0.10103594	0.09746776
9	0.09644601	0.09449299	0.09257447	0.09069008	0.08883946	0.08702224	0.08348647
10	0.08524137	0.08329094	0.08137882	0.07950457	0.07766777	0.07586796	0.07237750
11	0.07609197	0.07414904	0.07224818	0.07038889	0.06857065	0.06679294	0.06335690
12	0.06848395	0.06655217	0.06466619	0.06282541	0.06102923	0.05927703	0.05590199
13	0.06206157	0.06014373	0.05827535	0.05645577	0.05468426	0.05296011	0.04965085
14	0.05657073	0.05466897	0.05282032	0.05102397	0.04927912	0.04758491	0.04434494
15	0.05182507	0.04994110	0.04811381	0.04634229	0.04462560	0.04296276	0.03979462
16	0.04768483	0.04582000	0.04401537	0.04226991	0.04058254	0.03895214	0.03585765
17	0.04404313	0.04219852	0.04041758	0.03869914	0.03704197	0.03544480	0.03242519
18	0.04081684	0.03899333	0.03723690	0.03554622	0.03391992	0.03235654	0.02941260
19	0.03794033	0.03613862	0.03440734	0.03274501	0.03115006	0.02962086	0.02675301
20	0.03536108	0.03358175	0.03187614	0.03024259	0.02867933	0.02718456	0.02439293
21	0.03303659	0.03128011	0.02960057	0.02799611	0.02646478	0.02500455	0.02228900
22	0.03093207	0.02919881	0.02754565	0.02597051	0.02447123	0.02304557	0.02040577
23	0.02901880	0.02730906	0.02568249	0.02413682	0.02266965	0.02127848	0.01871393
24	0.02727283	0.02558683	0.02398703	0.02247090	0.02103580	0.01967900	0.01718902
30	0.01937133	0.01783010	0.01639154	0.01505144	0.01380539	0.01264891	0.01058640

Periods	14%	15%	16%	17%	18%	19%	20%
1	1.00000000	1.00000000	1.00000000	1.00000000	1.00000000	1.00000000	1.00000000
2	0.46728972	0.46511628	0.46296296	0.46082949	0.45871560	0.45662100	0.45454545
3	0.29073148	0.28797696	0.28525787	0.28257368	0.27992386	0.27730789	0.27472527
4	0.20320478	0.20026535	0.19737507	0.19453311	0.19173867	0.18899094	0.18628912
5	0.15128355	0.14831555	0.14540938	0.14256386	0.13977784	0.13705017	0.13437970
6	0.11715750	0.11423691	0.11138987	0.10861480	0.10591013	0.10327429	0.10070575
7	0.09319238	0.09036036	0.08761268	0.08494724	0.08236200	0.07985490	0.07742393
8	0.07557002	0.07285009	0.07022426	0.06768989	0.06524436	0.06288506	0.06060942
9	0.06216838	0.05957402	0.05708249	0.05469051	0.05239482	0.05019220	0.04807946
10	0.05171354	0.04925206	0.04690108	0.04465660	0.04251464	0.04047131	0.03852276
11	0.04339427	0.04106898	0.03886075	0.03676479	0.03477639	0.03289090	0.03110379
12	0.03666933	0.03448078	0.03241473	0.03046558	0.02862781	0.02689602	0.02526496
13	0.03116366	0.02911046	0.02718411	0.02537814	0.02368621	0.02210215	0.02062000
14	0.02660914	0.02468849	0.02289797	0.02123022	0.01967806	0.01823456	0.01689306
15	0.02280896	0.02101705	0.01935752	0.01782209	0.01640278	0.01509191	0.01388212
16	0.01961540	0.01794769	0.01641362	0.01500401	0.01371008	0.01252345	0.01143614
17	0.01691544	0.01536686	0.01395225	0.01266157	0.01148527	0.01041431	0.00944015
18	0.01462115	0.01318629	0.01188485	0.01070600	0.00963946	0.00867559	0.00780539
19	0.01266316	0.01133635	0.01014166	0.00906745	0.00810284	0.00723765	0.00646245
20	0.01098600	0.00976147	0.00866703	0.00769036	0.00681998	0.00604529	0.00535653
21	0.00954486	0.00841679	0.00741617	0.00653004	0.00574643	0.00505440	0.00444394
22	0.00830317	0.00726577	0.00635264	0.00555025	0.00484626	0.00422943	0.00368962
23	0.00723081	0.00627839	0.00544658	0.00472141	0.00409020	0.00354156	0.00306526
24	0.00630284	0.00542983	0.00467339	0.00401917	0.00345430	0.00296727	0.00254787
30	0.00280279	0.00230020	0.00188568	0.00154455	0.00126431	0.00103443	0.00084611

ASSIGNMENT Chapter 13

Name _____ **Date** _____ **Score** _____

Skill Problems

True or False. Circle the best answer.

T F   **1.** A present value problem involves a series of investments, usually of an equal amount and at regular intervals.

T F   **2.** The money set aside in the form of payments and its accumulation is termed an annuity due.

T F   **3.** To calculate compound interest, the interest from the previous period is subtracted from the principal amount.

T F   **4.** The total amount returned to the investor is termed the investment amount.

T F   **5.** An annuity due requires that the payment be made at the end of the period.

T F   **6.** Money has no "time value."

T F   **7.** The current value of the future amount is termed the future value.

T F   **8.** In an ordinary annuity, the payment is made at the end of the period.

T F   **9.** The computer has made compounding interest inexpensive for the banking industry.

T F   **10.** When a company has plans to replace worn equipment, they often establish a trust.

Compute the investment balance when interest is compounded using the charts in the chapter.

Amount Invested	Annual Interest Rate	Term of Investment (Years)	Compound Frequency	Interest Rate Per Period	Number of Periods Paid	Compounded Amount
$ 4,500	6%	5	Semi.	**11.** _____	**12.** _____	**13.** _____
14,500	4	7½	Qtr.	**14.** _____	**15.** _____	**16.** _____
58,700	5	6	Annually	**17.** _____	**18.** _____	**19.** _____
75,840	10	3	Qtr.	**20.** _____	**21.** _____	**22.** _____
6,000	9	2	Monthly	**23.** _____	**24.** _____	**25.** _____
42,000	8	7	Semi.	**26.** _____	**27.** _____	**28.** _____

Compute the compounded quarterly balance in the following eighteen-month savings account investment.

Quarter Ending Date	Beginning Balance	5% Annual Interest Pmt. for Quarter	Balance at End of Quarter
June. 30, 01	$4,000.00	**29.** _____	**30.** _____
Sep. 30, 01	**31.** _____	**32.** _____	**33.** _____
Dec. 31, 01	**34.** _____	**35.** _____	**36.** _____
Mar. 31, 02	**37.** _____	**38.** _____	**39.** _____
June 30, 02	**40.** _____	**41.** _____	**42.** _____
Sep. 30, 02	**43.** _____	**44.** _____	**45.** _____

Complete the following problems to find the present value.

	Future Value	Annual Interest Rate	Compound Frequency	Term before Receipt	Effective Rate Per Period	Number of Periods Paid	Present Value
46.	$41,000	10%	Qtr.	3 years	_____	_____	_____
47.	54,500	6	Semi.	5	_____	_____	_____
48.	16,750	5	Annually	7	_____	_____	_____
49.	20,000	8	Semi.	15	_____	_____	_____
50.	1,000	4	Qtr.	6	_____	_____	_____
51.	16,000	18	Monthly	2	_____	_____	_____

Fill in the missing information in problems 52 through 61. Use the compound interest and present value tables.

	Present Value	Annual Rate of Interest	Term of Investment	Compound Frequency	Future Value
52.	$23,500.00	6%	6 years	Semi.	_____
53.	41,000.00	4	6	Qtr.	_____
54.	_____	8	5	Semi.	12,500.00
55.	_____	4	9	Annual	23,000.00
56.	5,860.00	10	2	Qtr.	_____
57.	_____	5	4	Semi.	8,000.00
58.	26,000.00	9	6	Annual	_____
59.	_____	5	10	Annual	6,000.00
60.	4,000.00	10	3	Qtr.	_____
61.	61,000.00	4	4	Qtr.	_____

Complete the following problems to find the cash value of the annuity.

	Investment Amount	Payment Frequency	Annuity Term	Type of Annuity	Interest Rate	Effective Rate Per Period	Number of Periods Paid	Future Value
62.	$700	Semi.	10 years	due	4%	_____	_____	_____
63.	640	Qtr.	7$\frac{1}{2}$	ordinary	6	_____	_____	_____
64.	50	Monthly	2	ordinary	9	_____	_____	_____
65.	654	Semi.	8	due	4	_____	_____	_____
66.	3,600	Annual	28	ordinary	6	_____	_____	_____
67.	9,500	Annual	12	due	5	_____	_____	_____

Fill in the missing information in problems 68 through 77. Use the ordinary annuity table or sinking funds table.

Present Value—Investment Amount	Payment Frequency	Annuity Term	Sinking Fund or Type of Annuity	Interest Rate	Cash Value—Future Value
68. $1,000.00	Annual	10 years	Ordin.	5%	_____
69. 5,600.00	Semi.	7	Due	6	_____
70. _____	Semi.	5	Sink. fund	9	12,500.00
71. _____	Annual	12	Sink. fund	5	14,000.00
72. 6,250.00	Annual	3	Due	3	_____
73. _____	Semi.	4	Sink. fund	4	67,000.00
74. 1,250.00	Semi.	6	Ordin.	6	_____
75. 1,250.00	Annual	9	Ordin.	5	_____
76. _____	Annual	13	Sink. fund	6	190,400.00
77. _____	Annual	11	Sink. fund	3	53,000.00

Using the sinking funds table, determine the payment amount in the following problems.

	Future Value	Payment Frequency	Sinking Fund Term	Interest Rate	Effective Rate Per Period	Number of Periods Paid	Investment Amount
78.	$15,400	Annual	13 years	8%	_____	_____	_____
79.	40,000	Semi.	5	7	_____	_____	_____
80.	78,000	Qtr.	4	12	_____	_____	_____
81.	$170,000	Annual	8	5$\frac{1}{2}$	_____	_____	_____
82.	11,500	Semi.	2	9	_____	_____	_____
83.	8,000	Annual	10	11	_____	_____	_____

Business Application Problems

1. Hampe Associates purchased a building for $95,000, giving the seller a note earning 8 percent compounded quarterly and full repayment after three years. What is the amount of the payoff at the end of three years?

2. The Wellar Corporation must pay a debt of $121,000 in five years. If the corporation knows that other investment opportunities pay 6 percent compounded semiannually, what is the current value of the debt?

3. What amount should Wellar Corporation, in problem 2, deposit in a 6 percent sinking fund at the end of each quarter to be able to pay off their debt?

4. Bill just made the sixth annual payment of $500 to an ordinary annuity that pays 7 percent compounded annually. What is the value of the annuity?

5. Bill, in problem 4, makes no more payments to the annuity for three years. He will let the current balance grow at 7 percent compounded annually. What will be the new balance in the account?

6. Bill, in problem 5, withdrew the balance of the annuity at the end of three years of no deposits. What is the amount of interest earned over the nine-year life of the annuity?

7. Suli Ahmed has two certificates of deposit that pay 6 percent semiannually. He opened the first CD for $4,000 three years ago and the second CD for $1,000 one year ago. What is the value of his CDs in total?

8. An ordinary annuity was purchased five years ago. The annuity pays 6 percent compounded quarterly. The quarterly payments have been $500. What is the amount of interest earned on the annuity to date?

9. Alicia Hemez placed $162,000 she inherited in a savings account earning 3 percent compounded quarterly. She needs $200,000 to enroll in law school. She intends to enroll in two years. What will be the balance in the savings account in two years?

10. In problem 9, what is the additional amount that Alicia will need to deposit at the end of each quarter in a 6 percent quarterly compounded investment to reach her $200,000 goal for law school?

11. Kimpton and Company placed $5,000 in a CD for a business deal five and a half years ago. This CD paid 6 percent, compounded quarterly. What would be the value of the CD today?

12. The Dalton Company wishes to pay off a promissory note due in eight years. The firm decides to establish a sinking fund and invest money at 11 percent compounded annually. The payments will be made once a year. The maturity value of the promissory note is $140,000. What will be the payment amount?

13. A promissory note for $16,000 including interest is due in two years to the Bulster Company. What is the present value of the note if money can be invested at 6 percent compounded semiannually?

14. A 6 percent contingent ordinary annuity was purchased by James D. Armour seven years ago. The semiannual payment was $600. What is the cash value of the annuity today?

15. The parents of Jamez Hotz wished to save some money for her college education. They felt that if they could save $20,000 by the time she became 18 years old the balance of the education cost could be earned by part-time jobs, possible scholarships, etc. Jamez will soon have her twelfth birthday. What will be the payment amount on an ordinary annuity at a 6 percent rate if payments are made annually?

16. Phil Orel deposited $1,700 in his savings account on July 1, 1986. The bank paid 4 percent compounded semiannually. On June 30, 1992, he deposited an additional $3,100 in his account. The bank began compounding deposits quarterly on January 1, 1996. On June 30, 1997, Phil withdrew the total amount in his savings account to take a trip to Europe. What was the value of the withdrawal?

17. Genny Portan was recently widowed. She and her husband had made twelve annual payments of $2,500 on an annuity due at 6 percent. What is the value of the annuity before the next payment is due?

18. The P.H. Rickton Corporation borrowed $67,500 from Villance Trust Company with a promissory note at 6 percent for five years. The Rickton Corporation negotiated an ordinary annuity at 4 percent on the same date. Payments will be made at six-month intervals. What will the payment amount be in order to repay the maturity value of the note to Villance Trust Company in five years? (Use the sinking funds table.)

19. Benzol Products will need to retire (pay off) a $165,000 debt in eight years. In order to prepare for the payoff, the firm has decided to set aside some amount at the end of each year. The insurance firm they have been speaking to said that the amount set aside would earn 7 percent. What is the amount that will need to be set aside each year?

20. Jebson and Associates invested $60,000 five years ago at 8 percent compounded semiannually. What is the compounded amount?

21. The Wycoff Corporation will need to retire a bond debt that is due to mature in six years. The bond issue has a total face value of $6,300,000. The treasurer has recommended that a sinking fund be set up to retire the debt. The funds can be invested at 7 percent compounded semiannually. What will be the amount of the semiannual payment to the sinking fund?

22. Write a complete sentence in answer to this question. Lester placed $5,000 in a cash account with his stockbroker. The account pays an annual rate of 9 percent compounded monthly. After five months, he decided to purchase 300 shares of a stock at a price of $15.50 per share. The broker charged a fee of $\frac{1}{2}$ percent of the total purchase price. Lester used the money from his cash account to pay for the stock. How much was left in the account after the shares were bought and the broker had been paid?

23. Write a complete sentence in answer to this question. A savings and loan association is paying depositors 4 percent interest on savings deposits, and compounds the interest every six months. The funds are then lent to home buyers at 9 percent for home mortgages (monthly repayments). This savings and loan association has deposits and loans totaling $51,000,000. The overhead for office operations is maintained at 1 percent of deposits. What is the amount of profit for a one-year period?

★ Challenge Problem

Silvia Ryan is a widow. Her husband left $20,000 for her living expenses. She placed the $20,000 in a savings account that paid 4 percent compounded semiannually. At the beginning of each year, she transferred $3,000 from her savings account to her checking account. Each month, she wrote herself a check for $250. How many monthly $250 checks will she be able to use before the $20,000 is used up? She made the first transfer at the end of the first year.

Name _____ Date _____ Score _____

Complete the following problems. Use the appropriate tables:

Present Value	Annual Rate of Interest	Term of Investment	Compound Frequency	Future Value
1. $ 2,000.00	8%	4 years	Qtrly.	$ _____
2. _____	9	6	Annual	5,000.00
3. 19,000.00	10	7	Semi.	_____
4. _____	8	3	Qtrly.	9,500.00

Investment Amount	Payment Frequency	Annuity Term	Type of Annuity or Sinking Fund	Interest Rate	Cash Value— Future Value
5. $ 400.00	Annual	7 years	Due	9%	$ _____
6. _____	Semi.	5	Sink. Fund	8	30,000.00
7. 375.00	Semi.	6	Ordinary	10	_____
8. _____	Annual	2	Sink. Fund	6	50,000.00

9. Dolores Heine placed $300 in a new savings account that paid 5 percent interest compounded semiannually. Her money has been on deposit for $4\frac{1}{2}$ years. What is the value of Ms. Heine's account at the present time if no other deposits or withdrawals have been made?

10. The Blackburn Corp. purchased a contingent annuity due eight years ago with payments of $750 annually. The rate of interest on the annuity was 7 percent. The treasurer of the firm decided to terminate the annuity and purchase bonds with the cash value. What was the cash value of the annuity?

DISCOUNTS

14

SPORT-A-MERICA

In the new store, the additional space will allow for expansion of the merchandise now carried by Sport-A-Merica. Kerry knows that in order to continue to grow, the store should include new items. When the sales agent for Tourist Bicycles calls on her, she sees the possibility of adding racing models to the current line. The addition of a fine line of racing bicycles and related equipment would mean that Sport-A-Merica could sponsor road races. The store could use the two delivery vehicles as escorts for the cyclists, gain some media exposure for the sponsorship of the event, and promote public relations through the cyclists and their friends. Most importantly, the addition of the racing line would add to the profits of the firm.

The price offered by Tourist Bicycles, Inc. is dependent on the volume of sales and the shipping arrangements. Specifically, the agent for Tourist Bicycles offered a volume discount of 20 percent off the list price and another 5 percent for the handling of the freight. The agent also specified terms of trade of 2/10, n/30. Kerry considers placing an order for 20 bicycles with a list price of $1,295 each.

The discount would make a substantial difference in the net price to Sport-A-Merica and, therefore, in the amount of profit. Kerry decides to discuss this with Hal. It would be important to make this addition to the product line before the beginning of the next selling season for bicycles.

Trade Discounts
Objective 1

Manufacturers, wholesalers, and other middlemen often offer their customers discounts based on the buyer's providing some marketing functions. These discounts are termed **trade discounts.** The seller offers this price reduction if the buyer purchases in large quantities, transports or stores the merchandise for the seller, promotes the product to the final consumer, gathers sales information for the seller, or simply negotiates successfully. The number of marketing functions performed will determine the size of the trade discounts offered. The size of the discounts varies from seller to seller.

Finding the Selling Price

list price

selling price

The trade discount is computed from a **list price** established by the seller, and given in a catalog or on a price sheet. Since not all buyers will perform the same marketing functions, the trade discounts are not listed. The trade discount is subtracted from the list price in order to find the **selling price.** This is the amount that the seller is willing to accept for the merchandise when a buyer provides the agreed-upon marketing functions.

Example 1

A piece of office equipment is listed at $1,460. The seller agrees to offer the buyer a 15 percent trade discount if the buyer transports and installs the equipment. What is the selling price of the equipment?

$1,460 = List price (base)
  ×.15 = Trade discount percent (rate)
 $219 = Trade discount amount (percentage)

$1,460 = List Price
 −219 = Trade discount amount
$1,241 = Selling price

(1,460 × .15 =)

The same problem can be solved by using another shorter approach: Subtract the trade discount percent from 100 percent and then multiply.

Example 2

$1,460 = List price
  ×.85 = (100% − 15% trade discount)
$1,241 = Selling price

(1,460 × .85 =)

Example 3

If the selling firm in the previous example were to offer an additional 10 percent trade discount for allowing on-site inspections of the equipment by other prospective customers, the new selling price would be calculated as follows.

$1,460.00 = List price
 −219.00 = Trade discount amount (15%)
$1,241.00 = (Base)
     ×.10 = Trade discount percent (rate)
  $124.10 = Trade discount amount (10%) (percentage)

$1,241.00
 −124.10 = Trade discount amount (10%)
$1,116.90 = Selling price

(1,241 × .10 =)

When two or more trade discounts are offered, it is termed a *chain discount,* or *series of trade discounts*. Notice that the series of trade discounts, 15 percent and 10 percent, is *not* the same as a 25 percent discount.

$1,460 $1,460
  ×.25   −365
 $365 $1,095

Direct Method

The **direct method** is a much quicker and effective way to find the selling price. Each discount is subtracted from 100 percent, and each successive difference is then used to multiply the successively discounted prices, beginning with the list price, to find the selling price.

Example 4

The 20 percent and 5 percent trade discounts offered to Sport-A-Merica by Tourist Bicycles' agent on a group of 20 bicycles would have a significant effect on the selling price to Sport-A-Merica. The list price to the retailer of each bicycle is $1,295. What is the selling price of the group of 20 bicycles?

$1,295 Each
   ×20
$25,900 Total

100%	100%
−20%	−5%
80%	95%

$25,900 (list price) × .80 = $20,720
$20,720 × .95 = $19,684 (selling price)

Note: This method will be used to determine the answers to the problems in this chapter.

(1,295 × 20 =   25,900 × .80 =   20,720 × .95 =)

Example 5

Monson Distributing purchased 50 computers that would be connected with the existing server. The list price per computer was $600. J. Bonner, the distributor, agreed to allow the buyer a series of 10, 5, and 20 percent discounts for quantity, promotional considerations, and user feedback reasons. What will be the selling price of the lot of 50 computers?

$600
×50
$30,000 Total list price

$30,000 × .90 = $27,000
$27,000 × .95 = $25,650
$25,650 × .80 = $20,520 selling price

(600 × .50 =   30,000 × .90 =   27,000 × .95 =   25,650 × .80 =)

Exercise A

Find the total selling price in each of the following problems.

	List Price per Unit	Units Purchased	Trade Discount Offered	Total Selling Price
1.	$ 64.00	100	20% (6,400 × .80 =)	$5,120.00
2.	1,200.00	40	15	_____
3.	59.00	17	19	_____
4.	96.00	60	25	_____
5.	189.00	12	10	_____
6.	48.00	12	25	_____
7.	120.00	144	10	_____
8.	18.00	50	5 and 20	_____
9.	185.00	3	10 and 10	_____
10.	69.00	100	20 and 10	_____

Objective 2
equivalent trade discount

There is a method that can be used to find one single discount percent equal to a series of trade discounts. This is termed an **equivalent trade discount.** It is found by subtracting the product of two trade discounts from their sum. This method is useful when the same series of discounts are used over and over.

Example 6

An 8 percent and a 20 percent discount are offered to the A. J. Rislone Corporation for purchasing 1,500 vertical hydrators listed at $46 each. A & K Distributing, the seller, offered the discounts because of the size of the purchase and because Rislone will be penetrating a new market area. The single equivalent trade discount and the selling price are computed as follows:

$$20\% \qquad\qquad\qquad 20\% = .20$$
$$+8\% \qquad\qquad\qquad \times 8\% = \times.08$$
$$\overline{28\%} = \text{Sum of trade discounts} \qquad\qquad \overline{.0160} = 1.6\% \text{ product of trade discounts}$$
$$-1.6\% = \text{Product of trade discounts } (.20 \times .08)$$
$$\overline{26.4\%} = \text{Single equivalent discount } (20\% \text{ and } 8\%)$$

The single equivalent discount can be used to find the total trade discount amount.

$$\$69,000 \quad (1,500 \times \$46) = \text{Total list price}$$
$$\underline{\times .264} = \text{Equivalent trade discount percent}$$
$$\$18,216 = \text{Total trade discount amount}$$

By subtracting the single equivalent trade discount from 100%, the difference can be multiplied by the list price to find the selling price.

$$100\% \qquad \$69,000 = \text{Total list price}$$
$$\underline{-26.4\%} \qquad \underline{\times.736}$$
$$73.6\% \qquad \$50,784 = \text{Total selling price}$$

$$(.20 + .08 = \qquad .20 \times .08 =)$$

When there are three or more trade discounts offered, a single equivalent trade discount can still be found. The equivalent discount of two of the trade discounts with the third discount will provide an equivalent of all three discounts.

Exercise B

Find the total selling price of the following problems.

	List Price per Unit	Units Purchased	Trade Discount Offered	Total Selling Price	
1.	$45.00	200	5 and 8%	$7,866.00	(9,000 × .874 =)
2.	60.00	100	10 and 15	_____	
3.	25.00	25	5 and 12	_____	
4.	525.00	144	10 and 20	_____	
5.	1,200.00	65	20 and 5	_____	
6.	80.00	100	7, 5, and 10	_____	
7.	42.00	500	10, 10, and 20	_____	
8.	130.00	25	5, 8, and 30	_____	
9.	125.00	20	5, 10, and 10	_____	
10.	7.50	50	10, 5, and 17	_____	

Cash Discounts
Objective 3

terms of trade

Manufacturers, wholesalers, and other middlemen also offer their customers **cash discounts.** These discounts are not based on the customer's providing marketing functions, but are designed to motivate the customer to pay for the merchandise promptly. The motivating tool is a reduction in the selling price if the payment is made at an early date. The date (or dates) is specified in the **terms of trade** by the seller. Examples of terms of trade are 3/10,n/30 and 2/30,n/60; and these are shown on the invoice or bill. In each of

○ *Cash discounts are offered to firms by other firms, not to final consumers.*

the previous examples, the first number represents a percent discount. The second number following the slash mark represents the last day after the invoice or billing date that the discount is available.

Example 7

Terms of trade 1/10,n/60 mean that a 1 percent cash discount off the selling price of the merchandise is offered if the invoice is paid within the 10 days following the invoice date. The n/60 means that the customer has 60 days following the invoice date to pay the bill: In the first 10 days, a 1 percent discount is offered; in the next 50 days, no discount is offered. Sixty days after the invoice date, the unpaid bill may be placed in the hands of a collection agent. Firms that have their accounts placed in the hands of a collection agent will incur a financial penalty and the loss of a favorable status with the seller. The seller probably will no longer extend credit terms to the buyer, therefore requiring that future purchases be made on a cash basis (C.O.D.). Most new firms begin on a cash purchase basis until the seller is convinced that the firm is creditworthy.

○ *A cash discount of $31.39 to WBC Systems, Inc., for prompt payment, based on the terms of 2/10,n30.*

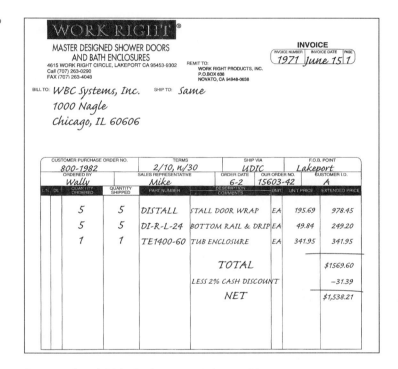

Courtesy of Work Right Products, Inc., Lakeport, CA.

Example 8

An invoice listing goods purchased on May 26, with terms of trade 3/5,2/30,n/60, would give the buyer until May 31 to receive a 3 percent discount from the selling price, until June 25 to receive a 2 percent discount from the selling price, and until July 25 to receive no discount with no penalty on the selling price.

Finding the Net Price

net price

When the cash discount is subtracted from the selling price, the balance is termed the **net price.** If there is no cash discount available in the terms of the sale, the selling price may also be called the net price. The net price paid to the seller will complete the financial obligation.

Example 9 On October 17, Sport-A-Merica purchased merchandise from Pratt Body Builders, Inc., at a selling price of $1,650 with terms of trade of 2/10,1/30,n/90. What is the net price if payment is made on November 6?

$$\begin{array}{r} \$1,650 \\ \times .01 \\ \hline \$16.50 \end{array}$$

$$\begin{array}{r} \$1,650.00 = \text{Selling price} \\ -16.50 = \text{Cash discount} \\ \hline \$1,633.50 = \text{Net price} \end{array}$$

or

$$\begin{array}{r} \$1,650 = \text{Selling price} \\ \times .99 = \text{Percent of net price } (100\% - 1\%) \\ \hline \$1,633.50 = \text{Net price} \end{array}$$

Note that Sport-A-Merica did not pay for the merchandise by October 27 and therefore did not qualify for the 2 percent cash discount.

(1,650 × .99 =)

Cash discounts are an important financial consideration, even though the discount rates sound small. Terms of trade of 2/10,n/30 equal a potential 36 percent annual discount rate. The proof is shown by the following: 2 percent discount can be taken if payment is made within ten days. By not paying on the tenth day, the money is in effect *borrowed* for the next 20-day period of the year. There are at least 18 such 20-day periods in a full year. Eighteen periods times 2 percent equals 36 percent annually (if the process were to repeat itself for a full year).

○ *What would the net price be to Haeck & Co. if payment was made by November 9th?*

invoice   **carbonless**

INVOICE

Home Town Video   INVOICE NO.
9535

1 - 9
Mon - Fri   Movies
4 - 9   Sega &
Sunday   Super
                   Nintendo

                   Haeck & Company

120 W Grand River, Laingsburg   80 State Road
Phone 651-5975 for Reservations   Pacer, OH

CUSTOMER'S ORDER	SALESMAN	TERMS	SHIPPED VIA	F.O.B.	DATE
H-52-6804	Tim	2/10, n/45	St. Clair National	Pacer	Nov. 1st
32	Video tapes			19 —	608 —
14	Games			31 50	441 —
3	Display tables w/wiring			146 50	439 50
			Total		1,488 50

carbonless

Courtesy of Home Town Video, Laingsburg, MI.

Exercise C

Find the net price in the following problems. Note the dates.

	Selling Price	Terms of Trade	Invoice Date	Payment Date	Net Price (Payment Amount)
1.	$675.00	3/15,n/30	July 26	Aug. 8	$654.75 (675 × .97 =)
2.	450.00	2/10,n/30	Nov. 10	Nov. 17	

Exercise C Continued

	Selling Price	Terms of Trade	Invoice Date	Payment Date	Net Price (Payment Amount)
3.	$ 200.00	1/10,n/90	Sept. 16	Oct. 30	_____
4.	65.00	2/10,1/30,n/60	Aug. 20	Sept. 12	_____
5.	175.00	1/30,n/60	May 13	July 1	_____
6.	1,400.00	n/30	Sept. 15	Oct. 10	_____
7.	70.00	2/15,1/60,n/90	Jan. 3	Feb. 16	_____
8.	500.00	2/10,n/120	June 30	July 2	_____
9.	120.00	1/20,n/45	Aug. 1	Aug. 13	_____
10.	89.00	1/30,n/60	Dec. 2	Dec. 30	_____

End of Month (E.O.M.)   A variation in terms of trade is the end-of-month (**E.O.M.**) stipulation. Terms of trade of 2/10,n/60,E.O.M. indicate that the buyer computes the days of discount beginning at the end of the month in which the purchase was made. If the purchase is made after the twenty-fifth of the month, then an additional month is allowed in the discount time period.

Example 10   A $595 (selling price) purchase of merchandise was made on June 12, with terms of trade of 2/10,n/60,E.O.M. What is the last date that a cash discount is available?

In this example, E.O.M. means the end of the month of June (month in which the purchase is made). Ten days after the end of the month of June is July 10. The 2 percent discount is offered until July 10.

Example 11   An invoice for $69.80 is dated March 28 with terms of 3/20,n/30,E.O.M. What is the last date that a cash discount is available?

Here, E.O.M. would normally mean the end of the month of March. However, because the purchase was made after the twenty-fifth of the month, an additional month is added (April). Twenty days after the end of the month of April is May 20. The 3 percent discount is offered until May 20.

If payment is made before May 20, the net price on this purchase would be $67.71.

(69.80 × .97 =)

Exercise D

Find the net price in the following problems.

	Selling Price	Terms of Trade	Invoice Date	Payment Date	Net Price (Payment Amount)
1.	$ 190.00	2/10,n/30,E.O.M.	Oct. 15	Nov. 20	$190.00 (NA)
2.	150.00	1/20,n/60,E.O.M.	July 20	Sept. 26	_____
3.	60.00	1/10,n/30,E.O.M.	April 30	June 10	_____
4.	750.00	2/20,E.O.M.	Feb. 28	April 16	_____
5.	600.00	1/15,n/30,E.O.M.	May 10	June 30	_____
6.	260.00	3/10,n/60,E.O.M.	Aug. 31	Oct. 7	_____
7.	550.00	2/10,n/20,E.O.M.	Jan. 16	Mar. 28	_____
8.	1,530.00	1/30,E.O.M.	Dec. 23	Jan. 25	_____
9.	32.50	4/10,2/30,E.O.M.	June 30	Aug. 3	_____
10.	700.00	2/20,1/30,E.O.M.	July 3	Aug. 25	_____

Receipt of Goods
(R.O.G.)

Another variation in terms of trade is the receipt-of-goods (**R.O.G.**) stipulation. This, like E.O.M., also specifies an extension of the period during which the buyer can still receive the cash discount. Terms of trade of 3/10,n/30,R.O.G. indicate that the buyer computes the days of discount beginning with the date that the merchandise is actually received by the buyer. The R.O.G. stipulation is often used when goods are not available for immediate shipment by the seller. By specifying R.O.G., the buyer also has an opportunity to inspect the goods received before making payment.

Example 12

A $1,385 purchase of equipment on August 20 is received on October 6. The terms of trade are 2/5,n/60,R.O.G. What is the latest date that a discount is available and what is the net price?

R.O.G. means that the discount period will be computed starting on the date that the goods are received: October 6 plus five days equals October 11. The net price equals:

$1,385 = Selling price (base)
×.02 = Cash discount rate (rate)
$27.70 = Cash discount amount (percentage)

$1,385.00 = Selling price
−27.70 = Cash discount amount
$1,357.30 = Net price, due on October 11

(1,385 × .02 =)

Exercise E

Find the net price in the following problems.

	Selling Price	Terms of Trade	Invoice Date	Date Received	Payment Date	Net Price (Payment Amount)
1.	$620.00	3/10,n/30,R.O.G.	Aug.6	Sept. 10	Sept. 17	$601.40 (620 × .97 =)
2.	85.00	2/5,n/10,R.O.G.	June 25	July 20	July 22	
3.	120.00	1/10,n/60,R.O.G.	Jan. 16	Feb. 1	Feb. 13	
4.	450.00	2/5,n/30,R.O.G.	April 3	May 30	June 16	
5.	327.00	1/10,n/20,R.O.G.	Sept. 1	Sept. 3	Sept. 11	
6.	700.00	1/30,n/60,R.O.G.	July 5	July 27	Aug. 13	
7.	175.00	2/10,n/45,R.O.G.	Oct. 26	Nov. 30	Dec. 7	
8.	78.00	2/20,n/30,R.O.G.	May 1	June 16	July 3	
9.	325.00	1/20,R.O.G.	Feb. 28	March 26	March 31	
10.	802.00	1/5,n/30,R.O.G.	Nov. 2	Nov. 16	Nov. 30	

Combination of Cash and Trade Discounts

When both cash and trade discounts are offered, the trade discount is subtracted before the cash discount is computed. The model below shows the relationship.

list price
− trade discounts
selling price
− cash discount
net price

Example 13

Sport-A-Merica purchased 125 baseball mitts from the O and W Manufacturing Company at a list price of $32 each. Sport-A-Merica was offered a series of trade discounts of 10 percent and 20 percent, on terms of trade of 3/10,n/60,E.O.M. The purchase was made on July 12, delivery was made on July 18, and payment was made on July 21. What was the net price on the purchase?

$32 = Price per unit at list price
×125 = Units purchased
$4,000 = Total list price
×.28 = Single equivalent discount (10% and 20%)
$1,120 = Trade discount amount

$4,000 = Total list price
−1,120 = Trade discount amount
$2,880 = Total selling price
×.03 = Cash discount rate
$86.40 = Cash discount amount

$2,880.00 = Total selling price
−86.40 = Cash discount
$2,793.60 = Net price

(4,000 × .28 =   2,880 × .03 =)

Exercise F

Find the net price in the following problems.

	Total List Price	Trade Discounts	Terms of Trade	Invoice Date	Date Received	Payment Date	Net Price
1.	$ 582	10 and 5%	1/10,n/30,E.O.M.	July 5	July 7	July 12	$492.63
							(582 × .855 = × .99 =)
2.	1,600	10 and 7	2/10,n/60	May 3	May 6	June 5	
3.	270	5 and 7	2/10,n/60,R.O.G.	March 1	March 19	March 25	
4.	950	15 and 10	3/5,n/60,E.O.M.	July 29	Aug. 17	Sept. 3	
5.	2,800	10 and 10	1/10,n/90	Dec. 2	Dec. 9	Dec. 11	

Transportation Charges
Objective 4

The price of merchandise varies due to discounts made available by the seller. Much of the discount variation depends on the negotiating ability of the two parties. Another factor included in the negotiations that can influence the final price is the transportation charges.

Whatever the method of delivery, there is a transportation cost that must be paid by one of the parties. Either the seller will include the transportation cost in the quoted price, or the buyer will have to pay the cost in addition to the negotiated price. The transportation cost is variable; it may be as high as the cost of the merchandise itself, depending on the distance traveled, the means of delivery, and the weight. The businessperson should consider this cost as an important price factor.

F.O.B.

F.O.B. is an abbreviation for "free on board." It means that the seller will include in the quoted price the service of placing the merchandise on the transportation device—truck, boat, plane, train, etc.—for "free." From a cost standpoint, F.O.B. does not mean very much. The transportation device will be located near the seller's place of business. The meaningful portion of F.O.B. follows the abbreviation by indicating the location to which the merchandise is free of transportation costs.

F.O.B. Destination

When the buyer and seller agree that the merchandise will be shipped to a location specified by the buyer, and all of the transportation costs will be paid by the seller, the term **F.O.B. destination** is used. This means that the seller will ship the merchandise free to the buyer's place of business. The price per unit quotation will always include transportation. The buyer will usually not be aware of the transportation costs.

Example 14

Downson & Patrick sold 12 pairs of women's slacks to the Tops & Bottoms Shop in Palo Alto, California. The price of each pair of slacks was $26. The merchandise will be shipped via truck by Baldwin Transportation Co. for a charge of $43.72. Downson & Patrick agreed to ship F.O.B. Palo Alto (the destination) from New York. How much will the Tops & Bottoms Shop pay the seller (terms of 2/10,n/60) if payment is within the discount period?

$26   Price per pair of slacks
×12   Pairs of slacks
$312   Total selling price
×.02   (2% cash discount)
$6.24

$312.00 = Total selling price
−6.24 = Cash discount
$305.76 = Net price

The net price of $305.76 will be paid by Tops & Bottoms to Downson & Patrick for the merchandise, including transportation. Downson & Patrick owes Baldwin Transportation Co. $43.72 for the transportation.

(26 × 12 =   312 × .02 =)

Sometimes the buyer specifies several locations if the merchandise is to be shipped to all of the buyer's plants, retail locations, or warehouses. The result is the same if *F.O.B. destination* is specified. The seller pays the freight costs to the multiple locations that the buyer specifies.

F.O.B. Shipping Point

○ *F.O.B. charges can amount to a large sum of money. Be sure you know who is paying when you negotiate price!*

When the buyer and seller agree that the price of the merchandise does not include the transportation charges, **F.O.B. shipping point** is specified. Under these terms, the buyer is responsible for all of the costs associated with moving the merchandise from the seller's location. In this case, F.O.B. indicates that the seller will place the merchandise on the transportation device free. From that point on, the costs are covered by the buyer. This allows the buyer to negotiate the transportation method and cost separately. This may aid in reducing costs and speeding up delivery. Each cost can be identified using F.O.B. shipping point. Transportation charges will not be part of the seller's invoice amount.

Example 15 Sport-A-Merica purchased 42 exercise sets for their stock. Each set was priced at $268, F.O.B. shipping point, 3/10, n/30. Built Well, Inc., the seller, indicated that the best method of shipment would be by truck at $38 a set. What would be the total cost of the 42 sets delivered if payment is made within the discount period?

$268 = Price per set
+38 = Transportation cost per set
$306 = Total cost per set delivered
×42 = Sets
$12,852 = Total cost/42 sets delivered

The amount paid to Built Well, Inc. would be:

$268 = Selling price per set
×42 = Sets
$11,256 = Selling price of 42 sets
×.03
$337.68 = Cash discount $11,256.00 = Selling price
−337.68 = Cash discount
$10,918.32 = Net price/42 sets

The amount paid to the common carrier would be:

$38 = Price of delivery per set
×42 = Sets
$1,596 = Total cost of delivery/42 sets

$10,918.32 = Cost of 42 sets
+1,596.00 = Cost of delivery
$12,514.32 = Total cost of 42 sets delivered

Note that the 3/10, n/30 terms apply to the merchandise only. The common carrier did not offer a cash discount on the cost of the transportation.

(268 + 38 =   306 × 42 =)   (268 × 42=   11,256 × .03 =   38 × 42 =)

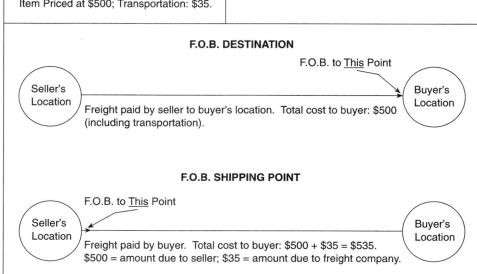

F.O.B. Cost Comparison

Item Priced at $500; Transportation: $35.

F.O.B. DESTINATION

F.O.B. to This Point

Seller's Location   Buyer's Location

Freight paid by seller to buyer's location. Total cost to buyer: $500 (including transportation).

F.O.B. SHIPPING POINT

F.O.B. to This Point

Seller's Location   Buyer's Location

Freight paid by buyer. Total cost to buyer: $500 + $35 = $535.
$500 = amount due to seller; $35 = amount due to freight company.

Exercise G

Fill in the missing amount in the problems below. Assume that the payment was made within the discount period and that the terms of trade apply to the sales price of the merchandise.

	Sales Price per Unit	Units Purchased	Terms of Trade	Freight Cost F.O.B. Destination	Cost to Buyer (total)
1.	$14.65	144	2/10,n/30	$ 16.46	$2,067.41
					(14.65 × 144 = × .98 =)
2.	_____	90	n/60	135.68	765.00
3.	18.00	150	3/10,n/30	72.00	_____
					Amount Due Seller
4.	6.00	30	—	27.00	_____
5.	60.00	_____	1/10,n/60	32.00	1,782.00

Exercise H

Fill in the missing amount in the following problems. Assume that payment was made within the discount period and that the terms of trade apply to the sales price of the merchandise.

	Sales Price per Unit	Units Purchased	Freight Cost F.O.B. Shipping Point	Terms of Trade	Cost to Buyer (total)
1.	$ 1.44	1,500	$ 162.50	3/10,n/20	$2,257.70
					(1.44 × 1,500 = × .97 = +162.50 =)
2.	15.00	80	2,359.00	n/60	_____
3.	1,950.00	12	493.51	1/10,n/30	_____
					Amount Due Seller
4.	84.90	_____	125.00	n/60	$12,735.00
5.	_____	71	792.00	3/10,n/45	5,234.12

C.O.D.

C.O.D. means cash on delivery. The buyer must pay for the merchandise in cash upon delivery, or it is returned to the seller's location.

C.O.D. prices tend to be higher than F.O.B. destination prices, because the seller is risking that the buyer will pay for the merchandise at delivery. If the buyer refuses to pay, the seller must pay the transportation cost of both directions and lose the sale. This type of pricing is used when selling impulse consumer merchandise, such as that sold through Internet or television merchandise programs, and when dealing with middlemen with poor or new credit ratings. The merchandise price will be quoted "C.O.D." or "plus C.O.D. charges."

Example 16

The Palderson Company sold Mrs. Alma Dickinson four bolts of material for her new upholstery business. The bolts were quoted at $268 per bolt C.O.D. What is the amount that the delivery firm will collect when the merchandise is accepted by Mrs. Dickinson?

$268 = Price per bolt/C.O.D.
  ×4 = Bolts
$1,072 = Total price for 4 bolts

(268 × 4 =)

Example 17

The Watson family was watching television and noticed an announcement that stated an item could be purchased for $29.95 plus C.O.D. charges. The family decided to order one. When the delivery man came to the home, he indicated that the cost of delivery was $4.50. How much is due the delivery man?

$29.95 = Price of merchandise
 +4.50 = C.O.D. charges
$34.45 = Total amount due

(29.95 + 4.50 =)

ASSIGNMENT Chapter 14

Name _____ Date _____ Score _____

Skill Problems

Find the total selling price in each of the following problems

	List Price per Unit	Units Purchased	Trade Discount Offered	Total Selling Price
1.	$64.50	200	15%	_____
2.	995.00	31	26	_____
3.	5.45	1,412	10	_____
4.	39.95	120	10	_____
5.	629.00	50	15	_____
6.	95.00	30	10	_____
7.	65.00	144	10 and 20	_____
8.	125.00	25	5	_____
9.	65.00	30	5 and 15	_____
10.	41.50	144	7, 20, and 5	_____

Find the total selling price in each of the following problems using the direct method.

	List Price per Unit	Units Purchased	Trade Discount Offered	Total Selling Price
11.	$145,100.00	2	5 and 8%	_____
12.	367.00	47	10 and 15	_____
13.	19.95	651	10, 5, and 3	_____
14.	10,483.00	8	20 and 10	_____
15.	150.00	245	5 and 12	_____

Find the total selling price in each of the following problems using the equivalent trade discount.

	List Price per Unit	Units Purchased	Trade Discount Offered	Equivalent Trade Discount	Total Selling Price
16.	$ 842.00	411	10 and 10%	_____	_____
17.	8,164.00	203	17 and 5	_____	
18.	29.60	764	15, 9, and 5	_____	
19.	8.45	120	27 and 20	_____	
20.	60.00	200	14 and 5	_____	

Find the last date that the cash discount is available in the following problems.

	Terms of Trade	Invoice Date	Last Date Discount Available
21.	3/10,n/30 E.O.M.	June 1	_____
22.	1/20,n/45 E.O.M.	Nov. 10	_____
23.	2/10,n/120 E.O.M.	Dec. 27	_____
24.	3/15,n/90 E.O.M.	July 18	_____

	Terms of Trade	Invoice Date	Date Goods Received	Last Date Discount Available
25.	2/10,n/30,R.O.G.	June 20	Nov. 6	_____
26.	3/15,n/60,R.O.G.	Sept. 1	Oct. 29	_____
27.	1/5,n/45,R.O.G.	Jan. 1	Apr. 9	_____
28.	1/30,n/90,R.O.G.	Oct. 25	Nov. 26	_____

Find the net price in the following problems. Note the dates.

	Selling Price	Terms of Trade	Invoice Date	Payment Date	Net Price
29.	$939.00	1/30,n/60	June 30	July 16	_____
30.	1,237.00	2/10,n/120	Nov. 14	Nov. 30	_____
31.	429.95	1/20,n/45	Dec. 9	Dec. 24	_____
32.	40.00	2/15,1/60,n/90	May 4	July 1	_____
33.	1,930.00	n/30	March 1	March 2	_____

Find the net price in the following problems.

	Total List Price	Trade Discounts	Terms of Trade	Invoice Date	Date Received	Payment Date	Net Price
34.	$900	10 and 7%	1/10,n/90	Mar. 3	Apr. 2	Mar. 11	_____
35.	295	5 and 10	3/5,n/60,E.O.M.	Feb. 7	Mar. 9	Mar. 10	_____
36.	940	10 and 10	2/10,n/30,R.O.G.	Oct. 4	Nov. 10	Nov. 15	_____
37.	300	5 and 8	1/10,n/90	Sep. 8	Nov. 22	Sep. 16	_____

Fill in the missing amount in the following problems. Assume that payment was made within the discount period and that the terms apply to the sales price for the merchandise.

	Selling Price	Terms of Trade	Invoice Date	Date Received	Payment Date	Net Price
38.	$259.00	3/10,n/20	July 12	July 15	July 20	_____
39.	890.00	1/20,n/90,E.O.M.	Aug. 29	Sept. 2	Oct. 15	_____
40.	59.50	2/15,n/60,R.O.G.	Oct. 21	Nov. 3	Nov. 16	_____
41.	120.00	2/10,n/60,E.O.M.	Feb. 12	March 1	March 7	_____
42.	856.00	n/30	April 22	April 26	May 10	_____

	Total List Price	Trade Discounts	Terms of Trade	Invoice Date	Date Received	Payment Date	Net Price
43.	$3,260.00	5 and 10%	2/10,n/30	July 6	July 9	July 20	_____
44.	695.00	12	2/5,n/30,R.O.G.	Jan. 5	Feb. 16	Feb. 20	_____

Sales Price Per Unit	Units Bought	Trade Discounts	Terms of Trade	Transportation Costs	Total Cost to Buyer	Amount Due Seller
$19.95	122	5 and 10%	3/10,n/20	$95.00 F.O.B. destination	45. _____	46. _____
9.70	148	12	2/15,n/60, E.O.G.	850.00 F.O.B. shipping point	47. _____	48. _____
5.05	35	10 and 10	n/30	65.00 F.O.B. shipping point	49. _____	50. _____
87.45	48	5, 4, and 10	2/5,n/90, R.O.G.	44.50 F.O.B. destination	51. _____	52. _____
6.82	912	—	3/10,n/90	78.75 F.O.B. shipping point	53. _____	54. _____
7.49	74	15 and 10	—	29.99 C.O.D.	55. _____	56. _____
55.00	35	20 and 8	5/30,n/90	64.95 F.O.B. destination	57. _____	58. _____
184.25	341	4 and 7	1/10,n/30 R.O.G.	984.44 F.O.B. shipping point	59. _____	60. _____

Determine the missing amount in the following problems. Assume that the payment was made within the discount period and that the terms of trade apply to the invoice amount.

	Amount of Invoice	Terms of Trade	Transportation Agreement	Transportation Costs	Amount Due Seller
61.	$ 453.80	2/10,n/60	F.O.B. destination	$32.80	_____
62.	980.00	—	C.O.D. included	62.68	_____
63.	1,698.00	1/10,n/45	F.O.B. shipping point	—	_____

Business Application Problems

1. Anter Company sold 140 units to Kass Furniture at a list price of $345 each. The invoice indicated a quantity discount of 15 percent. What is the selling price for the total sale?

2. Junior Devins purchased 50 TVs for the retirement complex his company is building. Each TV is priced at $439. The invoice indicates terms of 2/15, n/30. If payment is made on the tenth day following the invoice date, what is the total net amount due?

3. A multistation exercise unit was purchased by Realife Body at a list price of $1,250. A trade discount of 17 percent was negotiated. The terms on the invoice indicated 2/10,n/40. The purchase was made on July 14th. Payment was made on July 20th. What is the amount of the trade discount?

4. In problem 3, what is the last day that the cash discount is available to Realife Body?

5. In problem 3, what is the amount that Realife Body will need to remit to pay off the invoice in full on July 20th?

6. Thurston's, a retail store, ordered 300 gas grills listed at $239 each from Mentrail Manufacturing on April 3. The seller agreed to a 10 percent discount due to the size of the order. They also agreed to a 15 percent discount because Thurston's was arranging and paying for shipping. The seller also agreed to terms of 1/15,n/30, R.O.G. The goods were shipped on April 12 and received on April 16. The invoice was paid on April 28. What was the total list price of the order?

7. What is the total selling price of the order in problem 6?

8. What is the total net price of the order in problem 6?

9. If the shipping cost in problem 6 were $450, what is the total cost of each unit to Thurston's?

10. Kelsey, Inc. sold three sets of office furniture to Jambor Rhe Associates for $1,649 per set, list, F.O.B. destination. The agreement stipulated a trade discount of 12 percent and terms of 2/10,E.O.M. The purchase was made on March 12. What is the amount that Kelsey should expect in full payment on April 20?

11. The Sales Furniture Store ordered 20 dining sets at a list price of $230 each. The manufacturer offered a 12 percent discount for the size of the order. What is the sales price of the total order?

12. Billings and Alpert Company received a price quotation from two firms for 900 units of a processing tool used in the canning business. A representative of the Bernwalther Corporation quoted a price of $3.91 per unit F.O.B. destination, with terms of trade of 2/10,n/60. A spokesperson for the Altron Digital Corporation quoted a price of $3.78 per unit F.O.B. shipping point, with terms of trade 3/15,n/45. Total shipping costs were expected to be $93.60 for the 900 units. Which price quotation should be accepted, considering the price of the merchandise, transportation costs, and cash discounts?

13. Kelly bought 50 units of raw materials for her manufacturing firm at a list price of $17 per unit. Terms of trade were 3/10,n/20,E.O.M. The purchase was made on March 12, the goods were delivered on April 6, and payment was made on April 16. What was the amount of the check sent by Kelly's company?

14. Karen Liscomb was instructed to pay for a purchase made on September 1 for 30 pieces of office furniture that were listed at $174 each. A trade discount of 20 percent was offered, as well as terms of trade of 1/10,n/30. What is the net price if payment is made on September 9?

15. The Aikman Office Furniture Company ordered three desks with matching chairs for their inventory. The items were priced at $1,900 for each set (desk and chair). The manufacturer offered a 10 and a 15 percent discount if the Aikman Office Furniture Company would handle and ship the merchandise themselves. What is the selling price of a set to the buyer?

16. The Fredrick Insurance Agency purchased six new computers listed at $480 each. They received a 10 percent discount because they purchased more than five, and they received an additional 5 percent discount because they took immediate delivery. Terms of trade were 2/5,n/30. What was the net price if payment was made within the discount period?

17. A check for the net amount of $84.28 is received from a customer who was extended terms of 2/10,n/30. The check is past the discount period. You must contact the customer and indicate the correct amount due. What is the amount due?

18. Sharon Sadlowski, payments clerk and secretary for the Tog and Slumber Shop, made out a check to the P. K. Worth Company for a purchase made on July 14. The purchase was for 30 travel wear sets priced at $31 per set. The P. K. Worth Company extended terms of 1/10,n/45, E.O.M. The price quoted was F.O.B. shipping point. Freight on the purchase was $21.80. The check will have the date of August 5. What should be the amount of the check to P. K. Worth Company?

19. Peter Frank ordered 20 men's jackets at the Apparel Center in Chicago for his men's shop. The jackets were priced at $90 each, with terms of trade of 2/20,1/30,n/45,R.O.G. The purchase was made on March 12 for fall delivery. The merchandise was received on August 5 and payment was made on August 22. What was the net price of the lot of 20 jackets?

20. Using the invoice below, show the purchase of seven refrigerators from Wilton's by Petti and Barret Construction Company on July 17, 2001. Each refrigerator will cost $722 after a trade discount has been considered. There will be a $184 C.O.D. charge added to the invoice. Find the total amount due at delivery.

THE BEST SERVICE COSTS NO MORE

WILTON'S
TV & APPLIANCE , INC.

2703 PINEGROVE
(810) 982-9549
PORT HURON, MICHIGAN 48060

Customer's Order No.	43807	Date	7/17	20	

Sold To   Petti & Barret Construction Co.
Address   1800 North
City   Warren, MI   48213

SOLD BY	CASH	C.O.D.	CHARGE	ON ACCT.	MDSE.RETD.	PAID OUT
Irwin		✓				

Quantity	DESCRIPTION	Price	Amount
7	GE 19 cu. ft. frost free refrigerators - white	$722	
	C.O.D. charges	$184. —	
	TAX		
	TOTAL		

THANK YOU –Please keep this copy for reference.

ALL claims and returned goods MUST be accompanied by this bill.
082280   Received by

Bee Line Business Systems - Mt. Clemens, Michigan 48043

Courtesy of Wilton's TV & Appliance, Inc., Port Huron, MI.

21. Pulver and Company sold three video systems to Tec-Rec, Inc. The price was $1,680 per system F.O.B. the buyer's locations in St. Paul, Boston, and Atlanta. Terms of 3/15,n/90 were offered. The transportation costs totaled $567.80. How much did Pulver and Company receive for the systems after the cash discount and transportation costs were considered?

22. Patio Encounters paid $382.50 for a table-and-four-chair set after the supplier provided a 15 percent trade discount. What was the list price of the set?

23. Write a complete sentence in answer to this question. Randy Marsh paid an invoice with a check for $509.25. The vendor had agreed to a 25 percent trade discount and terms of 3/10,n/60. Randy paid the invoice within the first ten days. What was the list price of the goods before the trade and cash discounts?

24. Write a complete sentence in answer to this question. A check for $1,162.80 was sent on August 30 for merchandise purchased from Eron Corporation. Eron Corporation allowed a series of 15 percent, 10 percent, and 5 percent trade discounts, as well as terms of 3/10,E.O.M. What was the list price of the merchandise? The invoice was dated July 12.

25. Write a complete sentence in answer to this question. The Jackson and Fulmer Company will need to borrow the money necessary to pay the $56,850 invoice from Acme Corporation. The terms of trade are 2/10,n/30. The company will receive enough revenue on the twenty-fifth day after the invoice date to repay the lender. The lender will charge 17 percent for the short-term loan. Will the company save any money by borrowing the payment amount for the 15 days? How much? Use exact interest.

★ **Challenge Problem**

Deanna purchased 1,200 pounds of a food additive for the Ideal Food Company at a list price of $9 per pound. The purchase was made on February 12. Delivery was made on March 31. Deanna negotiated a series of trade discounts of 10 percent, 10 percent, and 7 percent. Terms of trade were 3/10,1/30,n/45,R.O.G. Because of an overpurchase, $1,300 of the additive was returned to the supplier on April 16 for credit at list price. A partial payment of $500 was made on May 7, and the balance was paid on May 28. What was the amount of the payment on May 28? The supplier uses E.O.M.

Name _____ **Date** _____ **Score** _____

Select the best phrase or term.

Phrases/Terms

_____ 1. Equivalent trade
discount

_____ 2. Cash discount

_____ 3. Trade discount

_____ 4. Seller pays
transportation costs

_____ 5. Chain discount

_____ 6. Buyer pays the
transportation costs to
the seller, who pays
the transportation firm

_____ 7. Terms of trade

_____ 8. R.O.G.

_____ 9. Net price

_____ 10. Marketing functions

_____ 11. Buyer pays the
transportation costs to
the transportation firm

_____ 12. E.O.M.

_____ 13. List price

_____ 14. Selling price

A. Discount period begins at end
of the month
B. Three times per year
C. C.O.D.
D. Abbreviation of cash discount
agreement
E. Offered for buyer aiding the
seller
F. List price less trade discounts
G. Selling, storing, gathering information
H. F.O.B. destination
I. F.O.B. shipping point
J. A reduction in the selling price
K. Usually found in a catalogue or
price sheet
L. Equal to two or more trade
discounts
M. Discount period begins when
the goods are received
N. Amount due after cash discount
is subtracted
O. More than one trade discount

Compute the selling price in the following problems.

List Price	Trade Discounts Offered	Selling Price
15. $600	18%	_____
16. 890	10 and 20%	_____

Compute the net price in the following problems. Assume discounts are taken.

Amount of Invoice	Transportation Agreement	Terms of Trade	Transportation Costs	Amount Due Seller
17. $900	F.O.B. destination	1/10,n/60	$12.00	_____
18. 748	F.O.B. shipping point	2/10,n/30,R.O.G.	—	_____

Compute the net price in the following problems. Assume discounts are taken.

	Total List Price	Trade Discounts	Terms of Trade	Invoice Date	Payment Date	Date Rec'd	Net Price
19.	$ 800	10 and 20%	2/10,n/30,R.O.G.	Dec. 3	Jan. 5	Dec. 28	_____
20.	1,000	5, 5, and 10%	1/20,n/60	May 10	May 31	May 12	_____
21.	650	5, 10, and 8%	2/15,n/60,E.O.M.	July 16	Aug. 12	July 20	_____

22. The accounts payable clerk of the A-K Marketing Company made out a check for the amount due on the purchase of ten file cabinets. The list price of the cabinets was $92 each. The firm had received an 18 percent trade discount on terms of 2/30,n/60. What is the amount of the check if payment is made within the discount period?

23. What is the amount due on a $600 purchase of merchandise bought on April 22? There is a 10 and 12 percent trade discount offered on terms of 2/5,1/20,n/60,R.O.G. The goods were received on May 2 and the payment is made on May 18.

MARKUP

OBJECTIVES

After mastering the material in this chapter, you will be able to:

1. Calculate sales based on cost or sales. p. 286

2. Calculate cost based on cost or sales. p. 289

3. Calculate markup based on cost or sales. p. 289

4. Calculate the markdown or new selling price. p. 290

5. Convert a markup on cost to a markup on sales. p. 291

6. Convert a markup on sales to a markup on cost. p. 291

7. Determine wholesale catalog prices when a trade discount and a markup are to be considered. p. 292

8. Understand and use the following terms:
   Markup
   Cost Price
   Selling Price
   List Price

SPORT-A-MERICA

 Hal has ordered a new line of exercise equipment for Sport-A-Merica. The items, including rowing machines, stationary walking machines, and electronic stationary bikes, will arrive at the end of the month. Using the feedback he receives from customers and information from the trade association, he knows the area is a good market for the equipment. Sales of this equipment will offset the sluggish sales during the slower months when the weather turns bad. The major buyers of this equipment are expected to be either senior citizens who want to stay inside where it is safe and close to a phone, or exercise enthusiasts who want to set up an exercise area of their own to avoid joining a gym. Though these items will not sell in the volume Sport-A-Merica is used to selling, say, tennis balls, the fact that they are big-ticket items will allow for substantial profits.

Hal knows that the price marked on the item must cover the cost of the item *and* a portion of the overhead, just to break even. The exercise machine that he plans to purchase for $420 will be marked up by 20 percent, for example. That markup will cover his overhead and provide a profit, especially when he sells all of the units that he plans to order.

At the same time, he has some items in the tennis area that are not moving very fast, including a few of the remaining tennis shirts that he purchased in the spring. He is considering a 40 percent markdown just to see if he can move the inventory. In order to make room for the newest products in the equipment area, he feels Sport-A-Merica should advertise a markdown in order to sell them quickly. All of this price concern is an ongoing question for the firm. It is a question that Hal continues to struggle with, recalculate, and adjust.

In order for a merchandising firm to stay in business, it must sell its merchandise for more than it pays for it. The merchant must increase the price enough to cover overhead, such as the cost of utilities, rent, taxes, wages, and insurance. There must be an incentive to risk the capital investment that is required to go into business. This incentive, or profit, is a return on the capital investment, a reward for the risk taken, and, in some cases, a compensation for hours worked by the owners. Overhead and profit are grouped together *markup* and termed **markup** by merchants. Markup is the difference between the amount that is paid for the merchandise and the amount that it is sold for by the merchant. Markup is used by all members of the channel of distribution, such as retailers, wholesalers, distributors, manufacturers, and independent salespeople. Any firm or person dealing with the flow of goods or materials through the channel of distribution comes into contact with markup. Members of the channel of distribution are often termed *middlemen*. This term is used because their activities are between the manufacturer of the product and its use or consumption by the final consumer.

cost price   The price the firm pays for an item is termed the **cost price.** Cash and trade discounts must be subtracted to find the cost price, which is the net price paid for merchandise.

selling price   The **selling price** is the amount for which merchandise is sold to the customers of the firm. The selling price is found by adding a markup to a cost price (i.e., cost price + markup = selling price).

Example 1

A merchant purchased a desk for $165 and marked it up by $70. What is the selling price on the desk?

$165 = Cost price
+70 = Markup
$235 = Selling price

(165 + 70 =)

Often the markup is not a specific dollar amount but a percent. The percent is determined by considering the amount of overhead and profit that the firm must earn to stay in business. The percent of markup must be based on one of the two factors in the formula: cost price + markup = selling price.

Markup on Cost

Middlemen often mark up goods based on cost. The cost price of the merchandise is increased by the markup rate. In this case, the markup rate is based on the cost price of the merchandise. The cost price is, therefore, 100 percent (the base). The selling price percent is the total of the cost price (100 percent) and the markup rate.

Objective 1

Example 2

A gift wholesaler purchased a carload of decorative radios for a price of $15 per item. The goods were priced to sell using the company's standard markup rate of 40 percent. What is the selling price?

○ *Consider using the percent model in markup problems.*

100%	+	40%	=	140%
cost price	+	markup	=	selling price
$15				

The selling price in this example is the amount. The 140 percent is the rate and the $15 cost price is base. The amount is found by:

A = B × R
A = $15 × 1.40 (100% + 40%)
A = $21 (selling price)

○ *Solving Algebraically*
S = Selling Price
S = 1.4 × C
S = 1.4 × $15
S = $21

(15 × 1.40 =)

Example 3

Sport-A-Merica marks up pool merchandise 55 percent on cost. A pool cleaner is purchased for $8. What is the selling price?

100%   +   55%   =   155%
cost price + markup = selling price
   $8

A = B × R
A = $8 × 1.55
A = $12.40 (selling price)

(8 × 1.55 =)

Exercise A

Determine the selling price and markup amount for the following cost prices. Round off your answer to the nearest cent.

	Cost Price	Markup	Markup Percent on Cost	Selling Price
1.	$ 40.00	$20.00	50%	$60.00 (40 × 1.50 =)
2.	60.00	_____	20	_____
3.	100.00	_____	35	_____
4.	125.00	_____	40	_____
5.	68.00	_____	23	_____

Markup on Sales

Goods can also be marked up based on selling price. If the markup is based on the selling price or sales, then the selling price is 100 percent (the base).

Example 4

Using the information in example 2, what is the selling price of an item that cost $15 if it will be marked up 40 percent on sales?

60%   +   40%   =   100%
cost price + markup = selling price
   $15

The selling price in this example is the base (100 percent). The 60 percent is the rate, and $15 is the amount. The base is found by:

B = A ÷ R
B = $15 ÷ .60
B = $25 (selling price)

(15 ÷ .60 =)

○ *Solving Algebraically*
 S = Selling Price
 S = C + .4S
 .6S = C
 S = C ÷ .6
 S = $15 ÷ 6
 S = $25

Notice that a problem that indicates a markup on sales is handled differently *and* results in a different answer than one that indicates markup on cost.

○ *A given percent markup on sales is different than a markup on cost. Which provides more dollars in markup?*

Example 5

Hal marked up a $420 piece of exercise equipment 20 percent on sales. What was the selling price that Sport-A-Merica will mark on the price tag?

80%   +   20%   =   100%
cost price + markup = selling price
   $420

Base = Amount ÷ Rate
Base = $420 ÷ .80
Base = $525 (selling price)

(420 ÷ .80 =)

Exercise B

Determine the selling price and markup amount. Round off your answer to the nearest cent.

	Cost Price	Markup	Markup Percent on Sales	Selling Price
1.	$150	$37.50	20%	$187.50 (150 ÷ .80 =)
2.	250	_____	30	_____
3.	80	_____	25	_____
4.	48	_____	70	_____
5.	13	_____	45	_____

Exercise C

Determine the selling price. Note the basis of the markup rate. Round off your answers to the nearest cent.

	Cost Price	Markup Rate	Markup on	Selling Price
1.	$90.00	10%	Cost Price	$99.00 (90 × 1.10 =)
2.	30.00	40	Cost Price	_____
3.	160.00	25	Selling Price	_____
4.	146.00	60	Selling Price	_____
5.	12.00	50	Selling Price	_____
6.	14.50	40	Cost Price	_____
7.	6.50	60	Cost Price	_____
8.	780.00	45	Cost Price	_____
9.	19.00	25	Selling Price	_____
10.	46.00	70	Selling Price	_____

Finding Cost, Sales, Markup, or Markdown

There are occasions when a middleman needs to know an answer other than the selling price in a markup problem. The information needed may be the cost price, the markup amount, the markup percent, the percent of cost to sales, etc. There are many possible questions. These questions require the student to understand the markup formula and the way the percent works. It would be difficult to memorize every procedure for every problem type. The following examples represent some types of markup problems. You should determine if the problem is a markup on cost or one on sales, then solve for either base, rate, or amount.

Notice that in all of the examples, a basic procedure is followed: (1) the markup formula is used, (2) the markup base is identified with 100 percent, (3) the dollar values available are included in the formula, (4) the percent values available are included in the formula, and (5) the unknown factor is determined. This procedure is an excellent method to solve markup problems. It is recommended that the student adopt this procedure until another method proves to be more workable.

Objective 2

Example 6 To find the cost price when the markup is on cost—What is the cost price of an article priced to sell at $40, if the merchant has marked it up 20 percent on cost?

100%   +   20%   =   120%
cost price + markup = selling price
$40

Base = Amount ÷ Rate
Base = $40 ÷ 1.20
Base = $33.33 (answer)

(40 ÷ 1.20 =)

Example 7 To find the cost price when the markup is on sales—What is the cost price of an item priced to sell at $65, if the markup is 30 percent on sales?

70%   +   30%   =   100%
cost price + markup = selling price
$65

Amount = Base × Rate
Amount = $65 × .70
Amount = $45.50 (answer)

(65 × .70 =)

Objective 3

Example 8 To find the markup when the markup percent is on sales—What is the markup on an item priced to sell at $50, if the markup percent is 20 percent on sales?

80%   +   20%   =   100%
cost price + markup = selling price
$50

Amount = Base × Rate
Amount = $50 × .20
Amount = $10 (answer)

(50 × .20 =)

Example 9 To find the markup when the markup percent is on cost—What is the markup on an item priced to sell at $80, if the markup percent is 30 percent on cost?

100%   +   30%   =   130%
cost price + markup = selling price
$80

Base = Amount ÷ Rate
Base = $80 ÷ 1.3
Base = $61.54 (cost price)

Amount = Base × Rate
Amount = $61.54 × .30
Amount = $18.46 (answer)

(80 ÷ 1.30 =   61.54 × .3 =)

Example 10

To find the percent of cost to sales—What is the percent of cost to sales of an item that has a 25 percent markup on cost? The item is priced to sell at $80.

$$100\% \ + \ 25\% \ = \ 125\%$$
cost price + markup = selling price
$$\$80$$

Base = Amount ÷ Rate
Base = $80 ÷ 1.25
Base = $64 (cost price)
Rate = Amount ÷ Base
Rate = $64 ÷ $80
Rate = 80% (answer)

(80 ÷ 1.25 =   64 ÷ 80 =)

Exercise D

Fill in the blank lines below. Note the markup base—cost or sales. Round off your answers to the nearest percent or cent.

	Cost Price	Selling Price	Markup Percent	Markup on	Markup
1.	$128.00	$160.00	20%	Sales	$32.00 (160 × .20 =)
2.	_____	50.00	30	Sales	_____
3.	40.00	_____	70	Cost	_____
4.	93.00	_____	20	Cost	_____
5.	_____	46.00	35	Sales	_____
6.	_____	47.50	30	Sales	_____
7.	22.00	39.95	_____	Cost	_____
8.	45.00	65.50	_____	Sales	_____
9.	_____	_____	25	Sales	57.00
10.	_____	_____	35	Cost	80.00
11.	20.00	48.00	_____	Cost	_____
12.	36.00	48.00	_____	Sales	_____
13.	126.00	200.00	_____	Sales	_____
14.	_____	56.00	20	Sales	_____
15.	42.00	_____	350	Cost	_____

Objective 4
Markdown or Loss

There are times that a retailer may decide to mark down the current inventory in order to move the merchandise. The retailer usually decides on a rate of markdown and then marks the price tags accordingly. The markdown is always based on the previous selling price.

Example 11

Hal has decided to mark down some of the tennis shirts that had been priced at $59.95. The markdown will be 40 percent. What will be the sale price?

$$60\% \ + \ 40\% \ = \ 100\%$$
sale price + markdown = selling price (previous)
$$\$59.95$$

Amount = Base × Rate
Amount = $59.95 × .60
Amount = $35.97 (sale price)

○ *How sale merchandise is marked!*

$59.95

(30 × .60 =)

When merchandise is sold at an actual loss (less than the cost price), the percent of loss must be calculated. The loss is always based on the cost price.

Example 12   Find the percent of loss of an item that cost $50 and was sold for $40.

100%
cost price + loss = selling price
$50   − $10 = $40

Rate = Amount ÷ Base
Rate = $10 ÷ $50
Rate = 20% (loss)

(10 ÷ 50 =)

Exercise E

Determine the answer in the exercises below. Round off your answer to the nearest percent or cent.

	Selling Price	Markdown	Sale Price
1.	$ 85	30%	$59.50 (85 × .70 =)
2.	17	40	
3.	40	35	
4.	17.95	20	
	Cost Price	**Selling Price**	**Loss (%)**
5.	$80	$60	
6.	35	20	
7.	80	68	
8.	129	89	

Conversion of Markup Rate
Objectives 5 & 6

A markup of 20 percent on sales is not the same amount as a markup of 20 percent on cost. When a merchant wishes to compare various markups, the markups must be stated using the same base. This will allow for a direct comparison of markup rates. The markup rate on one base can be changed to another base by using the same markup formula.

To change the markup rate on one base (cost or sales) to another base, divide the current markup percent by the new base percent.

Example 13   What is the markup percent on sales if the markup percent on cost is 30 percent?

100%   +   30%   =   130%
cost price + markup = selling price

Markup percent on sales = $\frac{30\% \text{ (current markup percent)}}{130\% \text{ (new base percent − sales)}}$

Markup percent on sales = 23% (answer)

The new markup percent on sales in the above example can be used in the markup formula as follows.

77%   +   23%   =   100%
cost price + markup = selling price

(Continued on next page)

Example 13
continued

The two markup values (30 percent on cost and 23 percent on sales) are equal. To prove this, use a cost price of $3 in each case to determine the selling price.

on Cost	*on Sales*
100% + 30% = 130%	77% + 23% = 100%
cost price + markup = selling price	cost price + markup = selling price
$3	$3
Amount = Base × Rate	Base = Amount ÷ Rate
Amount = $3 × 1.30	Base = $3 ÷ .77
Amount = $3.90 (the selling price)	Base = $3.90 (the selling price)

(.30 ÷ 1.30 =)

Example 14

What is the markup percent on cost if the markup percent on sales is 20 percent?

    80%   +   20%   =   100%
cost price + markup = selling price

$$\text{Markup percent on cost} = \frac{20\% \text{ (current markup percent)}}{80\% \text{ (new base percent} - \text{cost)}}$$

Markup percent on cost = 25% (answer)

(.20 ÷ .80 =)

In the previous example, a markup of 20 percent on sales is equal to a markup of 25 percent on cost. Proof: What is the selling price of a product that cost $85.60 (using 20 percent on sales and 25 percent on cost)?

on Cost	*on Sales*
100% + 25% = 125%	80% + 20% = 100%
cost + markup = selling	cost + markup = selling
price price	price price
$85.60	$85.60
Amount = Base × Rate	Base = Amount ÷ Rate
Amount = $85.60 × 1.25	Base = $85.60 ÷ .80
Amount = $107 (answer)	Base = $107 (answer)

Exercise F

Determine the markup percent on the new base in the exercises below. Round off your answers to the nearest percent.

Markup Percent on Cost	Markup Percent on Sales	Markup Percent on Cost	Markup Percent on Sales
1. 40%	29% (.40 ÷ 1.40 =)	2. 50	___
3. ___	40	4. ___	50
5. 70	___	6. 30	___
7. ___	40	8. 20	___
9. 45	___	10. ___	20

Catalog Pricing
Objective 7

list price

Merchandise sold by wholesalers or manufacturers is often priced in a catalog and may be subject to a discount (see chapter 14). Trade discounts are available for larger purchases and for the buyer performing marketing functions for the seller. The seller must determine the price to be shown in a catalog in order to allow for trade discounts. This price is known as the **list price.** Trade discounts are subtracted from the list price to find the selling price. The relationship of cost price, selling price, and list price is shown here.

(Cost Price + Markup) = Selling Price
Selling Price + Trade Discount = List Price

Example 15

What is the list price on an item that cost a wholesaler $17 if the wholesaler will mark it up 20 percent on sales and will allow for a 10 percent trade discount?

80%   +   20%   =   100%
cost price + markup = selling price
   $17

Base = Amount ÷ Rate
Base = $17 ÷ .80
Base = $21.25 (the selling price)

90%   +   10%   =   100%
selling price + trade discount = list price
   $21.25

Base = Amount ÷ Rate
Base = $21.25 ÷ .90
Base = $23.61 (list price)

(17 ÷ .80 =   21.25 ÷ .90 =)

Exercise G

Determine the list prices in the exercises below. Round off your answers to the nearest cent.

	Cost Price	Markup Percent	Markup on	Trade Discount	List Price
1.	$ 40	150%	Cost (40 × 2.50 =)	15% (100 ÷ .85 =)	$117.65
2.	25	35	Sales	10	_____
3.	163	40	Sales	15 and 10	_____
4.	47	60	Cost	21	_____
5.	80	55	Cost	17	_____
6.	65	7	Sales	20 and 15	_____
7.	72	42	Sales	15	_____
8.	1,530	80	Sales	10 and 15	_____
9.	96	12	Cost	10	_____
10.	120	17	Sales	20	_____

Additional Trade Discounts

When a middleman finds another source for a product at a lower price, the middleman may decide not to issue a new catalog immediately. The lower-cost price would provide a higher profit for the seller. However, in order to stay competitive, the middleman may decide to offer an additional trade discount, possibly through an E-mail communication, passing along to the customer the cost price reduction. The percent of additional trade discount that can be passed on is found by (1) determining the difference between the old and new cost prices, and (2) dividing the difference by the old cost price.

Example 16

The cost price of a toy robot is $190. A new supplier can reduce the cost price to $172. What will be the additional trade discount the seller can offer if the markup rate and trade discounts remain the same?

   $190 Old cost price
   −172 New cost price
   $18 Difference

$$\frac{\$18 \text{ difference}}{\$190 \text{ old cost price}} = 9\% \text{ additional trade discount}$$

(18 ÷ 190 =)

Exercise H

Determine the list price and additional trade discount. Round off your answer to the nearest percent or cent.

Cost Price	Markup Percent	Markup on	Trade Discount	List Price	New Cost Price	Additional Trade Discount Percent
1. $127.00	15%	Sales	10% (127 ÷ .85 =)	$166.01 (149.41 ÷ .90 =)	$120.00	6% (7 ÷ 127 =)
2. 427.00	35	Cost	20		360.70	
3. 320.00	20	Cost	10 and 15		268.00	
4. 95.50	38	Sales	$7\frac{1}{2}$		76.00	
5. 147.00	80	Sales	22		138.60	

ASSIGNMENT Chapter 15

Name _____ Date _____ Score _____

Skill Problems

Determine the selling price and markup amount. Round off your answer to the nearest cent.

Cost Price	Markup		Markup Percent on Cost	Selling Price	
$ 95.00	**1.** _____		55%	**2.** _____	
700.00	**3.** _____		35	**4.** _____	
55.40	**5.** _____		50	**6.** _____	
1,746.50	**7.** _____		23	**8.** _____	
49.00	**9.** _____		70	**10.** _____	

Cost Price	Markup		Markup Percent on Sales	Selling Price	
$ 80.00	**11.** _____		55%	**12.** _____	
120.00	**13.** _____		35	**14.** _____	
55.40	**15.** _____		50	**16.** _____	
1,746.50	**17.** _____		23	**18.** _____	
48.50	**19.** _____		70	**20.** _____	

Determine the selling price. Note the basis of the markup rate. Round off your answers to the nearest cent.

Cost Price	Markup Rate	Markup on	Selling Price
$600.00	10%	Cost Price	**21.** _____
369.00	25	Selling Price	**22.** _____
84.60	50	Selling Price	**23.** _____
10.95	70	Cost Price	**24.** _____
381.50	34	Selling Price	**25.** _____

Fill in the blank lines below. Note the markup base—cost or sales. Round off your answers to the nearest percent or cent.

Cost Price	Selling Price	Markup Percent	Markup on	Markup
26. _____	$184.50	30%	Sales	**27.** _____
50.00	**28.** _____	35	Cost	**29.** _____
30. _____	1,040.64	20	Sales	**31.** _____
648.12	**32.** _____	70	Sales	**33.** _____
25.00	40.00	**34.** _____	Cost	**35.** _____
$140.00	**36.** _____	35	Cost	**37.** _____
47.50	62.00	**38.** _____	Cost	**39.** _____
60.00	85.00	**40.** _____	Sales	**41.** _____
42. _____	188.00	20	Cost	**43.** _____
44. _____	**45.** _____	40	Cost	60.00
46. _____	**47.** _____	30	Sales	72.00
72.00	**48.** _____	60	Sales	**49.** _____
50. _____	27.50	40	Sales	**51.** _____
248.00	**52.** _____	30	Cost	**53.** _____
54. _____	12.80	120	Cost	**55.** _____

Determine the answer in the exercises below. Round off your answer to two decimal places.

Cost Price	Selling Price	Markdown	Sale Price		Loss (%)	
$74.00	$95.50	40%	**56.** _____		**57.** _____	
80.20	90.00	30	**58.** _____		**59.** _____	
46.81	53.83	30	**60.** _____		**61.** _____	
650.00	910.00	65	**62.** _____		**63.** _____	
891.65	936.23	25	**64.** _____		**65.** _____	

Determine the markup percent on the new base in the exercises below. Round off your answers to the nearest percent.

	Markup Percent on Cost	Markup Percent on Sales
66.	46%	_____
67.	_____	23
68.	80	_____
69.	_____	60
70.	125	_____

Complete the following problems. Round off your answers to the nearest cent.

Cost Price	Markup Percent	Markup on	Selling Price	Trade Discount	Catalog List Price
$30.00	40%	Sales	**71.** _____	15%	**72.** _____
73. _____	75	Cost	**74.** _____	15	147.50
46.00	30	Cost	**75.** _____	6	**76.** _____
25.50	30	Sales	**77.** _____	17	**78.** _____
79. _____	20	Sales	**80.** _____	20	60.00

Determine the list prices and the additional trade discount percent in the exercises below. Round off your answers to the nearest percent or cent.

Cost Price	Markup Percent	Markup on	Trade Discount	Catalog List Price	New Cost Price	Additional Trade Discount Percent
$50.00	125%	Cost	20%	**81.** _____	$ 45.00	**82.** _____
612.50	15	Sales	5	**83.** _____	600.00	**84.** _____
84.40	20	Cost	25	**85.** _____	75.50	**86.** _____
78.85	38	Cost	15	**87.** _____	75.35	**88.** _____
340.60	62	Sales	20	**89.** _____	295.00	**90.** _____
42.00	20	Cost	12	**91.** _____	37.50	**92.** _____
23.60	70	Cost	10	**93.** _____	21.90	**94.** _____
126.00	35	Sales	15 and 10	**95.** _____	93.70	**96.** _____

Business Application Problems

Round off your answer to nearest percent or cent.

1. Price Wise Stores purchased a lawn mower for $259. The store will need to make a markup of 40 percent on cost. At what price will the store sell the lawn mower?

2. In problem 1, what is the amount of the markup?

3. If Price Wise Stores, in problem 1, would have used a markup of 40 percent on the selling price, what would be the price of the lawn mower?

4. Professional Kitchens, Inc. bought a piece of used stainless steel preparation equipment for $275. They were able to resell the equipment for $400 a week later. What markup did they earn on cost?

5. What was the rate of markup on the selling price in problem 4?

6. All the merchandise at Steve's store is priced to sell at $49.95 or lower. Steve uses a markup of 35 percent on cost for all of his merchandise. If he is to buy items to sell in his store, what is the most he can pay for them and maintain the current maximum price level?

7. Kevin's Kernels sells popcorn for $4.75 a bag. The supplier indicated that Kevin could produce 45 popped bags from the box of kernels he purchased for $7. What is the potential amount of markup on the purchase?

8. What is the rate of markup on cost that Kevin could earn in problem 7?

9. A leaf blower sold for $189 at the Bayview Hardware Store. The unit was purchased for $83 through a supplier in Mexico. What is the rate of markup on the selling price?

10. Dawn Donuts produces the best pecan rolls in Medford. The total cost of ingredients for a batch of 100 rolls is $31.25. Each of the pecan rolls is sold for $1.25 by 10:00 a.m. each day. What is the rate of markup on the ingredients?

11. The Carnes Gift Shop purchased an antique for $40 and resold it for $90. To the nearest percent, what is the markup percent on sales?

12. Arrigo Dodge prices all repair parts to sell at a 60 percent markup based on cost. The counter-parts salesperson must price out the lamp assembly to be sold to Carl's Storage shown in the following invoice. The cost price listed on the video display unit is $17.90. Place the selling price in the list column on the invoice below.

14. Keeline Wilcox sold 20 Kentia plants to the Colwell Nursing Home, a nonprofit organization. The plants cost the firm $17.90 each. They were marked to sell for $22.90 each. Keeline Wilcox offered to give the buyer a discount as shown on the following invoice. How much markup was made on the sale?

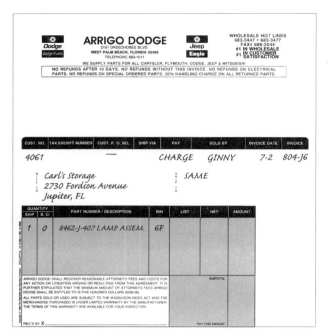

Courtesy of Arrigo Dodge, West Palm Beach, FL.

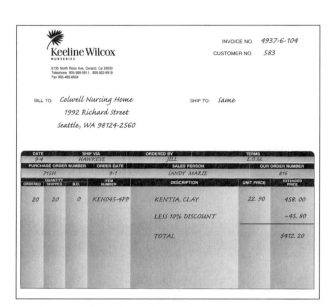

Courtesy of Keeline Wilcox, Oxnard, CA.

13. Sarah marked some out-of-season boots to sell for $31 a pair. The boots cost the store $56 a pair and had been marked up 23 percent on sales. Rita asked Sarah to determine the percent of markdown on the original selling price. Round off your answer to the nearest percent.

15. The Stork Shop realized a 42 percent markup on cost on all products sold last year. The merchandise bought for resale cost $37,860. There was $9,650 worth of overhead during the year, including utilities, rent, and insurance. What was the amount of the owner's profit for the year? (overhead + profit = markup)

16. Whitican & Associates, art dealers, planned to mark down a painting 50 percent. The painting was purchased from an artist for $40. They wish to maintain a 35 percent markup on sales after the markdown. What was the pre-marked-down price?

17. Linda's Natural Foods buys dried fruit for resale. A single package of dried apples costs $1.80. The package is marked up 60 percent on sales. How many packages must be sold for the store to realize a total markup of $240?

18. Janis has a markup of 100 percent on cost in her dress shop. A business associate, Cyndi, indicates that she does better than Janis because she earns a 60 percent markup on sales. Who earns more per sale, Janis or Cyndi?

19. For what price must a merchant sell a desk that cost him $300 in order to earn a markup of 30 percent on sales after a 10 percent trade discount has been offered to his customers?

20. The Cabinet Corporation bought a buffet for $600 from the Stillo Manufacturing Company on July 17. The Cabinet Corporation operates under a policy of a 40 percent markup on sales. The corporation allows a 10 percent trade discount to its customers. What will be the list price on the buffet that costs $600?

21. A dress marked "25 percent off" is on sale for $169. The dress was originally marked up 150 percent on cost. What was the cost of the dress?

22. A diamond ring is marked $890. The diamond cost the jeweler $200 and is marked up 250 percent on cost. The setting cost the jeweler $114. What is the percent of markup on cost on the setting? Round off your answer to the nearest percent.

23. Write a complete sentence in answer to this question. If the markup on food items at a convenience store is 12 percent on cost, how much merchandise must be sold to have total markups of $200,000 per month?

24. Write a complete sentence in answer to this question. Angelo sells popcorn at all of the athletic games at Bucken Stadium. His popcorn sells for $5 a bag. Angelo marks up the ingredients (popcorn, oil, salt, etc.) 1,000 percent on cost. What is the cost of the ingredients?

25. Write a complete sentence in answer to this question. Desmond and Associates have taken an inventory of the materials in their storage facility. The value of the items at cost is $853,980. The accountant said that these items will be sold at an average markup of 22 percent of the selling price. If Desmond and Associates sell the complete inventory, what will they receive in revenue?

★ Challenge Problem

The Cronton & Trion Company purchased 70 special timers at a list price of $37.80 each. The Burrs Corporation allowed a 15 percent trade discount, as well as terms of 2/15,n/30, E.O.M. The purchase was made on July 18. The timers were backordered and actually arrived on September 29. Four were found to be defective and were returned. The invoice was paid on October 17. The timers were sold through the Cronton & Trion catalog. The markup rate of 40 percent on cost was made.

A second purchase of 70 timers was made on November 2, C.O.D., at $31 each with no trade discount. They were marked up in the same way as the first purchase. What is the total dollar amount of markup realized by Cronton & Trion on the two purchases? Round your answer off to the nearest cent.

Name _____ **Date** _____ **Score** _____

Complete the problems below. Round off your answer to the nearest cent or percent.

1. Markup = 30% on cost
   Cost price = $50
   Selling = $ _____

2. Markup = 60% on sales
   Cost price = $38
   Selling price = $ _____

3. Markup = 40% on cost
   Selling price = $200
   Cost price = $ _____

4. Markup = 50% on sales
   Selling price = $47
   Cost price = $ _____

5. Selling price = $60
   Cost price = $25
   Markup = _____ % on cost

6. Selling price = $35
   Cost price = $20
   Markup = _____% on sales

7. Markup = 40% on sales
   Markup = _____% on cost

8. Markup = 50% on cost
   Markup = _____% on sales

9. Markup = $80
   Markup = 12% on cost
   Selling price = $ _____

10. Markup = $30
    Markup = 8% on sales
    Cost price = $ _____

11. Cost price = $120
    Markup = 38% on sales
    Trade discount = 25%
    List price = $ _____

12. List price = $380
    Trade discount = 15 and 22%
    Markup = 25% on sales
    Cost price = $ _____

13. Out-of-Doors, Inc. purchased a tent for $280. The tent was then marked up 40 percent on cost. What is the selling price of the tent?

14. The Shipe Ice Cream Parlour marks up half-gallon containers 20 percent on cost. What is the highest cost the firm can pay for the item and sell it for $2.29? Round off your answer to the nearest cent.

15. Richard Yake's men's store purchased suits for $105 each. They were marked to sell for $249. The last of the suits were put on sale for $89. What is the percent of loss on the last suits sold?

INVENTORY AND TURNOVER 16

OBJECTIVES

After mastering the material in this chapter, you will be able to:

1. Determine the value of ending inventory using the L.I.F.O., F.I.F.O., average cost, specific identification, or retail inventory estimating method. p. 304

2. Determine the cost of goods sold using the L.I.F.O., F.I.F.O., average cost, specific identification, or retail inventory estimating method. p. 304

3. Determine the rate of merchandise turnover for a firm. p. 313

4. Understand and use the following terms or concepts:
Beginning Inventory
Purchases
Goods Available for Sale
Ending Inventory
Cost of Goods Sold
L.I.F.O.
F.I.F.O.
Average Cost
Retail Inventory Estimating Method
Specific Identification Method
Turnover

SPORT-A-MERICA

 It's Saturday, and Kerry and Hal are sitting down to discuss how Sport-A-Merica has been doing in the recent past. It is important to focus on the overall financial picture at least once a month. They know that markup is an important part of how profits are generated; they also know that the more items they are able to sell over a period of time, the better the total financial picture. Some sports items in the store are sold every day. Things like tennis balls and running shoes are constantly in demand. Other items are "slow movers," but bring a high profit when they are sold. Things like sleeping bags and clay pigeon throwers are sold only occasionally. These items contribute a great deal to the year-end profit and to the image of a complete sports store; though they sit for some time and bring a limited profit, they do contribute to the bottom line for the store.

Both Kerry and Hal know that a comparison to a previous period of time would help them to determine how well the store is doing. The time of year the comparison is to be made will also make a difference. They know, for example, that the spring and the holiday seasons are busy for them. If a comparison is to be made, it should be made at the same time each year.

Hal has kept records of the stock amounts on hand since the store first opened. He thinks about the face mask item often. It sells well and must be re-ordered several times a year. Keeping track of how many are in stock, how many have been sold, and how many were received in the last order is very time-consuming. In addition, the price of this item keeps changing, and the records must reflect these changes. Each time he has ordered the face mask, the price has gone up one dollar. He must determine the current value of the stock that he has in the store. This will give them a head start in making comparisons with the most recent months.

Kerry and Hal proceed with the analysis. They will have a much better grasp on how Sport-A-Merica is doing by the end of the day.

Inventory Methods

For firms that buy merchandise for resale, the value of the merchandise on hand at tax time or inventory time is very important. Several methods of determining that value will be presented in this chapter. Each method has its own advantage and special purpose. Knowledge of these methods is useful for students studying for merchandising, distribution, or accounting positions.

beginning inventory

At the beginning of an accounting period, a firm has an inventory of merchandise on hand to sell. This amount is termed **beginning inventory,** the value of inventory that the firm begins with in a period. During that period, the firm will buy more merchandise to add to the inventory. The total value bought is termed **purchases.** Beginning inventory plus purchases equal **goods available for sale,** all of the merchandise that can be sold during this period; not necessarily all available at one time in that period.

purchases

goods available for sale

Objectives 1 & 2

ending inventory

cost of goods sold

The value of the goods on hand at the end of the period is termed **ending inventory.** By subtracting the ending inventory from the goods available for sale, the value of the goods that were sold during the period can be found. The cost price of the goods that were sold during the period is termed **cost of goods sold.** The word merchandise can be substituted for goods to create "merchandise available for sale" and "cost of merchandise sold." Note that (1) the period covered may be a year, six months, a month, or even one day; (2) all of the values in the example are at cost price, not selling price; (3) the ending inventory from one period becomes the beginning inventory for the next period; and (4) this inventory model repeats itself every period.

○ Purchase orders such as this one are used to replenish the stock in the firm. Work Right is ordering 144 pieces from Pace International, Inc., to add to their stock.

○ Bar coding [officially known as the Uniform Product Code (UPC)] on most packaged goods plus scanning devices provide a very efficient method to take a physical inventory.

Courtesy of Work Right Products, Inc., Lakeport, CA.

Example 1   Inventory model

    Beginning inventory
  + Purchases
    Goods available for sale
  − Ending inventory
    Cost of goods sold

Example 2   The Dosen Paper Company had 147,000 units of inventory on hand valued at $123 each on June 1. During the month, another 46,000 units were purchased at the same price. On June 30, the Accounting Department found that there were $15,922,350 of goods on hand. What is the value of the goods that were sold?

$18,081,000	Beginning inventory (147,000 units × $123/unit)—June 1
+5,658,000	Purchases (46,000 units × $123/unit)
$23,739,000	Goods available for sale
−15,922,350	Ending inventory—June 30
$7,816,650	Cost of goods sold—in the month of June

$$(147,000 \times 123 = \qquad + 5,658,000 = \qquad - 15,922,350 =)$$

In this example, the cost price per unit of beginning inventory and of purchases was the same amount.

For income tax purposes, the cost value of ending inventory or cost of goods sold is very important. A firm can legally increase the cost of the goods sold, thereby reducing gross income. Income taxes are reduced because the taxable base, net income before taxes, is reduced.

    Revenue from sales
  − Cost of goods sold
    Gross income
  − Operating and selling expenses
    Net income before taxes
  × Tax rate
    Taxes

Inventory Valuation

Valuation of inventory costs considers units of the same product being purchased at different unit-cost prices during the period. In order to determine the value of inventory, you must first determine the unit prices to be assigned to the items on hand at the end of the period (ending inventory). It is necessary to assume an arbitrary flow of costs of merchandise through the company. The three most frequently used methods in valuation of inventory are

1. L.I.F.O.
2. F.I.F.O.
3. Average Cost

L.I.F.O.

L.I.F.O. is an abbreviation for last in–first out. The **L.I.F.O.** concept considers that the most recently purchased goods (costs incurred) are sold first and their costs should be charged against revenue. The units remaining in the ending inventory come from the earliest purchases at their respective costs.

L.I.F.O
Flow of Costs

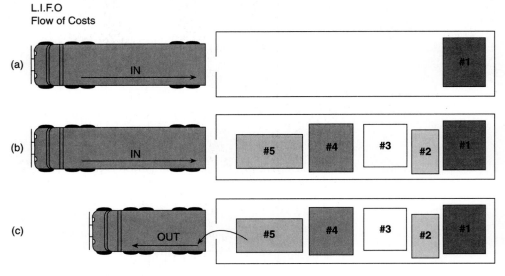

Last In–First Out (L.I.F.O.) = #5

The L.I.F.O. method will reduce gross income in times of rising prices (because the unit cost of the most recently acquired merchandise will be higher than the unit cost of the earlier purchases). This higher unit cost is used in calculating the cost of goods sold.

Example 3

Hal had made the following purchases of a face mask for Sport-A-Merica over the past year.

January	10	10 units at $5 each
March	3	20 units at $6 each
May	30	15 units at $7 each
September	3	5 units at $8 each

Hal has determined that there were 12 units of the item on hand on December 31.

A. What were the quantity and value of units available for sale?

	Units	Price per Unit	Total Price
	10	× $5	= $50
	20	× 6	= 120
	15	× 7	= 105
	+5	× 8	= +40

Total quantity and dollar value of units available for sale =   50 units (quantity) $315 (value)

○ *Remember, the assignment of costs is independent of the actual physical movement of merchandise.*

B. What was the value of ending inventory of the 12 units, using L.I.F.O.?

January	10	10 units at $5 each = $50
March	3	+2 units at $6 each = +12
Total		12 (quantity) $62 (value)

C. What is the value of the cost of goods sold?

Goods available for sale	50 units	$315
− Ending inventory	−12 units	− 62
Cost of goods sold	38 units	$253 (value)

or

September	3	5 units	at $8 each =	$40
May	30	15 units	at 7 each =	105
March	3	18 units	at 6 each =	108
Cost of goods sold		38 units		$253 (value)

Example 4   A new firm, Websters, Incorporated, made the following purchases during the first quarter year of operations.

January	16	450 units at $21.70 each
	21	600 units at 22.00 each
	28	250 units at 22.00 each
February	5	300 units at 21.90 each
	17	600 units at 22.10 each
	23	225 units at 22.10 each
March	6	730 units at 22.15 each
	11	400 units at 21.95 each
	16	350 units at 22.20 each
	28	600 units at 22.20 each
	31	400 units at 22.21 each

The firm sold 2,600 units during the quarter.

A. How many units were on hand at the end of the quarter?

Total purchases	4,905 units	
− Units sold	−2,600 units	(4,905 − 2,600 =)
Ending inventory	2,305 units	

Notice that the inventory model was used with alterations.

B. What was the value of ending inventory of 2,305 units, using L.I.F.O.?

Purchase of January 16 of 450 units at $21.70 each = $9,765.00
         (first purchase)

January	21	600 units at $22.00 each =	13,200.00
January	28	250 units at $22.00 each =	5,500.00
February	5	300 units at $21.90 each =	6,570.00
February	17	600 units at $22.10 each =	13,260.00
February	23	105 units at $22.10 each =	+ 2,320.50

(of the 225 units)

Value of ending inventory of 2,305 units, using L.I.F.O. = $50,615.50

C. What is the value of cost of goods sold at 2,600 units, using L.I.F.O.?

Purchase on March 31 of 400 units at $22.21 each = $8,884.00
         (most recent purchase)

March 28 of 600 units at $22.20 each =	13,320.00
March 16 of 350 units at $22.20 each =	7,770.00
March 11 of 400 units at $21.95 each =	8,780.00
March 6 of 730 units at $22.15 each =	16,169.50
February 23 of 120 units at $22.10 each =	2,652.00

(of the 225 units)

Value of cost of goods sold of 2,600 units, using L.I.F.O. = $57,575.50

Example 5   The Ulmac and Sterver Co. made the following purchases during the month of June.

June	4	120 units at $9.02 each
	16	200 units at $9.05 each
	21	310 units at $9.06 each
	30	150 units at $9.10 each

The beginning inventory on June 1 of 175 units was made up of 80 units purchased in May at $8.90 each and 95 units purchased in April at $8.55 each; 750 units were sold during June.

(Continued on next page)

Example 5
continued

A. How many units were on hand at the end of the month?

175 units—Beginning inventory (June 1)
+780 units—Purchases during June
955 units—Goods available for sale
−750 units—Sold during June
205 units—Ending inventory (June 30)

B. What is the value of ending inventory of 205 units, using L.I.F.O.?

Value of Ending Inventory, Using L.I.F.O.

95 units of beginning inventory purchased in April at $8.55 each	=	$812.25
80 units of beginning inventory purchased in May at $8.90 each	=	712.00
30 units purchased June 4 at $9.02 each	=	270.60
Value of ending inventory of 205 units, using L.I.F.O.	=	$1,794.85

C. What is the value of the cost of goods sold, 750 units, using L.I.F.O.?

Value of Cost of Goods Sold, Using L.I.F.O.

Purchase on June 30 of 150 units at $9.10 each = $1,365.00
(most recent purchase)
June 21   310 units at $9.06 each =   2,808.60
      16   200 units at $9.05 each =   1,810.00
        4   90 units at $9.02 each =   811.80
Value of cost of goods sold of 750 units, using L.I.F.O. = $6,795.40

F.I.F.O.

F.I.F.O is an abbreviation for first in–first out. The **F.I.F.O.** concept considers that the goods are sold in order of their purchase. Therefore, the units remaining in the ending inventory come from the most recent purchases at their respective costs.

○ *Consider introducing First in–first out, Last in–still there (FIFO-LIST) to your students to help them understand the FIFO concept.*

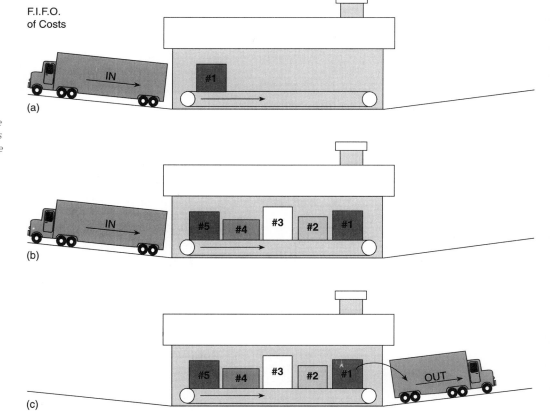

F.I.F.O.
of Costs

(a)

(b)

(c)

First In–First Out (F.I.F.O.) = #1

Example 6   Using the data from Sport-A-Merica in example 3:

A. What were the quantity and value of units available for sale? The answer to this question is the same as it would be in example 3.

Total quantity <u>50 units</u>   Dollar value <u>$315</u>

B. What was the value of ending inventory of the twelve units, using F.I.F.O.?

September 3   5 units at $8 each = $40
May   30   + 7 units at $7 each = +$49

Total   <u>12</u> quantity   <u>$89</u> value

C. What is the value of the cost of goods sold?

Goods available for sale   50 units $315
− Ending inventory   <u>−12 units</u>   <u>− 89</u>
    Cost of goods sold   38 units $226 value

or

January   10   10 units at $5 each = $50
March   3   20 units at $6 each =   120
May   30   <u>8 units at $7 each =   56</u>
Cost of goods sold   38 units $226 value

Example 7   Using the purchase data from Websters, Incorporated in example 4:

A. What was the value of ending inventory of 2,305 units using F.I.F.O.?

Value of Ending Inventory, Using F.I.F.O.

Purchase on March 31 of 400 units at $22.21 each = $ 8,884.00
        (most recent purchase)
        March 28 of 600 units at $22.20 each =   13,320.00
            16 of 350 units at $22.20 each =   7,770.00
            11 of 400 units at $21.95 each =   8,780.00
            6 of 555 units at $22.15 each = <u>+12,293.25</u>
            (of the 730 units)

Value of 2,305 units, using F.I.F.O. = $51,047.25

B. What is the value of the cost of goods sold, 2,600 units, using F.I.F.O.?

Value of Cost of Goods Sold, Using F.I.F.O.

Purchase on January 16 of 450 units at $21.70 each = $9,765.00
        (first purchase)
        January 21 of 600 units at $22.00 each = 13,200.00
            28 of 250 units at $22.00 each = 5,500.00
        February 5 of 300 units at $21.90 each = 6,570.00
            17 of 600 units at $22.10 each = 13,260.00
        February 23 of 225 units at $22.10 each = 4,972.50
            March 6 of 175 units at $22.15 each = <u>+3,876.25</u>
            (of the 730 units)

Value of cost of goods sold of 2,600 units, using F.I.F.O. = $57,143.75

Exercise A

Determine the answers in the exercises below.

	1.	2.	3.	4.
Beginning Inventory (units)	38	300	140	650
Cost per unit	$1.27	$4.75	$40.00	$12.50
Purchases				
January 16	600 at $1.28	75 at $4.90	20 at $35.00	130 at $10.00
February 3	350 at $1.28	200 at $4.88	35 at $34.60	150 at $10.00
15	90 at $1.30	150 at $4.95	60 at $34.70	275 at $9.59
22	400 at $1.30	200 at $4.95	40 at $35.60	300 at $10.05
March 2	250 at $1.35	225 at $5.00	75 at $36.10	500 at $10.25
Goods available for sale (units)	1,728	_____	_____	_____
Goods sold (units)	1,428	_____	_____	_____
Ending inventory (units)	300	350	180	732
Inventory method	L.I.F.O.	F.I.F.O.	L.I.F.O.	F.I.F.O.
Ending inventory	$383.62	_____	_____	_____
Cost of goods sold	$1,855.14	_____	_____	_____

$(38 \times 1.27 =)$
$(262 \times 1.28 =)$

Average Cost

Average cost is another method of determining inventory values. This method does not identify any specific units that were sold, but instead uses a weighted average of all units available for sale. The **average cost** is calculated by dividing the total dollar value of units available for sale by the total number of units available.

Example 8

Using the data from Sport-A-Merica in example 3:

A. What was the quantity and value of the units available for sale? The answer to this question is the same as it would be in example 3.

Total quantity: 50 units   Dollar value: $315

B. What was the value of ending inventory of the 12 units, using average cost?

$$\frac{\text{Total dollar value}}{\text{Units available}} = \frac{\$315}{50} \times 12 \text{ units} = \$75.60$$

Round off at final answer only.

C. What is the value of cost of goods sold?

Goods available for sale	50 units	$315.00
− Ending inventory	−12 units	− 75.60
Cost of goods sold	38 units	$239.40

or

$$\frac{\text{Goods available for sale}}{\text{Units available}} \quad \frac{\$315.00}{50} \times 38 \text{ (units sold)} = \$239.40$$

$(315 \div 50 = \quad \times 12 =)$ $(315 - 75.60 =)$

Example 9   Durst & Associates had 400 units of a file separator in stock at the end of the year. At the beginning of the year, there were 350 units in stock, at an average cost of 87 cents each. Two purchases were made during the year: one on July 16, for 800 units at 93 cents each; and one on November 12, for 250 units at 94 cents each.

A. How many units were sold during the year?

350	Beginning inventory (units)
+1,050	Purchases (800 + 250 units)
1,400	Goods available for sale (units)
−400	Ending inventory (units)
1,000	Cost of goods sold (units)

1,000 units were sold during the year.

B. What is the value of ending inventory, using average cost?

Value of ending inventory, using average cost

350	Units of beginning inventory	at 87¢ each =	$304.50
800	Units purchased	at 93¢ each =	744.00
+ 250	Units purchased	at 94¢ each = +	235.00
1,400	Total units available	Total cost =	$1,283.50

$$\frac{\$1,283.50}{1,400} = \frac{\text{Total cost}}{\text{Total units available}} \times \frac{400 \text{ units}}{\text{ending inventory}} = \frac{\$366.71 \text{ value of}}{\text{ending inventory}}$$

C. What is the value of the cost of goods sold, using average cost?

Value of cost of goods sold, using average cost

$$\frac{\$1,283.50}{1,400} = \frac{\text{Total cost}}{\text{Total units available}} \times \frac{1,000 \text{ units of}}{\text{goods sold}} = \frac{\$916.79 \text{ cost of}}{\text{goods sold}}$$

Exercise B

Notice that the average cost method considers the number of units each purchase, not the number of purchases. Determine the answers in the exercises below using the average cost method.

	1.	2.
Beginning inventory—units	720	300
	(18,136 ÷ 1,970 =)	
	(× 730 =)	
Beginning inventory (Cost per unit)	$9.80	$1.27
Purchases		
July 24	400 at $7.90/unit	7,200 at $1.30/unit
August 10	250 at $8.40/unit	4,000 at $1.31/unit
November 26	600 at $9.70/unit	5,500 at $1.36/unit
Ending inventory—units	730	200
Ending inventory value	$6,720.45	
Cost of goods sold value	$11,415.55	

(Continued on next page)

Exercise B Continued

	3.	4.
Beginning inventory—units	60	550
Beginning inventory (Cost per unit)	$32.00	$18.50
Purchases		
July 24	90 at $32.00/unit	650 at $19.00/unit
August 10	50 at $32.25/unit	400 at $19.20/unit
November 26	125 at $31.75/unit	750 at $18.70/unit
Ending inventory—units	50	400
Ending inventory value	_____	_____
Cost of goods sold value	_____	_____

Retail Inventory
Estimating Method

Occasionally, the manager may need only an estimate of the value of inventory. The **retail inventory estimating method** provides that estimate. This approach replaces the need for an actual physical inventory of the merchandise to estimate its value. In order for this method to be close to the actual value of inventory, the markup rate for all of the firm's products must be similar and consistent. The percent of cost to selling price is used with the sales for the period to estimate the cost price of goods sold and ending inventory.

The comparison of cost to sales is made from goods available for sale. All sales are made from goods available for sale. With a consistent markup rate, an estimate of inventory value at cost can be made. The information in the inventory model in example 1 must be available at selling price. Again, the inventory model has been altered to find ending inventory.

Example 10

	Cost Price	Selling Price
Beginning inventory	$62,580	$97,427
+ Purchases	+158,690	+229,352
Goods available for sale	$221,270	$326,779
− Sales		−261,408
Ending inventory		$65,371

Find the value of ending inventory at cost price from the above information.

$$\frac{\text{Percent of cost to sales of}}{\text{goods available for sale}} = \frac{\text{Cost price}}{\text{Selling price}} = \frac{\$221{,}270}{\$326{,}779} = 67.7\% \text{ (rounded)}$$

67.7 percent of the ending inventory of $65,371 at selling price must be the cost price.

$65,371—Ending inventory at selling price
×.677—Percent of cost to sales
$44,256.17—Ending inventory at cost price (estimate)

Exercise C

Determine the answers to the exercises using the retail inventory estimating method. Round off the percent of cost to sales to the nearest tenth of a percent.

	Cost Price	Selling Price
1. Beginning inventory	$4,762	$9,827
+ Purchases	+3,846	+7,842
Goods available for sale	$8,608	$17,669
− Sales		−$13,931
Ending inventory	$1,820.41 (8,608 ÷ 17,669 = × 3,738 =)	$3,738

	Cost Price	Selling Price
2. Beginning Inventory	$8,662	$10,421
+ Purchases	+37,356	+43,930
Goods available for sale	_____	_____
− Sales		−$35,567
Ending inventory		_____

	Cost Price	Selling Price
3. Beginning Inventory	42,618	$60,247
+ Purchases	+38,065	+57,348
Goods available for sale	_____	_____
− Sales		−$103,900
Ending inventory		_____

	Cost Price	Selling Price
4. Beginning Inventory	$3,421	$5,608
+ Purchases	+132,641	+215,648
Goods available for sale	_____	_____
− Sales		−$21,432
Ending inventory		_____

Specific Identification Method

When the number of sales are few, the individual merchandise items may be marked with the date and cost of the purchase. This method allows the middleman to identify which items were sold or are part of ending inventory. The **specific identification method** is used when the price of the merchandise is high or when only a few items are sold.

Example 11

Hal located the following units of a ball thrower in the Sport-A-Merica storage room at the end of last month.

3 units purchased on July 16 at $82 each
1 unit purchased on April 23 at $78.64 each
5 units purchased on September 7 at $77.90 each
2 units purchased on June 30 at $81.45 each

What is the value of ending inventory?

3 units × $82.00 each =	$246.00
1 unit × $78.64 each =	78.64
5 units × $77.90 each =	389.50
+ 2 units × $81.45 each =	+ 162.90
11 units purchased at	$877.04 = ending inventory value

Turnover
Objective 3

Turnover is the number of times the average inventory is sold during the year.

$$\text{Turnover} = \frac{\text{Sales at cost}}{\text{Average inventory}}$$

The average inventory is found by totaling the value of all inventories taken during the year and dividing that sum by the number of inventories taken. Both average inventory value and sales value must be at cost price. The answer is a number, not a dollar amount or a percent, although the number can be changed to a percent.

Example 12

A firm had sales of $66,000 during the year, with a markup of 20 percent on cost. Inventory taken at cost on January 1 was valued at $16,400; on June 30, it was $18,600; and on December 31, it was $13,000. What was the turnover?

○ *The answer to a turnover problem is a number, not a dollar amount or a percent.*

Average inventory

$16,400
18,600
+13,000
$48,000 ÷ 3 inventories = $16,000

$66,000
Cost price + markup = selling price
100%   +   20%   =   120%

Cost price = 1.20)$66,000 = $55,000 (sales at cost or cost or goods sold)

$$\text{Turnover} = \frac{\text{sales (at cost)}}{\text{average inventory (at cost)}} = \frac{\$55,000}{\$16,000} = 3.4 \text{ times}$$

Exercise D

Determine the turnover rate at cost price in the following problems. Round off answer to nearest tenth.

	1.	2.	3.	4.	5.
Inventories	$47,682.00	$23,050.00	$40,600.00	$2,800.00	$3,490.00
	53,802.00	65,400.00	42,400.00	4,380.00	6,660.00
	49,681.00	61,325.00	49,000.00	3,670.00	4,658.00
	57,304.00		50,600.00		5,360.00
Sales	$426,380.00	$117,430.00	$172,480.00	$24,000.00	$31,680.00
Markup %	23	30	40	35	50
Markup on	Cost	Cost	Sales	Cost	Sales
Cost of sales	$346,650.41				
Average inventory	$52,117.25				
Turnover	6.7				

(346,650.41 ÷ 52,117.25 =)

ASSIGNMENT Chapter 16

Name _____ Date _____ Score _____

Skill Problems

Select the best term for the definition below.

A. Average Cost   B. Beginning Inventory   C. Cost of Goods Sold   D. Ending Inventory
E. F.I.F.O.   F. Goods Available for Sale   G. L.I.F.O.   H. Purchases
I. Retail Inventory Estimating Method   J. Specific Identification Method
K. Turnover

1. _____ An accounting approach used to determine the value of inventory calculated by dividing the total dollar value of units available for sale by the total number of units available.

2. _____ The units in this ending inventory come from the most recent purchases at their respective costs.

3. _____ The number of times the average inventory is sold during the year.

4. _____ This approach replaces the need for an actual physical inventory of the merchandise.

Complete the following inventory model with the terms above.

Inventory Model:

Beginning inventory
+ _____ **5.** _____
= _____ **6.** _____
− _____ **7.** _____
= _____ **8.** _____

Determine the value of ending inventory in the problems below.

Beginning Inventory at Cost	Purchases at Cost	Net Sales	Markup on Cost	Cost of Goods Sold	Ending Inventory at Cost
$13,560	$47,390	$57,300	40%	**9.** _____	**10.** _____
2,658	12,256	13,560	35	**11.** _____	**12.** _____
91,420	51,200	63,420	45	**13.** _____	**14.** _____

Determine the answers in the exercises below.

Beginning inventory (units)	55	897
Cost per unit	$3.45	$29.95
Purchases:		
February 28	500 at 3.56	120 at 30.00
March 5	200 at 3.57	200 at 31.15
March 17	600 at 3.60	205 at 32.35
March 30	600 at 3.75	100 at 33.10
April 4	120 at 3.80	205 at 31.50
Goods available for sale (units)	**15.** _____	**19.** _____
Goods sold (units)	**16.** _____	**20.** _____
Ending inventory (units)	205	250
Inventory method	L.I.F.O.	F.I.F.O.
Ending inventory	**17.** _____	**21.** _____
Cost of goods sold	**18.** _____	**22.** _____

Beginning inventory		85 units at 2.25	191.25
Purchases	January 8	140 units at 2.35	329.00
	January 16	200 units at 2.10	420.00
	January 28	110 units at 2.45	269.50 (191.25 + 329 + 420 + 269.50 = 1,209.75)
Ending inventory		152 units	
Ending inventory F.I.F.O.	**23.** _____		
Cost of goods sold F.I.F.O.	**24.** _____		
Ending inventory L.I.F.O.	**25.** _____		
Cost of goods sold L.I.F.O.	**26.** _____		
Ending inventory Average costs (Do not round unit cost)	**27.** _____		
Cost of goods sold Average costs	**28.** _____		

Use for problems 29–34.

Beginning inventory = 420 units at $3.60 each

Purchases =
November   2   1,500 units at $3.70 each
           10   2,360 units at $3.75 each
           23   1,500 units at $3.72 each
           28   1,700 units at $3.76 each

Ending inventory = 360 units

Cost of goods sold

29. L.I.F.O. = _____

Ending inventory

30. L.I.F.O. = _____

Cost of goods sold

31. F.I.F.O. = _____

Ending inventory

32. F.I.F.O. = _____

Cost of goods sold

33. Average cost = _____

Ending inventory

34. Average cost = _____

Use for problems 35–40.

Beginning inventory = 47 units at $1.20 each

Purchases = April   3   150 units at $1.23 each
              26   250 units at $1.26 each
        May   7   300 units at $1.28 each
             30   200 units at $1.28 each
       June 13   350 units at $1.29 each
             19   200 units at $1.29 each

Ending inventory = 320 units

Cost of goods sold

35. F.I.F.O. = _____

Ending inventory

36. F.I.F.O. = _____

Cost of goods sold

37. L.I.F.O. = _____

Ending inventory

38. L.I.F.O. = _____

Cost of goods sold

39. Average cost = _____

Ending inventory

40. Average cost = _____

Use for problems 41–46.

Beginning inventory = 1,366 units at $14 each

Purchases =
January   5   600 units at $13.96 each
         18   4,500 units at $13.95 each
         22   3,100 units at $14.26 each
         26   8,450 units at $14.35 each

Ending inventory = 1,580 units

Cost of goods sold

41. Average cost = _____

Ending inventory

42. Average cost = _____

Cost of goods sold

43. L.I.F.O. = _____

Ending inventory

44. L.I.F.O. = _____

Cost of goods sold

45. F.I.F.O. = _____

Ending inventory

46. F.I.F.O. = _____

Determine the answers in the exercises below using the average cost method. Round off unit cost to nearest cent.

Beginning inventory (units)		840	400
Beginning inventory (cost per unit)		$10.75	$1.40
Purchases June 20		300 at 9.85	650 at 1.35
July 15		200 at 10.50	500 at 1.25
August 17		300 at 11.10	900 at 1.50
Ending inventory (units)		760	300
Ending inventory value	**47.** _____		**49.** _____
Cost of goods sold value	**48.** _____		**50.** _____

Determine the value of ending inventory at cost in problems 51 through 70 using the retail inventory estimating method. Use a cost to sales percent rounded off to the nearest tenth of a percent.

	Cost Price		Selling Price
Beginning inventory	$368.40		$605.00
Purchases	943.30		1,736.80
Goods available for sale	**51.** _____		**53.** _____
Sales			$2,141.60
Ending inventory	**52.** _____		**54.** _____

	Cost Price		Selling Price
Beginning inventory	$46,356.00		$51,840.00
Purchases	117,540.00		136,983.00
Goods available for sale	**55.** _____		**57.** _____
Sales			$102,421.00
Ending inventory	**56.** _____		**58.** _____

	Cost Price		Selling Price
Beginning inventory	$40,100.00		$108,560.00
Purchases	54,980.00		126,950.00
Goods available for sale	**59.** _____		**61.** _____
Sales			$180,670.00
Ending inventory	**60.** _____		**62.** _____

	Cost Price		Selling Price
Beginning inventory	$5,870.50		$8,218.70
+ Purchases	+5,604.40		+7,846.16
Goods available for sale	**63.** _____		**65.** _____
− Sales			−9,405.48
Ending inventory	**64.** _____		**66.** _____

	Cost Price		Selling Price
Beginning inventory	$987.40		$1,234.25
+ Purchases	+1,105.85		+1,382.31
Goods available for sale	**67.** _____		**69.** _____
− Sales			−1,744.37
Ending inventory	**68.** _____		**70.** _____

Determine the turnover rate at cost price in the following problems. Round off answers to nearest tenth.

Inventories			
		$4,085.50	$40,875.00
		3,045.60	65,087.00
		5,784.55	105,040.00
		4,563.40	
Sales		$15,487.55	$85,706.00
Markup %		20	50
Markup on		Cost	Sales
Cost of sales	**71.** _____		**74.** _____
Average inventory	**72.** _____		**75.** _____
Turnover	**73.** _____		**76.** _____

77. Inventories taken: July 1   = $13,560
October 1   =   14,388
December 31   =   13,840

Sales at cost   = $38,645

2. The average inventory value for Pedmix Company was $43,976 over the past 12-month period. The markup rate averages 27 percent of sales. If the turnover rate is 4.1, what is the value of sales for the 12-month period?

78. Inventories taken: January 1   = $68,542
March 31   =   72,840
June 15   =   65,473
August 20   =   69,568
October 31   =   73,869

Sales   = $630,470
Markup   = 35% on cost

3. The beginning inventory of Fredmore Associates was $36,720. Purchases for the period were $75,908. The value of the cost of merchandise sold was $40,860 for the period. What is the value of ending inventory?

4. In problem 3, what is the value of sales if the markup rate is 27 percent on sales?

79. Inventories taken: April 2   = $32,480
9   =   36,400
16   =   35,931
28   =   33,560

Sales   = $149,680
Markup   = 40% on sales

5. In problem 3, to the nearest tenth, what is the turnover rate?

Business Application Problems

1. The value of the cost of merchandise available for sale was $428,500 for the year. The average inventory was $68,400. To the nearest tenth, what was the turnover for the year?

6. The Balcom Corporation enjoyed a low inventory level of $52,930 at the end of the year. The cost value of sales was $340,817 for the year. The turnover rate was 4.2. What was the average inventory level for the year?

7. In problem 6, what was value of purchases for the period if the beginning inventory was $91,560?

8. In problem 6, the markup rate was 38 percent on cost. What is the value of sales for the period?

9. Hopson Corporation reduced their investment in inventory this year, while maintaining the same 40 percent level of markup on cost. Last year the average inventory level was $45,890. This year it was $41,984. Turnover was 6.5 last year and 7.2 this year. By what amount did Hopson Corporation increase or decrease sales this year?

10. Gulf Tire Company had a beginning inventory of $106,390. They ended the period with an inventory value of $89,600. During the period they made two purchases, one for $32,640 and one for $98,600. What is the value of the goods that were available for sale in the period?

11. Booth and Burtis, Inc., has a turnover rate of 5.4 and a markup of 40 percent on sales. Sales are $321,840 for the year. What is the average inventory that the firm carries in order to maintain this level of sales?

12. Arsenau and Co. found that the value of inventory on January 1 was $147,368; and on March 31, it was $153,820. Sales for the period were $413,875. Goods were marked up 40 percent on cost. What was the turnover to the nearest tenth?

13. The Jones Corporation had a beginning inventory for the year of 275 units valued at $14,307.50. At the end of the year, there were 265 units remaining. A purchase was made on March 16 for 360 units at $65 each, and on September 1 for 800 units at $71.40 each. What is the amount of sales for the period if the firm used a F.I.F.O. system and a 30 percent markup on cost?

14. Dawn's Discount enjoyed a fourth quarter sales record of $69,800 at a markup of 40 percent on sales. The purchases for the period were $32,590. The ending inventory was $21,730. What was the beginning inventory?

15. Barton Wholesalers had a fire that damaged a portion of their warehouse on June 23. However, inventory had been taken on June 1, and it totaled $85,642. Only one purchase for $38,650 was made in June. Sales for the first 23 days of the month were $95,391 at a 42 percent markup on sales. A physical inventory of the undamaged merchandise totaled $41,365. What is the approximate value of the loss?

$203,568. The firm marks up on sales, and they use a 40 percent rate. To the nearest tenth, what is the turnover for the month?

20. Write a complete sentence in answer to this question. Dalton Distributors uses a L.I.F.O. system for inventory. The firm had a beginning inventory of $190,000. Their purchases for the month totaled 8,500 units at $16.20 each. The purchases brought the total units available for sale to 20,120. The end-of-the-month inventory, as displayed on the computer printout, said there were 5,890 units available. A physical inventory of the warehouse showed there were only 5,530 units available. The accountant decided to write off the missing units as "stock shrink." What is the amount of the write-off?

16. Benson and Co. started the firm with a purchase of ten units at $14 each. During the month, they purchased an additional 20 units at $14.50 each. In the second month, they purchased 50 units at a total cost of $727.50. At the end of the second month, they had 17 units left in stock. If the firm elects to use a F.I.F.O. system, what is the value of ending inventory?

17. If Benson and Co. in problem 16 elects a L.I.F.O. system, what is the value of the cost of goods sold?

★ **Challenge Problem**

The Saranac Producers Association had a beginning inventory of 682 units of a feeder on June 13. The inventory had cost Saranac $947.30. During the next six months, the following purchases were made:

September 1 200 units at $1.47 each; terms 3/10,n/30
October   21 450 units at $1.51 each; terms 3/10,n/30
November 13 600 units at $1.52 each; terms 2/10,n/60

All discounts were taken.
  1,640 units were sold during the six months at $2.39 each.
  What is the value of ending inventory using L.I.F.O.?

18. In problem 16, Benson and Co. sells the merchandise at a markup of 30 percent on the selling price. What was the turnover for the two-month period using F.I.F.O.?

19. Write a complete sentence in answer to this question. The Sterling Corporation had sales of $589,630 for the month. At the beginning of the month, the inventory was $184,980. At the end of the month, the inventory was

Name _____ Date _____ Score _____

True or False

T F   **1.** Goods available for sale less ending inventory equals purchases.

T F   **2.** Ending inventory equals the previous period's beginning inventory.

T F   **3.** Turnover measures the number of times that the average inventory is sold.

T F   **4.** Specific identification is a method of determinimg the actual value of inventory.

Complete the following matrix for the missing values.

Beginning Inventory	Purchases	Ending Inventory	C.O.G.S.
5. $ 42,000	$175,000	$32,000	_____
6. _____	243,980	56,835	345,600
7. 39,800	156,000	_____	178,000

Determine the value of ending inventory in the problems below using the method indicated.

Beginning inventory = 890 units at $41.00 each
Purchases = January 17   400 units at $41.20 each
           May 30   650 units at $41.20 each
     September 16   300 units at $41.30 each
     November 3   650 units at $41.50 each
Ending inventory = 970

8. Ending Inventory using L.I.F.O. = _____

9. Ending Inventory using F.I.F.O. = _____

10. Ending Inventory using average cost = _____

11. Sales are made at a 25 percent markup on sales. What is the turnover in the problem just given if L.I.F.O. is used? Round off the answer to the nearest tenth.

12. Estimate the value of ending inventory at cost using the Retail Inventory Estimating Method.

	Cost Price	Selling Price
Beginning Inventory	$ 38,500	$ 84,600
Purchase	153,750	301,230
Goods Available for Sale	_____	
Sales		$260,900
Ending Inventory	_____	_____

13. Humbert and Associates found the value of inventory on July 1 to be $96,800. An additional inventory taken on August 1 totaled $42,690. The sales for the period were $120,970 at a markup of 40 percent on sales. What was the turnover to the nearest tenth?

14. The Woodard Company took inventory on January 1 and found that they had 345,587 meters of their wire product. The value was set at $251,873. The warehouse was broken into on January 12. All of the inventory was stolen. The value of the sales for the twelve days was $34,980 at a 40 percent markup on sales. What is the value of merchandise that was stolen? There were no purchases during the period.

DEPRECIATION

 17

OBJECTIVES

After mastering the material in this chapter, you will be able to:

1. Explain the concept of depreciation and depletion. p. 324

2. Calculate book value, accumulated depreciation, and depreciation using four methods. p. 324

3. Explain the advantages of each of the depreciation methods. p. 324

4. Calculate the depreciation amount for a partial year using three methods. p. 335

5. Understand and use the following terms and concepts:
Depreciation
Expected Life
Total Cost
Salvage Value
Total Depreciation
Accumulated Depreciation
Book Value
Market Value
Straight-Line Method
Depreciation Schedule
Units-of-Production Method
Sum-of-Years'-Digits (S.O.Y.D.) Method
Declining-Balance Method
Modified Accelerated Cost Recovery System (M.A.C.R.S.) Method
Depletion
Time Line

SPORT-A-MERICA

Hal is looking at the calendar after the Saturday session on the financial picture of Sport-A-Merica with Kerry. He realizes it is time to prepare the tax information for the accountant. This always means a lot of digging into the files. The records of the assets bought the past year will have to be reviewed for any changes that need to be brought to the attention of the accountant.

Another area that always needs reviewing is the schedule of depreciation on the fixed assets. Sport-A-Merica has purchased a new delivery vehicle this year. The cost of the vehicle, along with the cost of the necessary optional equipment, will need to be gathered. Hal knows the vehicle's useful life will be shorter than the dealer had estimated because Sport-A-Merica will be keeping the vehicle very busy with the increase in sales. Hal will talk to Kerry about which method she wishes to use to depreciate the vehicle. She had earlier suggested a method that would recognize the actual use of the vehicle instead of a method that involves merely the passage of time. Hal thinks that she is right on target. After all, the total cost of the vehicle was $28,400—no small amount! He will talk to her today so that he can be ready for the meeting with the accountant.

323

Objective 1   *depreciation*   **Depreciation** is an estimate of the amount of decrease in market value of an item over a period of time. A *depreciation expense* is allowed by taxing authorities, such as the federal government, for the decline in the item's value. The depreciation expense, in effect, reduces a firm's income tax. The Economic Recovery Tax Act of 1981 was seen in part as an incentive for business to invest in new machinery, plants, and equipment. These investments would then result in a favorable tax break for the business—a faster depreciation of the asset. This allows for a greater depreciation expense per year in the early years of the asset's life. Fixed assets, such as buildings, office equipment, trucks, and furniture, decline in value due to wear, an outdating of the style, or obsolescence. An office machine may have a useful life of only two or three years, whereas a building's life

expected life   may exceed fifty years. The useful life of an asset is termed the **expected life.** The expected life will be stated by the selling firm or provided through industry or government guidelines. Expected life is usually expressed in years or units of usefulness.

The amount of costs to be depreciated includes all of those costs necessary to make the item operational. Item costs, transportation costs, and installation costs make up the

total cost   **total cost** of an item. Consider an item such as a computer that was purchased at a price that included transportation and installation. If the buyer had agreed instead to purchase the computer without transportation or installation, those costs would still have been incurred by the buyer before the computer became operative. In the final analysis, the total costs would have been approximately the same.

All of the total costs are depreciable over the expected life, unless the firm feels that the item can be sold at the end of its expected life. The sales price at the end of the

salvage value   expected life of the item is termed the **salvage value.** Terms such as *scrap value, trade-in value, residual value,* and *end value* are also used. The selling firm can provide an estimate of the salvage value.

total depreciation   The **total depreciation,** or the amount that the item is depreciated over its expected life, is found by subtracting the salvage value from the total cost. A portion of the total depreciation is subtracted from the total cost each year. This amount is termed

accumulated depreciation   *depreciation*. By totaling the depreciation to date, the amount of **accumulated depreciation** can be determined. By subtracting the accumulated depreciation from the

book value   total cost, the amount of book value can be found. **Book value** is an estimate of an item's

○ *Book value is not limited to the value of a vehicle!*   market value. The book value is the amount at which the item is valued on the firm's accounting records.

Objective 2   *market value*   The **market value** is the amount that the item can be sold for at a given point in its expected life.

The terminology of depreciation is rather extensive; and at the same time, it is important to the understanding of depreciation.

Straight-Line Method
Objective 3

There are four basic methods that are used to determine depreciation. The **straight-line method** is the easiest to understand. This method establishes an equal amount of depreciation per year for an item. It is found by dividing the total depreciation by the years of expected life. Total depreciation is found by subtracting the salvage value from the total cost.

The following figures may help you to visualize these terms and the concepts they represent.

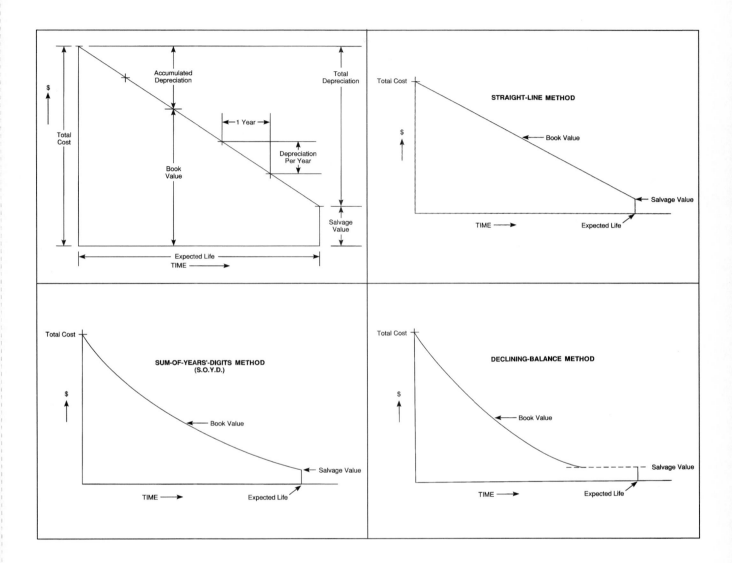

Example 1

A communications system purchased for $600 has an expected life of eight years and a salvage value of $100. What is the depreciation per year using the straight-line method?

$$\text{Depreciation} = \frac{\text{total depreciation}}{\text{expected life}} \text{ or } \frac{\text{total depreciation}}{n}$$

(n = the number of years of expected life)

$600 − $100 = $500 total depreciation

$$\text{Depreciation} = \frac{\$500}{8} = \$62.50 \text{ per year}$$

(500 ÷ 8 =)

The book value at the end of the second year in example 1 is found by subtracting the accumulated depreciation ($62.50 per year × 2 years = $125) from the total cost ($600). Book value = $600 − $125 = $475.

Example 2

Sport-A-Merica purchased an asset that originally cost $300. In addition, the firm paid $14 for transportation and $26 for installation. The asset is expected to have a useful life of ten years before it will be traded in for an estimated value of $60. What will the book value be at the end of six years?

$300 Item cost
+14 Transportation cost
+26 Installation cost
$340 Total Cost
−60 Salvage value
$280 Total depreciation

$$\text{Depreciation} = \frac{\$280 \text{ (total depreciation)}}{10 \text{ years expected life}} = \$28 \text{ per year}$$

$28 = Per year depreciation
×6 = Years
$168 = Accumulated depreciation at the end of the sixth year

$340 = Total cost
−168 = Accumulated depreciation
$172 = Book value, sixth year

(280 ÷ 10 =   × 6 =)

Depreciation Schedule

The **depreciation schedule** shows the effects of depreciation on the book value of the asset over the expected life. The depreciation schedule for example 2 would be as follows.

End of Year Number	Depreciation/Year	Accumulated Depreciation at the End of Year	Book Value (Cost − Accum. Depr.) at the End of Year
1	$28	$ 28	$312
2	28	56	284
3	28	84	256
4	28	112	228
5	28	140	200
6	28	168	172
7	28	196	144
8	28	224	116
9	28	252	88
10	28	280	60

Exercise A

Determine the missing values in the exercises below using the straight-line method.

	Item Cost	Transportation Cost	Installation Cost	Salvage Value	Expected Life (Years)	Year No.	Depreciation per Year	Accumulated Depreciation	Book Value
1.	$14,350	$95 (27,820 ÷ 12 =)	$16,585	$3,210	12	4	$2,318.33	$9,273.32	$21,756.68
2.	19,000	0	1,757	1,257	30	7			
3.	625	26	249	200	7	3			
4.	42,600	0	10,380	1,980	17	2			
5.	965	85	75	325	8	3			

Exercise B

Determine the missing values in the exercises below using the straight-line method and the formula

$$\text{Depreciation/year} = \frac{\text{Total depreciation}}{\text{Expected life}}.$$

	Total Cost	Expected Life (Years)	Salvage Value	Year No.	
1.	$70,000	10	$20,000	3	$55,000 book value (50,000 ÷ 10 =)
2.	10,000	8	2,000	—	$3,000 book value
3.	4,000	_____	1,000	6	$1,800 accumulated depreciation
4.	20,000	10	_____	2	$17,000 book value
5.	_____	7	6,000	3	$2,000 depreciation per year

Units-of-Production Method

In the straight-line method, the depreciation expense is the same amount every year. Some assets are not used an equal amount every year. Their use varies, and therefore, the depreciation amount should also vary. The **units-of-production method** acknowledges an unequal usage. The amount of depreciation is equal for each unit produced, but it is not necessarily the same total amount each year. This method is also termed the *hours-of-production method*.

Example 3

The delivery vehicle Sport-A-Merica purchased is expected to provide approximately 100,000 trouble-free miles over its life. The vehicle, including all of the add-ons the firm needed, cost $28,400. This year, Sport-A-Merica drove the vehicle 21,000 miles. The salvage value of the vehicle after 100,000 miles have been driven is estimated to be $2,400. What is the depreciation for this year?

$$\text{Depreciation} = \frac{\text{Total depreciation}}{\text{Expected total output capacity over life of asset}} \times \text{Output this period}$$

$$\text{Depreciation} = \frac{\$28,400 - \$2,400}{100,000 \text{ miles}} \times 21,000 \text{ miles} = \$5,460 \text{ for } 21,000 \text{ miles}$$

(26,000 ÷ 100,000 =   × 21,000 =)

The book value at the end of 49,000 miles would be as follows:

$$\frac{\$28,400 - \$2,400}{100,000 \text{ miles}} \times 49,000 \text{ miles} = \$12,740 \text{ for } 49,000 \text{ miles}$$

$\$ 28,400$ = Total cost
$\underline{-12,740}$ = Accumulated depreciation
$\$ 15,660$ = Book value after 49,000 miles

> A *depreciation schedule* for an asset depreciated by the units-of-production method could be created only if the usage rate could be known for the periods in the future.

Exercise C

Determine the missing values in the exercises below using the units-of-production method.

	Total Cost	Salvage Value	Expected Life (in Units)	Units This Period	Units to Date	Depreciation This Period	Book Value
1.	$43,000 (41,500 ÷ 140,000 =)	$1,500	140,000	1,980	35,923	$586.93	$32,351.40
2.	975	65	1,400	350	895	_____	_____
3.	23,680	1,180	30,000	3,500	19,600	_____	_____
4.	3,600	250	16,000	950	2,860	_____	_____
5.	19,780	430	21,000	610	13,548	_____	_____

Sum-of-Years'-Digits (S.O.Y.D.) Method

The market values of some assets do not decline by an equal amount each year. In order to develop a book value that approximates the market value of an item, the **sum-of-years'-digits (S.O.Y.D.) method** is sometimes used. This method accelerates the depreciation by allowing for the greater decline in the market value in the first years of an asset's life and a lesser decline in the later years of its life. It is similar to the straight-line method: The total depreciation is multiplied by a fraction to determine the depreciation for the year. However, each year, the fraction becomes smaller so that the depreciation amount also becomes smaller. The fraction's denominator is the sum of the years of expected life, and its numerator is the number of useful years remaining.

Example 4

An asset has an expected life of eight years. What is the fraction used to determine the depreciation for the second year using the S.O.Y.D. method?

Years of Expected Life	Year Number
8	1
7 ←	2 ←
6	3
5	4
4	5
3	6
2	7
$\underline{+1}$	8
36 = Sum of the years of expected life	

Beginning year 2 the asset had seven years of useful life left. The fraction would be $\frac{7}{36}$.

Example 5

An asset has an expected life of twelve years. What is the fraction used to determine the depreciation for the fifth year using the S.O.Y.D. method?

Years of Expected Life	Year Number
12	1
11	2
10	3
9	4
8 ←	5 ←
7	6
6	7
5	8
4	9
3	10
2	11
+1	12

Beginning year 5 the asset had eight years of useful life left. The fraction would be $\frac{8}{78}$.

78 = Sum of the years of expected life

The sum of the years can also be found as follows:

$$\frac{n \times (n + 1)}{2}$$ n = the number of years of expected life

Example 6

The expected life of an asset is 30 years. What is the sum of the years' digits?

$$\frac{n \times (n + 1)}{2} = \frac{30 \times (30 + 1)}{2} = 465$$

(30 × 31 =   ÷ 2 =)

This formula is very useful for assets with long life expectancies.

Exercise D

Determine the sum-of-years' fraction to be used in the following exercises.

Years of Expected Life	Sum of Years	Year Number	Fraction to be Used
1. 14 (14 × 15 = ÷ 2 =)	105	3	$\frac{12}{105}$
2. 20	_____	2	_____
3. 13	_____	4	_____
4. 15	_____	6	_____
5. 10	_____	10	_____

To determine the depreciation amount for a year, the fraction is multiplied by the total depreciation. Each year the fraction will change, and the new fraction will provide a different depreciation amount.

Example 7 A building purchased for $42,000 five years ago has an expected life of 20 years. The salvage value is estimated to be $3,000. What is the depreciation for each of the five years and what is the book value using the sum-of-years' method?

$42,000 − $3,000 = $39,000 (total depreciation)

	Years of Expected Life	Year Number
	20	1
(sum of values	19	2
for first five 90	18	3
years)	17	4
	16	5

$\frac{+1}{210}$ = Sum of the years of expected life

Depreciation

Year 1 = $39,000 × $\frac{20}{210}$ = $3,714.29

Year 2 = $39,000 × $\frac{19}{210}$ =   3,528.57

Year 3 = $39,000 × $\frac{18}{210}$ =   3,342.86   or $\frac{90}{210}$ × $39,000 = $16,714.29

Year 4 = $39,000 × $\frac{17}{210}$ =   3,157.14

Year 5 = $39,000 × $\frac{16}{210}$ =   2,971.43

Accumulated depreciation = $16,714.29 years 1–5
Book value = $42,000 − $16,714.29 = $25,285.71

(20 × 21 =   ÷ 2 =)

The *depreciation schedule* for the first five years of the building in example 7 using the S.O.Y.D. method would be as follows.

End of Year Number	Depreciation/Year	Accumulated Depreciation at the End of Year	Book Value at the End of Year
1	$3,714.29	$ 3,714.29	$38,285.71
2	3,528.57	7,242.86	34,757.14
3	3,342.86	10,585.72	31,414.28
4	3,157.14	13,742.86	28,257.14
5	2,971.43	16,714.29	25,285.71

Exercise E

Determine the missing values in the exercises below using the S.O.Y.D. method.

	Total Cost	Salvage Value	Expected Life (Years)	Year Number	Depreciation	Accumulated Depreciation	Book Value
1.	$ 48,500	$ 4,000	20	10	$ 2,330.95	$32,845.24	$15,654.76
							(44,500 × 11 = ÷ 210 =)
2.	21,000	600	14	4			

Exercise E Continued

Determine the missing values in the exercises below using the S.O.Y.D. method.

	Total Cost	Salvage Value	Expected Life (Years)	Year Number	Depreciation	Accumulated Depreciation	Book Value
3.	135,500	10,000	15	6	_____	_____	_____
4.	8,900	1,000	10	9	_____	_____	_____
5.	7,600	200	20	15	_____	_____	_____

Declining-Balance Method

The **declining-balance method** is similar to the sum-of-years'-digits method because both provide an accelerated amount of depreciation in the early years of an asset's life. This method is used more frequently and is even more accelerated than the sum-of-years'-digits method. The term *declining-balance* is very descriptive of the procedure used to determine the depreciation amount.

This method also has some similarity to the straight-line method. In the declining-balance method, the straight-line method rate of depreciation is doubled. This rate is then multiplied by the current book value. The depreciation amount is subtracted from the previous period book value to find the current year book value, or declined balance. The book value must not drop below the salvage value.

Note that the total cost the first year and the book value or declined value in subsequent years, not the total depreciation, are used to determine the amount of depreciation for the current year.

Example 8

A shop machine cost $840 and is expected to be used for an expected life of eight years. The salvage value is $150. What is the depreciation for each of the eight years using the declining-balance method?

$$\frac{1}{8 \text{ years of expected life}} \times 2 \text{ (double)} = \frac{2}{8} = 25\%$$

$840 = Total cost $\times$ .25 = $210 depreciation for first year
$\underline{-210}$ = Depreciation for first year
$630 = Book value year 1 or declined value
$\underline{\times .25}$
$157.50 = Depreciation for second year

$630.00 = Book value year 1
$\underline{-157.50}$ = Depreciation year 2
$472.50 = Book value year 2 $\times$ .25 = $118.13

$472.50 = Book value year 2
$\underline{-118.13}$ = Depreciation for third year
$354.37 = Book value year 3

$354.37 = Book value year 3 $\times$ .25 = $88.59
$\underline{-88.59}$ = Depreciation for year 4
$265.78 = Book value year 4 $\times$ .25 = $66.45
$\underline{-66.45}$ = Depreciation for year 5
$199.33 = Book value year 5 $\times$ .25 = $49.83

If $49.83 was subtracted from $199.33, the book value would fall below $150. The depreciation for year six is, therefore, $49.33, the difference between book value year 5 and the salvage value ($199.33 − $150.00 = 49.33). The depreciation amount for years 7 and 8 is zero.

The *depreciation schedule* for the eight years of the machine in example 8 using the declining-balance method would be as follows.

End of Year Number	Depreciation/Year	Accumulated Depreciation at the End of Year	Book Value at the End of Year
1	$210.00	$210.00	$630.00
2	157.50	367.50	472.50
3	118.13	485.63	354.37
4	88.59	574.22	265.78
5	66.45	640.67	199.33
6	49.33	690.00	150.00
7	0	690.00	150.00
8	0	690.00	150.00

Exercise F

Determine the missing values in the exercises below using the declining-balance method. Use a factor rounded off to three decimal places.

Total Cost	Salvage Value	Expected Life (Years)	Year Number	Depreciation	Accumulated Depreciation	Book Value
1. $3,800	$ 500	4 (3,800 × .5 =)	2 (1,900 × .5 =)	$950.00	$2,850.00	$950.00
2. 45,000	0	5	2			
3. 16,900	600	8	3			
4. 10,000	1,000	6	4			
5. 85,400	3,400	10	5			

Modified Accelerated Cost Recovery System (M.A.C.R.S.) Method

To improve the economy, the Economic Recovery Tax Act of 1981 included an incentive for business to invest in new equipment and other capital assets. This incentive was modified in 1986, and is now titled the **Modified Accelerated Cost Recovery System (M.A.C.R.S.)** (also known as the General Depreciation System, GDS). The M.A.C.R.S. method must be used for tax purposes and shortens the expected life of items placed in service after 1986. The salvage value is not considered in this method of computing depreciation. The shorter life and the elimination of scrap value combine to provide a larger depreciation amount, thereby reducing taxes for business. Items in service before 1986 must be depreciated for tax purposes by the A.C.R.S. methods (for years 1981 through 1986) or by the methods explained earlier in the chapter. Furthermore, a firm may not switch to the M.A.C.R.S. method on older items. Depreciable items may be classified by government regulation into eight categories.

Property Categories

Three-year items include over-the-road tractor units and race horses over two years old when placed in service.

Five-year items include autos, taxis, buses, trucks, computers and peripheral equipment, office machines, property used in research and experimentation, breeding cattle, and dairy cattle.

Seven-year items include office furniture and fixtures, and property that has not been designated in another class.

Ten-year items include vessels, barges, tugs, similar water transportation equipment, single-purpose agricultural or horticultural structure, and any tree or vine bearing fruit or nuts.

Fifteen-year property includes improvements made directly to land or added to it; items such as shrubbery, fences, roads, and bridges.

Twenty-year property includes farm buildings other than agricultural or horticultural structures.

Twenty-seven and one-half-year property includes residential real property.

Thirty-one and one-half-year items include nonresidential real property.

The M.A.C.R.S. method stipulates a 200 percent declining-balance method for items in the three-, five-, seven-, and ten-year categories, and a 150 percent declining-balance method for the fifteen- and twenty-year categories. The fifteen- and twenty-year categories use a straight-line method for real property.

The M.A.C.R.S. method also specifies that an item put into service in any of the categories from three to twenty years will use one-half of the normal depreciation for the first year.

Items in the last two categories (real property) are allowed a one-half month deduction for the month the item is put into service.

3-Year, 5-Year, 7-Year, 10-Year, 15-Year, and 20-Year Property (Half-year Convention)

	Depreciation Rate for Recovery Period					
Year	**3-Year**	**5-Year**	**7-Year**	**10-Year**	**15-Year**	**20-Year**
1	33.33%	20.00%	14.29%	10.00%	5.00%	3.750%
2	44.45	32.00	24.49	18.00	9.50	7.219
3	14.81	19.20	17.49	14.40	8.55	6.677
4	7.41	11.52	12.49	11.52	7.70	6.177
5		11.52	8.93	9.22	6.93	5.713
6		5.76	8.92	7.37	6.23	5.285
7			8.93	6.55	5.90	4.888
8			4.46	6.55	5.90	4.522
9				6.56	5.91	4.462
10				6.55	5.90	4.461
11				3.28	5.91	4.462
12					5.90	4.461
13					5.91	4.462
14					5.90	4.461
15					5.91	4.462
16					2.95	4.461
17						4.462
18						4.461
19						4.462
20						4.461
21						2.231

Residential Rental Property (27.5-Year) (Mid-month Convention)

	Use the Column for the Month of Taxable Year Placed in Service											
Year	**1**	**2**	**3**	**4**	**5**	**6**	**7**	**8**	**9**	**10**	**11**	**12**
1	3.485	3.182	2.879	2.576	2.273	1.970	1.667	1.364	1.061	0.758	0.455	0.152%
2–9	3.636	3.636	3.636	3.636	3.636	3.636	3.636	3.636	3.636	3.636	3.636	3.636%

**Nonresidential Real
Property (31.5-Year)
(Mid-month Convention)**

Use the Column for the Month of Taxable Year Placed in Service

Year	1	2	3	4	5	6	7	8	9	10	11	12
1	3.042	2.778	2.513	2.249	1.984	1.720	1.455	1.190	0.926	0.661	0.397	0.132%
2–7	3.175	3.175	3.175	3.175	3.175	3.175	3.175	3.175	3.175	3.175	3.175	3.175%

The depreciation amount is found by multiplying the cost by the percent in the table.
To find the book value, the accumulated depreciation is subtracted from the cost.

Example 9

Allen Bayes noted that the Muskegon Corporation purchased a research machine two years ago for $42,600. The machine will have a scrap value of about $750 at the end of its expected life. The material furnished by the I.R.S. indicates that this asset can be depreciated over a five-year period. What is the depreciation for this year and what is the book value?

$42,600 × .32 (for second year) = $13,632 depreciation for the second year

$42,600 × .20 (for first year) = $8,520

$42,600 Cost
−8,520 Depreciation for first year
−13,632 Depreciation for second year
$20,448 Book value
(Scrap value is not used.)

The *depreciation schedule* for the machine in example 9 for the first six years using the M.A.C.R.S. method would be as follows.

End of Year Number	Depreciation/Year	Accumulated Depreciation at the End of Year	Book Value at the End of Year
1	$ 8,520.00	$ 8,520.00	$34,080.00
2	13,632.00	22,152.00	20,448.00
3	8,179.20	30,331.20	12,268.80
4	4,907.52	35,238.72	7,361.28
5	4,907.52	40,146.24	2,453.76
6	2,453.76	42,600.00	0.00

Exercise G

Determine the missing values in the following exercises using the M.A.C.R.S. method.

	Cost	Category (Years)	Year	Depreciation	Accumulated Depreciation	Book Value
1.	$ 58,000	10	3	$8,352.00	$24,592.00	$33,408.00
2.	8,740	5	1	_____	_____	_____
3.	32,844	31.5	1, 3rd mo.	_____	_____	_____
4.	198,500	15	2	_____	_____	_____
5.	3,900	3	2	_____	_____	_____

Depletion

In the previous examples, the items were assets in a usable state tangible assets, such as machinery, buildings, and equipment. Other assets, such as natural resources, are not usually in a usable state. The conversion of these natural resources, termed *wasting* by the owner, is accounted for by a **depletion** allowance similar to depreciation.

The conversion of assets such as coal, timber, oil, natural gas, and minerals (when pumped, mined, or cut) reduces the balance of the value of the asset. The amount of depletion is determined by estimating the percent of the asset that has been used or "wasted" at cost.

Example 10

An oil field was purchased for $650,000. The field has an estimated 320,000 barrels of oil production possible. If 60,000 barrels were pumped in a year, what would be the depletion for the oil field?

$$\frac{60,000 = \text{Barrels pumped}}{320,000 = \text{Barrels available}} = 18.75\%$$

$$
\begin{array}{ll}
\$650,000 & = \text{Cost of oil field} \\
\underline{\times.1875} & = \text{Percent of depletion} \\
\$121,875 & = \text{Depletion amount}
\end{array}
$$

(60,000 ÷ 320,000 =   × 650,000 =)

Exercise H

Determine the amount of depletion in the following exercises. Round off the factor to the nearest hundredth of a percent.

Cost	Total Available	Wasted This Period	Depletion This Period
1. $450,800 (15,000 ÷ 70,000 =) (× 450,800 =)	70,000 tons	15,000 tons	$96,606.44
2. $1,600,000	80,000,000 cubic feet	90,000 cubic feet	_____
3. 680,000	6,000,000 board feet	800,000 board feet	_____
4. 230,000	50,000 barrels	8,000 barrels	_____
5. 750,000	1,840,000 tons	450,000 tons	_____

Partial Year
Objective 4

A firm must use the M.A.C.R.S. method of depreciation for tax purposes for assets put into service after 1981–1986. A firm may elect to use one of the other methods of depreciation for their *internal* records, as it may be felt that another method more closely fits real values. When the M.A.C.R.S. method is used, the owner may use a full year's depreciation no matter when the asset was purchased during the year. The other systems require that an adjustment be made to the amount of depreciation for the period. Not all assets are purchased at the beginning or end of a fiscal year; this concerns all assets whose asset life overlaps the fiscal year. There is little problem when the straight-line method is used, because the depreciation amount is the same each year. The S.O.Y.D. and declining-balance methods have a changing amount of depreciation per year and must be adjusted accordingly.

When an asset is put into service on or before the fifteenth of the month, the complete month is used to compute the amount of depreciation for the period. When it is put into service after the fifteenth, no depreciation is claimed for that month.

time line   A **time line** is useful for determining the correct month and year in partial-year depreciation problems. It will help you visualize the relationship of the calendar to the fiscal year of the firm, and to the placed-in-service date of the asset. This basic tool can be very beneficial in selecting the correct number of months of the correct year to use in solving the problem. It only takes a few moments to construct the time line and place the facts on it.

Example 11

Time line

○ *The time line can save you time (and will keep the facts straight!).*

Asset life

_____

Fiscal year

Example 12

A desk was purchased on August 1, 2000, for $870. The expected life is five years and the salvage value is $150. The firm has a fiscal year of June 1 to May 31.

What is the depreciation amount for the first two fiscal years ending May 31, 2002, using the straight-line method?

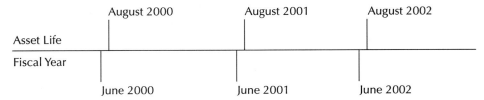

	August 2000	August 2001	August 2002
Asset Life			
Fiscal Year			
	June 2000	June 2001	June 2002

There are ten months in the first fiscal year (from August 2000 to June 2001), and a full year in the second fiscal year (June 2001 to June 2002).

The full-year depreciation for any year is $\dfrac{\$720}{5} = \$144$.

The first fiscal year depreciation would be $\dfrac{10}{12} \times \$144 = \$120$.

The depreciation for the first two fiscal years would total $264 ($144 + $120) using the straight-line method.

What is the depreciation amount for the first two fiscal years ending May 31, 2002, using the S.O.Y.D. method?

Years of Expected Life	Year Number
5 ←—————————	1
4 ←—————————	2
3 ←—————————	3
2	
1	

The S.O.Y.D. method requires that the asset life depreciation amounts be proportioned to the fiscal year of the firm.

	August 2000	August 2001	August 2002
Asset Life	Year #1	Year #2	Year #3
Fiscal Year	Year #1	Year #2	Year #3
	June 2000	June 2001	June 2002

The first fiscal year (ending June 2001) would be computed as shown below.

$$\$720 \times \frac{5}{15} \times \frac{10}{12} = \$200$$

The second fiscal year (ending June 2002) would be computed as shown below.

$$\$720 \times \frac{5}{15} \times \frac{2}{12} = \$40 \text{ plus } \$720 \times \frac{4}{15} \times \frac{10}{12} = \$160$$

The depreciation for the first two full fiscal years would be

$200 + $40 + $160 = $400 using the S.O.Y.D. method

Notice that the partial year of the first year is carried over into the computations of the second year. This process would be continued if the question were to ask for the depreciation for the third year, etc. In that case, the portion of the second year would be carried over into the computations of the third year.

Example 12
continued

What is the depreciation amount for the first two fiscal years ending May 31, 2002, using the declining-balance method?

The declining-balance method requires that the ten months of the first fiscal year be used to adjust the balance for the next year. After the first partial year, the succeeding years are computed as full years (without any adjustment for ten months/two months, etc.).

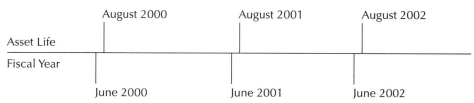

$870 total cost $\times \frac{2}{5}$ = $348 (full-year depreciation)
$348 $\times \frac{10}{12}$ = $290 for the fiscal year ending June 2001

The depreciation for the second year would be computed using the declined-balance and a full-year factor of $\frac{2}{5}$.

$870 − $290 = $580 (book value June 2001)
$580 $\times \frac{2}{5}$ = $232 for the year ending June 2002

The depreciation for the first two full fiscal years would be $290 + $232 = $522 using the declining-balance method.

Another example of the S.O.Y.D. and declining-balance methods should clear up any questions that might remain.

Example 13

A sign was purchased on July 10, 1999, for $7,100. The expected life is ten years, with a $1,200 salvage value. What is the depreciation for the fiscal year ending March 31, 2002, if the firm uses the S.O.Y.D. method?

	July 1999		July 2000	July 2001	July 2002
Asset Life		Year #1	Year #2	Year #3	Year #4
Fiscal Year					
		April 2000	April 2001	April 2002	

Years of Expected Life	*Year Number*	
10	1	
9 ←	2	
8 ←	3	
7	4	
6	5	
5	6	$7,100 Total cost
4	7	−1,200 Salvage value
3	8	$5,900 Total depreciation
2	9	
+1	10	
55		

$5,900 $\times \dfrac{9}{55} \times \dfrac{3}{12}$ = $241.36 (from April 2001 to July 2001)

$5,900 $\times \dfrac{8}{55} \times \dfrac{9}{12}$ = $643.64 (from July 2001 to April 2002)

$885.00 depreciation for the fiscal year ending March 31, 2002

Example 14 If the firm in example 13 had used the declining-balance method, what would be the amount of depreciation for the fiscal year ending March 31, 2002?

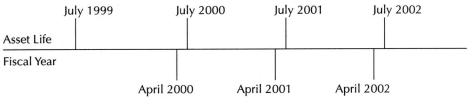

$7,100 total cost $\times \dfrac{2}{10} \times \dfrac{9}{12} = $1,065$ for the year ending March 31, 2000

$7,100 - $1,065 = $6,035$ book value April 2000

$6,035 \times \dfrac{2}{10} = $1,207$ $6,035 - $1,207 = $4,828$ book value April 2001

$4,828 \times \dfrac{2}{10} = 965.60 depreciation for the year ending March 31, 2002

Exercise I

Determine the depreciation amount in the exercises below. Use the depreciation method indicated.

	Total Cost	Expected Life (Years)	Salvage Value	Purchase Date	Fiscal Year Ended	Depreciation Method	Depreciation Amount
1.	$650	4	$50	August 15, 1999	January 15, 2001	S.O.Y.D. $\left(600 \times \dfrac{4}{10} \times \dfrac{7}{12} =\right)$ $\left(600 \times \dfrac{3}{10} \times \dfrac{5}{12} =\right)$	$215.00
2.	1,300	8	100	June 1, 1999	January 1, 2002	Declining-Balance	
3.	900	4	300	September 9, 1999	January 1, 2000	S.O.Y.D.	
4.	700	5	100	May 1, 1997	August 1, 2001	S.O.Y.D.	
5.	630	4	0	June 1, 2000	March 1, 2002	Declining-Balance	
6.	2,800	5	300	September 23, 2000	January 1, 2001	Straight-Line	
7.	750	10	50	July 3, 2000	March 1, 2002	S.O.Y.D.	
8.	3,260	5	250	January 12, 2002	July 1, 2002	Declining-Balance	
9.	4,500	4	300	August 3, 2000	June 1, 2002	Declining-Balance	
10.	700	8	50	July 8, 1998	January 1, 2002	Straight-Line	

ASSIGNMENT Chapter 17

Name _____ **Date** _____ **Score** _____

Skill Problems

Determine the missing values in the exercises below using the straight-line method and the following formula:

Depreciation/year = Total depreciation/Expected life

	Total Cost	Expected Life (yr)	Salvage Value	Year No.		
1.	$80,000	20	$10,000	4	_____	book value
2.	4,000	_____	800	6	1,920	accumulated depreciation
3.	_____	7	500	4	2,070	depreciation per year

Determine the missing values in the exercises below using the straight-line method.

Item Cost	Trans. Cost	Install. Cost	Salvage Value	Expected Life (yr)	Year No.	Deprec. per Yr	Accum. Deprec.	Book Value
$15,600	$95	$7,050	$3,400	10	8	**4.** _____	**5.** _____	**6.** _____
20,050	0	1,640	0	15	12	**7.** _____	**8.** _____	**9.** _____
750	48	150	150	17	8	**10.** _____	**11.** _____	**12.** _____
15,700	0	0	2,000	8	6	**13.** _____	**14.** _____	**15.** _____
894	105	355	300	4	3	**16.** _____	**17.** _____	**18.** _____

Determine the missing values in the exercises below using the units-of-production method. Use a four-decimal-place factor.

	Total Cost	Salvage Value	Expected Life (in units)	Units This Period	Depreciation This Period
19.	$ 1,950	$ 50	1,400	60	_____
20.	1,650	350	90,000	4,000	_____
21.	5,300	300	19,500	900	_____
22.	3,950	200	80,000	650	_____
23.	41,600	0	156,000	3,500	_____

Total Cost	Salvage Value	Expected Life (in units)	Units This Period	Units to Date	Deprec. This Period	Book Value
$45,000	2,500	150,000	2,870	45,972	**24.** _____	**25.** _____
1,000	0	25,500	840	22,451	**26.** _____	**27.** _____
15,540	2,500	200,000	10,500	190,000	**28.** _____	**29.** _____

Determine the sum-of-years fraction to be used in the following exercises.

Years of Expected Life	Sum of Years		Year Number	Fraction to be Used
16	**30.** _____		4	**31.** _____
15	**32.** _____		8	**33.** _____
42	**34.** _____		21	**35.** _____
12	**36.** _____		7	**37.** _____

Determine the missing values in the exercises below using the S.O.Y.D. method.

Total Cost	Salvage Value	Expected Life (yr)	Year No.	Deprec.	Accum. Deprec.	Book Value
$57,400	$ 2,000	12	8	**38.** _____	**39.** _____	**40.** _____
20,843	1,500	25	6	**41.** _____	**42.** _____	**43.** _____
164,800	20,000	20	4	**44.** _____	**45.** _____	**46.** _____
8,400	0	7	5	**47.** _____	**48.** _____	**49.** _____

Determine the missing values in the exercises below using the declining-balance method. Use factor rounded off to three decimal places.

Total Cost	Salvage Value	Expected Life (yr)	Year No.	Deprec.	Accum. Deprec.	Book Value
$ 4,800	$200	5	2	**50.** _____	**51.** _____	**52.** _____
9,870	500	8	3	**53.** _____	**54.** _____	**55.** _____
8,425	0	7	2	**56.** _____	**57.** _____	**58.** _____
95,200	4,400	10	4	**59.** _____	**60.** _____	**61.** _____

Complete the problems below. Use the method indicated. Round off declining-balance factor to the nearest tenth of a percent.

Total Cost	Salvage Value	Expected Life (Years)	Year Number	Depreciation	Accumulated Depreciation	Book Value	Method
$7,000	$500	20	8	62. _____	63. _____	64. _____	Straight-Line
8,600	600	8	4	65. _____	66. _____	67. _____	S.O.Y.D.
35,700	1,000	15	3	68. _____	69. _____	70. _____	Declining-Balance
1,250	50	20	3	71. _____	72. _____	73. _____	Declining-Balance
9,800	500	15	2	74. _____	75. _____	76. _____	S.O.Y.D.
950	150	10	3	77. _____	78. _____	79. _____	Straight-Line
5,600	0	10	4	80. _____	81. _____	82. _____	Declining-Balance
695	95	12	7	83. _____	84. _____	85. _____	Straight-Line
1,350	350	5	2	86. _____	87. _____	88. _____	S.O.Y.D.
856	56	15	3	89. _____	90. _____	91. _____	S.O.Y.D.

Complete the following depreciation schedule for a 4-year-old race horse purchased for $25,000. Use the M.A.C.R.S. method and charts provided in your chapter.

End of Year No.	Deprec. Rate (% from chart)	Deprec. per Year	Accum. Deprec. at the End of Year	Book Value at the End of Year
1	92. _____	93. _____	94. _____	95. _____
2	96. _____	97. _____	98. _____	99. _____
3	100. _____	101. _____	102. _____	103. _____

Determine the missing values in the following exercises using the M.A.C.R.S. method.

Cost	Category (years)	Year	Deprec.	Accum. Deprec.	Book Value
$ 80,600	20	2	104. _____	105. _____	106. _____
9,462	7	3	107. _____	108. _____	109. _____
174,600	31.5	1, 4 mo	110. _____	111. _____	112. _____
1,348	15	1	113. _____	114. _____	115. _____
9,000	3	2	116. _____	117. _____	118. _____
15,900	5	3	119. _____	120. _____	121. _____
85,000	10	1	122. _____	123. _____	124. _____

Determine the amount of depletion in the following exercises.

	Cost	Total Available	Wasted This Period	Depletion This Period
125.	$640,600	100,000 tons	14,000 tons	_____
126.	2,460,000	8,700 cubic ft	1,940 cu. ft	_____
127.	245,400	40,000 barrels	10,500 barrels	_____

Determine the partial year depreciation amount as indicated in the exercises below. Use the depreciation method indicated. Use a three-decimal-place factor.

	Total Cost	Expected Life (yr)	Salvage Value	Purchase Date	Fiscal Year Ended	Deprec. Method	Deprec. Amount
128.	$2,000	5 years	$250	5/1/01	7/1/03	S.O.Y.D.	_____
129.	1,680	4	300	9/23/00	3/1/02	Declining-Balance	_____
130.	840	8	100	8/4/99	1/31/02	S.O.Y.D.	_____
131.	2,500	4	0	9/9/99	6/30/01	Declining-Balance	_____
132.	900	5	50	5/23/00	12/31/02	Straight-Line	_____
133.	6,405	10	600	7/8/02	9/30/03	Declining-Balance	_____
134.	3,460	3	200	5/1/01	12/31/01	S.O.Y.D.	_____
135.	680	15	50	4/5/02	7/1/02	Straight-Line	_____
136.	2,500	6	100	1/31/03	6/30/04	Declining-Balance	_____

Business Application Problems

1. A machine purchased five years ago is fully depreciated and will be sold for the salvage value of $500. The original cost was $6,500. What was the depreciation per year using the straight-line method?

2. In problem 1, what would have been the depreciation for the second year using the S.O.Y.D. method?

3. The electronic component of an installation has an expected life of 5,000 hours. The component, installed, cost $8,000 and has a salvage value of $200. The component was used 800 hours in the most recent period. What is the amount of depreciation for the period?

4. What is the depreciation cost per hour in problem 3?

5. Kendron Corporation bought a wall cleaning unit for $6,800. The unit will be used for eight years and then salvaged for $400. The straight-line method will be used to determine depreciation. What is the depreciation per year?

6. In problem 5, what would be the book value of the cleaning unit at the end of three years?

7. If the Kendron Corporation in problem 5 would have used the declining-balance method to determine depreciation, what would have been the amount of depreciation for the second year?

8. A delivery trailer was purchased for $6,200. The trailer will be used for a period of ten years. At the end of the ten-year period it will be sold for $1,000. The firm will use the straight-line method to determine depreciation. What will be the accumulated depreciation amount at the end of the first three years?

9. Verton Academy bought a 15-passenger van for $26,000. They will depreciate the van using the M.A.C.R.S. system as a five-year item. The salvage value will be $850. What is the depreciation amount for the third year?

10. In problem 9, what is the accumulated depreciation at the end of the third year?

Blecher and Associates purchased a building for $139,000 for office and storage space. The building is expected to have a useful life of 20 years. The salvage value at the end of that period is estimated to be $20,000. The owner of the firm has asked that a comparison be made at the end of the tenth year using three methods of depreciation. What is the depreciation for the tenth year?

11. Straight-line method: _____

12. S.O.Y.D. method: _____

13. Declining-balance method: _____

14. Demmer Manufacturing uses a shear to cut thick metal pieces. The shear cost $129,000 when it was new 15 years ago. The shear is expected to have a useful life of 35,000 hours. It was used 1,500 hours this year. If the scrap value is expected to be $5,000, what is the depreciation amount for this year?

15. A heating system was purchased by the Fossum Import Company of New York. The installed system cost $89,000 when it was purchased 12 years ago. What is the difference in the amount of depreciation for this year if the system will have a scrap value of only $3,000 at the end of 15 years? Compare the straight-line and S.O.Y.D. methods.

16. If an item is depreciated at the rate of 5 percent per year using the straight-line method, what was the original cost if the depreciation amount per year is $265 and the salvage value is $800?

17. The Felder Co. purchased an office building on March 1, 1992, for $60,000. The building is expected to last for 20 years. On March 1, 2003, a real estate appraiser stated the building's value was probably $46,000. Mr. Felder, company president, has asked that the appraisal be compared to the book value. Which method of depreciation would come closest to the appraisal?

18. An asset was purchased two years ago for $4,000. The expected life is five years with an expected salvage value of $500. The firm will depreciate the asset using the sum-of-the-years'-digits method. What is the book value of the asset?

19. The Morrison Company has a dispute with the I.R.S. over the classification of a piece of property. Morrison feels that the item should be categorized (M.A.C.R.S.) as a five-year item, and the I.R.S. indicates that it should be a ten-year item. The dispute has been going on for two full years. The item cost $8,900 and will scrap out at $1,000 at the end of the expected life. What is the difference in the book values using the five- and ten-year categories?

20. The Forsythe Interdynamics Corporation purchased a specialized computer for $5,800 two years ago. The computer is in the five-year M.A.C.R.S. category. The firm is now interested in trading in the computer for a new unit that costs $3,790. The firm will receive a $2,000 allowance for the trade-in. What is the difference in the trade-in value and the book value of the computer?

21. Dr. Monroe purchased a new light-signaling system for the series of patient exam rooms in her new office building. The system will cost $14,780. The expected life of the system is ten years. The system will be junked at the end of its expected life. The doctor is unsure of which depreciation system to use. At the end of $3\frac{1}{2}$ years, what is the difference in the book value between the straight-line and S.O.Y.D. methods?

22. Write a complete sentence in answer to this question. The depreciation expense for the year was $900 on a piece of equipment that has now been used for the second full year. The item is to be depreciated over a six-year period and has a salvage value of $3,000. The item is depreciated using the S.O.Y.D. method. What was the original price of the item?

23. Write a complete sentence in answer to this question. If the item in problem 22 were depreciated over a five-year period using the straight-line method, what would be the accumulated depreciation at the end of the second year?

★ **Challenge Problem**

The Bayer Corp. purchased a hydraulic press for $76,587, F.O.B. Cincinnati. Bayer Corp. is located in Wattona, South Carolina. The transportation charges were $942. The press was purchased on terms of 2/10,n/30 on July 16, 2000. It was delivered on July 23, and the invoice was paid on July 24. Trade discounts of 10 percent and 5 percent were granted by the selling firm, Gray Iron, Inc. Installation costs were $15,321. The salvage value is estimated to be $5,000 at the end of its expected life of 25 years.

Bayer uses the declining-balance method to determine the depreciation expense. The first full (fiscal) year ended on April 15, 2002.

You are asked to determine the book value of the press on April 15, 2003.

Name _____ **Date** _____ **Score** _____

Select the best term for the explanations.

_____ **1.** Depletion

_____ **2.** S.O.Y.D. Method

_____ **3.** Straight-Line Method

_____ **4.** Declining-Balance Method

_____ **5.** Book Value

_____ **6.** Salvage Value

_____ **7.** M.A.C.R.S. Method

_____ **8.** Accumulated Depreciation

A. An equal amount of depreciation per year.

B. Made up of all costs incurred to make the item operable.

C. All properties are categorized to set the percent.

D. Depreciation per year × the number of years to date (straight-line).

E. Total cost less total depreciation: the residual value.

F. Allowance for a used portion of natural resource.

G. A method using $\frac{2}{15}$ as a fraction to multiply by salvage value.

H. A method considering the amount of operations performed. The fraction is multiplied by total cost to find book value.

I. A method using total cost or book value multiplied by a rate to accelerate the depreciation amount in the early years..

J. A method that totals the years of expected life. This number becomes the denominator in a fraction.

K. Total cost less the depreciation to date.

Calculate the missing values in each of the following problems. Use the depreciation method indicated.

	Total Cost	Salvage Value	Expected Life (years)	Year Number	Depreciation Amount	Accumulated Depreciation	Book Value	Method
9.	$21,000	$1,000	10	4	_____	_____	_____	S.O.Y.D
10.	35,900	1,900	8	2	_____	_____	_____	Declining-Balance
11.	12,600	50	20	17	_____	_____	_____	Straight-Line

	Total Cost	Salvage Value	Expected Life (Units)	Units this Period	Depreciation Amount	Method
12.	$42,000	$2,000	100,000	20,000	_____	Units-of-Production

13. Market Services Company purchased a new copier one year ago. The machine cost $5,800 and will be able to command a trade-in value of $500 when sold. The machine is categorized for a five-year life. What is the book value at the end of the first year using the M.A.C.R.S. method?

14. Hoffman Associates purchased a piece of land for $150,000 with an estimated 50,000 barrels of oil that could be produced from it. The firm pumped 2,000 barrels of oil in the third year of ownership. What is the depletion allowance for the third year?

15. A building was purchased ten years ago for $260,000. The expected life was established at thirty years. The firm decided to use a straight-line method to depreciate the property. The salvage value of the property will be $30,000. What is the book value of the property?

16. Flash Movers, Inc. purchased an airplane for $74,000 on October 17, 2000. The plane will have a salvage value of $3,000. The firm has decided to depreciate the property over a period of twenty years using the declining-balance method. Gwen Kimberlin, company accountant, must determine the amount of depreciation for the fiscal year ended March 31, 2003. What is the amount of depreciation for the period?

CALCULATOR SELECTION AND OPERATION

Thirty years ago most students and business people solved business math problems by using the skills they had learned and by putting their problems on paper. The only other option was an expensive, large, semiportable desk-model calculator. Today, the calculator, a small, highly portable battery-operated device, usually priced under $5, is available. This offers the student and business person a much better means to solve business math problems. Through the advances of technology, the student is able to take a large amount of the long and tedious efforts out of the "arithmetic" aspect of business math. More effort can be placed on understanding the "why" of business math.

The calculator has earned a place in the business world and is being used more in the classroom. Some cautions should be stated, however. Calculators should not be used for problems in the first three chapters of this book. The skills covered in these chapters are necessary for an understanding of the operation of the calculator and for working problems when a calculator is not available.

Selection

When selecting a calculator for business math, the student should look for a model that has (1) the four mathematical functions, (2) a floating decimal point, and (3) an eight-place readout. Other features that may interest the student are memory functions, especially the accumulating memory. This device can be purchased for $5 or less.

Operation

The fundamental operations on the calculator are performed just as you would state the problem. Press the keys on the calculator as you say the problem.

3 plus 9 equals 12

84 divided by 2 equals 42

With the floating decimal feature, the decimal point is automatically placed by the machine.

42.07 times .36 equals 15.1452

54.06 minus 18.561 equals 35.499

Fractions can be added and subtracted by finding the decimal equivalent of each fraction and then performing the addition or subtraction.

$\frac{3}{4}$ minus $\frac{5}{8}$ equals

| 3 | ÷ | 4 | = | **.75** |

| 5 | ÷ | 8 | = | **.625** |

| . | 7 | 5 | − | . | 6 | 2 | 5 | = | **.125** |

Fractions in problems with whole numbers can be done by using the following example.

$$147 \times \frac{3}{8} = 147 \times 3 = 441 \div 8 = 55.125$$

| 1 | 4 | 7 | × | 3 | = | **441** | ÷ | 8 | = | **55.125** |

Multiple-function problems are also performed the way they are read.

$$\frac{\left(3.18 \times \frac{1}{4}\right) + 27}{1.6} =$$

| 3 | . | 1 | 8 | × | 1 | = | **3.18** | ÷ | 4 | = | **.795** | + | 2 | 7 | = |

| **27.795** | ÷ | 1 | . | 6 | = | **17.371875** |

Memory

A memory feature can allow the operator to do a part of a problem, store the answer, and then retrieve the stored answer later. This is helpful when solved information is needed in conjunction with another part of the problem. The memory portion of the calculator usually has four keys: an M+ key for adding a value to memory, an M− key for subtracting a value from memory, an RM key for memory recall, and a CM key for clearing the memory. The following problems will show the advantages of the memory feature.

$$\frac{47 \times 16.5}{8.3 + 9.6} =$$

| 8 | . | 3 | + | 9 | . | 6 | = | **17.9** | M+ |

| 4 | 7 | × | 1 | 6 | . | 5 | = | **775.5** | ÷ | MR | = | **43.324022** |

The memory function saves writing down partial answers, errors in copying, and time!

$$(75.67 - 8.06) - (85 \times .1687)$$

| 8 | 5 | × | . | 1 | 6 | 8 | 7 | = | **14.3395** | M+ |

| 7 | 5 | . | 6 | 7 | − | 8 | . | 0 | 6 | = | **67.61** | − | RM | = | **53.2705** |

The operation of calculators differs slightly from one to another. You should consult the instruction pamphlet accompanying your unit or the instructions on the back panel of the unit.

Use the problems above to check the operation of your own calculator.

MATH SHORTCUTS

Interest-Time Shortcuts

When the interest rate and the time factor are considered together, a shortcut method for computing interest can be used. This method is often termed the 6 percent/60-day or 4 percent/90-day method. It is especially useful when a calculator is not available.

If a promissory note is negotiated for $1,868 at 6 percent for 60 days, the rate and time factors $(.06 \times \frac{60}{360})$ can be reduced to .01 $\left(\text{i.e., } \frac{.06}{1} \times \frac{60}{360} = .01\right)$. This factor multiplied by the principal will yield the interest amount. In this problem, the interest would be $18.68. By moving the decimal point in the principal amount two places to the left, we have determined the interest amount.

The same approach is used with a problem involving 4 percent and 90 days.

$$I = P \times R \times T$$

$$I = \$16,520 \times .04 \times \frac{90}{360} = \$165.20 \text{ interest}$$

Not all promissory notes are at the stated rates or time periods as those in the previous problems. The shortcut can still be used with some adjustments.

Problem: $I = P \times R \times T$

$$I = \$1,568 \times .06 \times \frac{90}{360}$$

$\left(.06 \times \frac{60}{360}\right) = .01$ 60 days = $15.68 interest
30 days = <u> 7.84</u> interest one-half the time period =
one-half the interest
90 days = $23.52 total interest

The same type of thinking can be applied to similar problems.

Problem: $I = P \times R \times T$

$$I = \$450 \times .08 \times \frac{90}{360}$$

$\left(.04 \times \frac{90}{360}\right) = .01$

    (4%/90 days) = $4.50 —interest
<u>+ (4%/90 days) =   4.50 —interest</u>
Interest   (8%/90 days) = $9.00

The application can become even more interesting, as well as challenging to the problem solver.

Problem: I = P × R × T

$$I = \$4,840 \times .075 \times \frac{90}{360}$$

(6%/60 days) interest	$48.40
(6%/30 days) interest	+24.20
(6%/90 days) interest	$72.60

$$\left(1\frac{1}{2}\%/90 \text{ days} = \frac{1}{4} \text{ of } 6\%/90 \text{ days}\right) \quad +18.15$$

Interest $\left(7\frac{1}{2}\%/90 \text{ days}\right)$ $90.75

Problem: I = P × R × T

$$I = \$1,570 \times .08 \times \frac{135}{360}$$

(4%/90 days) interest	$15.70
(4%/90 days) interest	+15.70
(8%/90 days) interest	$31.40

$$\left(8\%/45 \text{ days} = \frac{1}{2} \text{ of } 8\%/90 \text{ days}\right) \quad +15.70$$

Interest   (8%/135 days) $47.10

Aliquot Parts

Another shortcut method used in business problems is termed *aliquot parts*. This method makes use of the fractional parts of 100. Knowledge of fractions and their decimal equivalents is very useful here. The table below may serve as a refresher.

Common Fraction	Decimal Equivalent	Common Fraction	Decimal Equivalent
$\frac{1}{25}$	.04	$\frac{1}{6}$	.166
$\frac{1}{20}$	.05	$\frac{1}{5}$	.2
$\frac{1}{12}$	.0833	$\frac{1}{4}$	.25
$\frac{1}{11}$	.091	$\frac{1}{2}$	.5
$\frac{1}{10}$	.1	$\frac{5}{6}$	.833
$\frac{1}{9}$	.111	$\frac{3}{8}$	.375
$\frac{1}{8}$	.125	$\frac{5}{8}$	.625
$\frac{1}{7}$	.14285	$\frac{7}{8}$	.875

The clear advantage of aliquot parts becomes evident when the student can visualize the shortcut's use and then do the problem in his or her head.

Problem: What is the total price of 680 units of a product to be sold at 25¢ each?

25¢ is equal to $\frac{1}{4}$ of $1.00; $\frac{1}{4}$ of 680 = 170 or $170

Problem: What is the cost per unit of 750 units produced at a total cost of $1,860?

750 units equals $\frac{3}{4}$ of 1,000; $\frac{3}{4}$ of $1,860 = $\frac{\$1,395}{1,000}$ or $1.40 each

Problem: A product is priced at $28. Twenty percent of the product price is packaging cost. What is the packaging cost?

20% = $\frac{1}{5}$ of $28 = $5.60

MATH OF COMPUTERS

C

Throughout this text, we have been using a numbering system termed the decimal system. This system has a base of 10. It uses the digits 0 through 9. Once ten units have been counted, the decimal system records that ten by placing a one in the column to the left. Each place value counts ten units of the position to the right. This system works well for most quantitative problem-solving methods, but not for the computer. The computer operates using only two digits, 0 and 1, because it has only "on-off" or "yes-no" capability. Hence, there is the need for only two digits. This number system is termed the binary system because of the use of two (bi) digits.

Below is a conversion table of decimal numbers to binary numbers.

Decimal Numbers (Base 10)	Binary Numbers (Base 2)	
0	0	
1	1	(2^0 power = 1)
2	10	(2^1 power = 2)
3	11	(2^1 power + 2^0 power = 3)
4	100	(2^2 power = 4)
5	101	(2^2 power + 2^0 power = 5)
6	110	
7	111	
8	1000	
9	1001	
10	1010	
11	1011	
12	1100	
13	1101	
14	1110	
15	1111	
16	10000	
17	10001	
18	10010	
19	10011	
20	10100	

Notice that each one is a power of two. Each position to the left increases the power by one.

Converting binary to decimal

Problem: 100000 = 2^5 power = 32

Problem: 10100 = 2^4 power (16) + 2^2 power (4) = 20

Problem: 1011000 = 2^6 power (64) + 2^4 power (16) + 2^3 power (8) = 88

Converting decimal to binary numbers

Problem: 23 = 10111 binary; that is, 2^4 power (16) + 2^2 power (4) + 2^1 power (2) + 2^0 power (1)

Problem: 17 = 10001 binary; that is, 2^4 power (16) + 2^0 power (1)

Problem: 67 = 1000011 binary; that is, 2^6 power (64) + 2^1 power (2) + 2^0 power (1)

Decimal numbers can be converted to binary numbers by dividing the decimal number by two until the dividend is zero. The remainder, placed in reverse order, is the binary number.

Problem: Convert 259 decimal into a binary number.

```
2)259              Remainder
2)129                 1
2)64                  1
2)32                  0
2)16                  0
2)8                   0
2)4                   0
2)2                   0
2)1                   0
   0                  1
```

259 decimal equals 100000011 binary.

Problem: Convert 123 decimal into binary.

```
2)123              Remainder
2)61                  1
2)30                  1
2)15                  0
2)7                   1
2)3                   1
2)1                   1
   0                  1
```

123 decimal equals 1111011 binary.

EMPLOYMENT MATH TEST

Business firms often use a mathematics test as a screening device in personnel selection. The skill of mathematics as well as logic ability is measured in a mathematics test. Some of the types of problems used on employment tests are shown below.

Arithmetic Progression

This type of problem tests logic more than it tests mathematical skills. The applicant is tested on his or her ability to see an arithmetic pattern and extend that pattern to the blank space.

Problem: 2, 4, 8, 16, 32, _____
The pattern in this problem is to double the previous number. The blank would be filled with a 64.

Problem: 5, 10, 50, 100, 500, 1,000, _____
The pattern in this problem is to multiply the first value by 2, the next value by 5, and then repeat this pattern. The blank should be filled with 5,000.

Problem: 60, 30, 15, 20, 10, 5, 10, _____
The pattern is ÷ by 2, ÷ by 2, add 5. The blank should be filled with a 5 (÷ by 2).

Problems:

1. 72, 70, 66, 64, 60, 58, 54, 52, _____

2. 80, 70, 61, 53, 46, 40, _____

3. 360, 180, 60, 30, _____

4. 4,000, 800, 200, 40, 10, _____

5. 6,720, 840, 120, 20, 4, _____

6. 68, 61, 63, 56, 58, 51, 53, _____

7. 600, 300, 100, 25, _____

8. 72, 76, 73, 75, 79, 76, 78, 82, 79, _____

9. 8, 40, 10, 30, 15, _____

10. 3, 9, 27, 108, 432, _____

Find the answers below.

1.	48	**6.**	46
2.	35	**7.**	5
3.	10	**8.**	81
4.	2	**9.**	15
5.	1	**10.**	1,296

A Sample Employment Math Test

An employer may ask you to take a test to determine your math skills. This is legal if math will be part of your actual job performance (bona fide occupational skill). The following problems are a sampling of the type you may encounter on such a test. Test your skill. (See your instructor for answers to these problems.)

1. Add $\frac{1}{7}, \frac{1}{8}, \frac{1}{21}, \frac{1}{14}$

   (a) $\frac{57}{83}$ (b) $\frac{65}{168}$ (c) $\frac{15}{32}$ (d) $1\frac{13}{97}$

2. A photocopy machine can run 390 copies in one-half hour. How many copies can be run in $3\frac{1}{4}$ hours?

   (a) 2,535   (b) 1,268   (c) 1,860   (d) 2,350

3. To drive from Billerston to Dillworth takes $15\frac{1}{2}$ hours. The same trip by bus takes $27\frac{3}{4}$ hours. How much time is saved by driving?

   (a) $3\frac{3}{4}$ (b) $43\frac{1}{4}$ (c) $12\frac{3}{4}$ (d) $12\frac{1}{4}$

4. Sarah plans to save 10 percent of her salary each week. When she has saved enough she will buy a bond for $375. Her salary is $130 weekly. How many weeks will she need to work before she can buy her first bond?

   (a) 27   (b) 4,875   (c) 26   (d) 29

5. A women's apparel shop has 4,368 units in stock. Half of the stock is slacks, $\frac{1}{8}$ is sweaters, $\frac{1}{16}$ is belts and scarves. How many units are not included above?

   (a) 3,003   (b) 1,365   (c) 1,638   (d) 2,965

6. The receiving department uses 43 inspection forms each day. How many days will a six-gross inventory last?

   (a) 20   (b) 18   (c) 0   (d) 36

7. $\frac{1}{2}$ of $\frac{3}{8}$ of a stock of pens have been used in the past ten days. The total inventory was 862. What is the daily usage rate?

   (a) 162   (b) 32   (c) 46   (d) 16

8. $3\frac{5}{8} \times 17\frac{1}{4} =$

   (a) $62\frac{17}{32}$ (b) $13\frac{5}{8}$ (c) $50\frac{11}{32}$ (d) $73\frac{7}{16}$

9. 280 of the employees are in the painting department. They represent 40 percent of the total labor force. How many are in the labor force?

   (a) 112   (b) 11,200   (c) 700   (d) 168

10. 140 percent of x is 1,344. What is x?

    (a) 1,882   (b) 538   (c) 188   (d) 960

11. The Wilkum Company bought $630 of merchandise. They were able to deduct a 10 percent discount. What was the net amount due?

    (a) $693   (b) $63   (c) $567   (d) $630

12. A train averages 63 miles per hour. The departure time was 9:20 P.M. How far had it gone by midnight?

    (a) 168 miles   (b) 924 miles   (c) 147 miles   (d) 146 miles

13. A merchant purchased a lamp at $17 and marked it up by 30 percent over the cost. What was the amount of the markup?

    (a) $11.90   (b) $22.10   (c) $5.10   (d) $10.20

14. An employee, Jack Terril, earns a commission of $3\frac{1}{4}$ percent on sales. His sales were $43,680. What was the commission?

    (a) $1,419.60   (b) $1,310.40   (c) $1,683   (d) $15,387.68

15. If the interest on a 30-day promissory note for $500 is $6.67, what was the interest rate using ordinary interest?

    (a) 10.3%   (b) 5.8%   (c) 16%   (d) 12%

16. Bill earned a commission of $463 this week. This was a 30 percent increase over last week. How much did Bill earn last week?

    (a) $601.90   (b) $356.15   (c) $324.10   (d) $138.90

17. David word-processed 14 more letters today than did Karen. This represented a 20 percent increase over Karen's output. What was Karen's output?

    (a) 56   (b) 70   (c) 84   (d) 280

18. A farmer plowed a field at the rate of $2\frac{1}{4}$ acres per hour. How many acres will he plow in $3\frac{1}{2}$ hours?

(a) $5\frac{3}{4}$ (b) $7\frac{7}{8}$ (c) $8\frac{3}{4}$ (d) $9\frac{1}{8}$

19. Of the 465 students enrolled in a school, 93 were ill with the flu. What percent were ill?
(a) 20%   (b) 93%   (c) 50%   (d) 30%

20. If a 20 percent discount on an invoice saved a firm $79, what was the original amount of the invoice?
(a) $158   (b) $15.80   (c) $395   (d) $98.75

21. One hundred sixteen people registered for a workshop on energy conservation. One-fourth were late registrants. The early registrants showed a two-to-one male-to-female relationship. How many of the early registrants were males?
(a) 15   (b) 47   (c) 58   (d) 37

22. An $8.95 gallon of paint will cover 500 square feet. How much will it cost to cover the side walls of a room that measures 15 feet by 10 feet by 7 feet 6 inches high?
(a) $17.90   (b) $8.95   (c) $26.85   (d) $89.50

23. Multiply and round off to the nearest tenth: 8.63×5.0473.
(a) 43.6   (b) 43.5   (c) 13.7   (d) 37.8

24. A pool measuring 20 feet by 40 feet by 5 feet deep must be filled with water. How many hours will be required if it will fill at the rate of 900 gallons an hour? There are $7\frac{1}{2}$ gallons in a cubic foot.
(a) 33   (b) 4   (c) 23   (d) 31

25. Jim processes 40 reports a day, which is 25 percent more than Tom's output. How many does Tom process?
(a) 65   (b) 50   (c) 32   (d) 30

26. Mark had $3,957 in his savings account. He bought a new motorcycle for $1,680. What percent of his savings was left after the purchase?
(a) 58%   (b) 47%   (c) 42%   (d) 43%

27. If wheat sold for $3.60 a year ago, and it sells for $4.25 today, what is the percent of increase?
(a) 118%   (b) 18%   (c) 85%   (d) 15%

Complete the following using arithmetic progression.

28. 60, 30, 15, 20, 10, 5, 10, _____
(a) 10   (b) 0   (c) 5   (d) 15

29. 5, 10, 30, 120, _____
(a) 600   (b) 240   (c) 60   (d) 480

30. 72, 76, 73, 75, 79, 76, 78, 82, 79, _____
(a) 76   (b) 83   (c) 81   (d) 77

READING MARKET QUOTATIONS

The price of stocks and bonds are printed in the business section of most newspapers. The price, as well as its activity in the market, are of interest to investors.

A sample of a stock exchange report found in a newspaper is shown below.

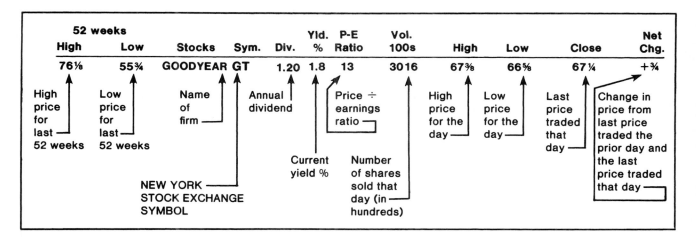

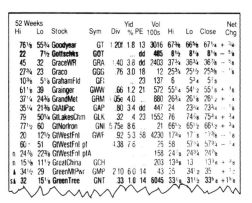

Bond prices are quoted in a similar fashion in the newspaper each business day. A

bond   sample of a **bond** exchange report is shown below.

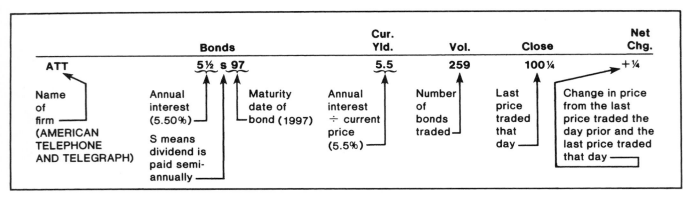

	Bonds	Cur. Yld.	Vol.	Close	Net Chg.
ATT	5½ s 97	5.5	259	100¼	+¼

Name of firm (AMERICAN TELEPHONE AND TELEGRAPH)

Annual interest (5.50%)

S means dividend is paid semi-annually

Maturity date of bond (1997)

Annual interest ÷ current price (5.5%)

Number of bonds traded

Last price traded that day

Change in price from the last price traded the day prior and the last price traded that day

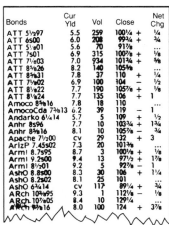

Bonds	Cur Yld	Vol	Close		Net Chg
ATT 5½97	5.5	259	100¼	+	¼
ATT 6s00	6.0	208	99¾	+	¾
ATT 5⅛01	5.6	70	91⅞		...
ATT 7s01	6.9	315	100⅞	+	⅛
ATT 7⅛03	7.0	934	101¾	+	⅝
ATT 8⅜26	8.2	140	105⅝		...
ATT 8⅝31	7.8	37	110	+	¼
ATT 7⅛02	6.9	100	104	−	½
ATT 8⅛22	7.7	190	105⅞	+	⅛
ATT 8⅛24	7.7	135	106	+	1
Amoco 8⅝16	7.8	18	110		...
AmocoCda 7⅜13	6.2	39	119	−	1
Andarko 6¼14	5.7	5	109	+	½
Anhr 8s96	7.7	10	103¾	+	1¾
Anhr 8⅜16	8.1	10	105⅞	−	¾
Apache 7½00	cv	29	132	+	3
ArizP 7.45s02	7.3	20	101⅜		...
ArmI 8.7s95	8.7	3	100⅛	+	⅛
ArmI 9.2s00	9.4	13	97½	+	1⅞
ArmI 8½01	9.2	5	92⅞	−	1
AshO 8.8s00	8.3	30	106	+	1¼
AshO 8.2s02	8.1	25	101		
AshO 6¾14	cv	117	89¼	+	¾
ARch 10⅜95	9.3	1	112⅛	−	⅛
ARch 10⅞05	8.4	10	129¼		...
ARch 9⅝16	8.0	100	124	+	3⅞

Mutual fund values are also reported in the newspaper each business day. A mutual fund represents ownership of several bonds, stocks, or a combination of both. The daily report lists each organization followed by their group of funds. The report below shows the Capital Appreciation Fund of Fidelity Investments.

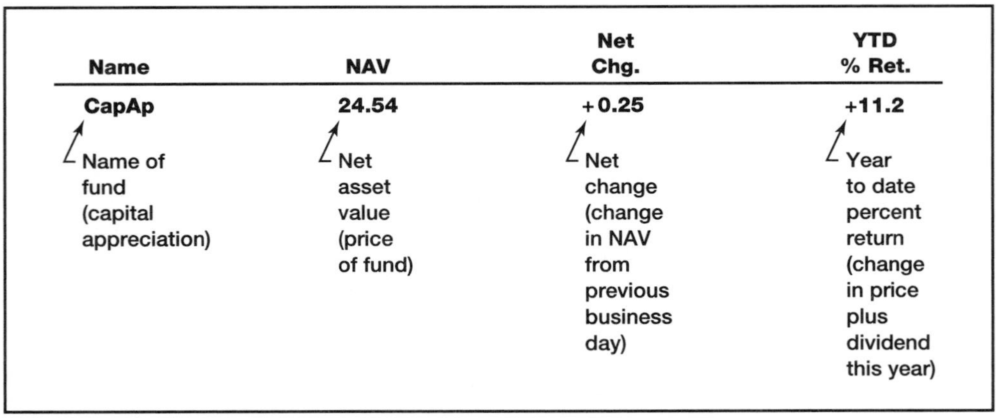

Name	NAV	Net Chg.	YTD % Ret.
CapAp	24.54	+0.25	+11.2

Name of fund (capital appreciation)

Net asset value (price of fund)

Net change (change in NAV from previous business day)

Year to date percent return (change in price plus dividend this year)

Name	NAV	Net Chg	YTD %ret
Fidelity Invest:			
A Mgr.	17.65	+0.13	+ 2.4
AggrGr	38.01	+0.80	+ 22.3
AMgrGr	19.24	+0.22	+ 3.0
AMgrIn	12.28	+0.04	+ 1.3
Balanc	17.68	+0.11	+ 8.7
BluCh	51.54	+0.75	+ 2.3
Canad r	14.73	+0.15	+ 5.3
CapAp	24.54	+0.25	+ 11.2
CpInc r	10.05	+0.01	+ 11.4
Cngs	393.32	+3.29	+ 4.8
Contra	59.55	+0.53	+ 5.8

Glossary

A

Accumulated Depreciation   The amount that an asset has depreciated to date.

Ad Valorem Duty   A type of customs duty that is a percent of the market value of the item.

Adjusted Balance   The actual amount available in the checking account at a point in time.

Amortizing   Paying off a loan by making monthly payments.

Amount   The result of taking a number of hundredths of the base.

Annual Percentage Rate (APR)   The interest rate on a loan to a consumer, stated as prescribed by Regulation Z of the Truth-in-Lending Act.

Annuity   A savings instrument that usually includes a series of investments that earn interest and accumulate to a desired amount by a specific future date.

Annuity Due   A type of annuity that requires that the investments be made at the beginning of the investment period.

Assessed Valuation   The base on which property taxes are computed.

Assets   Items that are owned by a firm or individual.

Automatic Teller Machine (ATM)   A machine available on a 24-hour basis to accept deposits and provide cash withdrawals from deposit accounts.

Average Cost Method   An accounting approach used to determine the value of inventory.

B

Balance Sheet   A summary of the financial condition of a firm at a particular point in time.

Balloon Payment   A single payment that is due on a loan (usually a home mortgage) at the end of one to five years, although the amount of regular payments is based on a longer term.

Bank Card   A type of credit card. This type of card is the most versatile.

Bank Statement   An analysis by the banking institution of the transactions and current balance of an account based on available information.

Bar Graph   A visual method of displaying numerical information in a two-dimensional form.

Base   The basis of comparison; the number of which so many hundredths is taken.

Bearer Bonds   Bonds owned by whoever (physically) holds them. The interest coupon is detached for payment.

Beginning Inventory   The value of the merchandise a firm has at the beginning of an accounting period such as a month or a year.

Beneficiary   The person or organization that will receive the financial benefits of a life insurance policy.

Blank Endorsement   The simplest form of endorsement. This endorsement requires only the endorser's signature; the endorser places no restrictions on the document.

Bond   A debt instrument of a firm or a government organization.

Book Value   The total cost of an asset less the accumulated depreciation.

Borrowing   Used in the subtraction function. Using a larger place value to provide enough to subtract from.

C

Calendar   The number of days in each month of a year.

Cash Discount   A reduction of the price of the merchandise offered to the buyer for early payment.

Cash Surrender Value   The money that will be received if the insurance policy holder gives up the claims of the policy.

Check   A demand by a depositor on his or her bank to pay the payee the amount specified.

Check Register   A record of the checking account balance from the viewpoint of the depositor.

Check Stub   A record of information about a check that is completed by the depositor of a checking account.

Circle Graph   A visual method to display numerical information. A circle is divided from the center into the portions that represent the data.

Closing Costs   The amount, paid at the closing meeting, for escrow, points, and fees.

C.O.D. (Cash on Delivery)   The merchandise delivered by this method must be paid for when it is received.

Coinsurance   The policy holder and the insurance company share in the financial loss of property damage.

Commission Formula   A combination of ever-increasing rates of pay for sales

designed to motivate an employee to sell more.

Common Denominator   A value that can be used to establish a common label for the fractions under consideration, for purposes of addition or subtraction.

Common Fraction   A type of fraction in which the numerator is smaller than the denominator.

Common Stock   A form of equity securities that provides a vote in the decisions of the firm.

Comparative Balance Sheet   A balance sheet that shows figures for two periods. A percent of change is shown and is determined by comparing the second period to the first period.

Compound Interest   A method of computing interest in which the interest earned in one period is added to the principal; interest for the next period is calculated on both interest and principal from the previous period.

Compound Rate   The combination of a specific rate and an ad valorem rate on imported goods; imposed by the federal government.

Compounded Amount   The amount returned to an investor at the end of the term of investment.

Cost of Goods Sold   The price that was paid for the merchandise that was sold during the period.

Cost of Investment   The fee that is paid to a middleman for facilitating the transaction.

Cost Price   The price that the retailer pays for goods.

Credit Card   A purchase device issued to those with a good credit history.

Cumulative   A stipulation on a preferred stock that allows the holder to receive dividends that were not paid in the past.

Currency Exchange   The rate at which money from one country trades for money from another country.

Customs Duty   A tax on imported goods; imposed by the federal government.

D

Data Set   A group of values that have something in common.

Debentured Bonds   Bonds that are issued by a firm without being tied to any physical equipment or other asset.

Decimal Point Placement Rule The answer in a multiplication problem should contain as many decimal places as the two values in the problem had.

Declining Balance Method An accelerated depreciation method that does not consider salvage value in the initial calculations; depreciation occurs at double the straight-line rate.

Deductions, Income Tax Filing Amounts that can be subtracted from the gross income amount for such items as interest expense on a home, contributions, and medical expenses.

Deductions, Payroll Amounts subtracted from gross pay for such items as taxes, contributions, and savings.

Denominator The number on the bottom of a fraction.

Dependents Those persons that are being supported by the wage earner and that can be claimed on the income tax return.

Deposit A sum of money that adds to the balance of a checking or savings account.

Deposit Slip A record device to verify a deposit.

Depositor The entity (an individual or a firm) that holds a checking or savings account.

Deposits in Transit Deposits of money, checks, and collections that have not caught up with the records of the bank.

Depreciation An adjustment in the value of an asset due to its decline in market value. An estimate.

Difference The result of subtraction.

Direct Method A fast method for computing the effect of trade discounts on the list price of an item.

Discount Amount The dollar amount of service charge that the purchaser of a note subtracts from the maturity value.

Discount Rate The percent of the note for service charge.

Dividend The number to be divided.

Dividends Profits distributed to shareholder owners of a corporation.

Divisor The number that is used to divide by. In a fraction, the number on the bottom.

Down Payment The amount that will be paid first and in addition to the amount financed in the purchase of an item.

E

Ending Inventory The value of the merchandise a firm has at the end of an accounting period such as a month or a year.

Endorsement The signature of the payee on the back of the check. There are several types of endorsements.

Endowment Life Policy A type of life insurance in which a large portion of the premium is directed toward a savings plan that may be used for college, retirement, etc.

E.O.M. (End of Month) One of the terms of trade for a cash discount; the buyer computes the days of discount beginning at the end of the month in which the purchase was made.

Equivalent Trade Discount A single percent equal to a series or chain of trade discounts.

Escrow Account An account kept by the lending institution into which the borrower pays in order to accumulate enough money to cover property taxes and property insurance when they come due.

Exact Interest An interest rate that uses a 365-day year. Used for personal loans only.

Exemption, Income Tax Filing The amount of earnings that the federal government allows a taxpayer to claim as untaxable income when filing the year-end tax return. This amount is based on the number of dependents that the taxpayer claims.

Exemptions, Payroll Legal dependents and the employee; can be claimed by the employee in order to reduce the amount of income tax withheld.

Expected Life The period of time or the number of cycles that an asset normally functions before it is disposed of.

Expenses Used up or expired assets.

F

Face Value The amount stated on a promissory note. The amount borrowed.

F.I.F.O (First In–First Out) A method of determining the value of ending inventory or cost of goods sold.

Fixed Payment Amount A payment amount on a home mortgage that is a set (fixed) amount for the term of the loan.

Fixed Rate An interest rate that remains the same (fixed) over the term of the loan. Usually on a home mortgage.

F.O.B. (Free on Board) Destination Transportation costs to the specified buyer's location are included in the price of the item.

F.O.B. (Free on Board) Shipping Point Transportation costs from the seller's location are to be paid by the buyer.

Form W-2 A federal government form that the employer completes indicating the earnings and deductions for the year.

Form W-4 A federal government form that the employee fills out to claim exemptions.

G

Gasoline Card A credit card issued by an oil company. Used primarily for credit purchases of gasoline.

Goods Available for Sale Beginning inventory plus net purchases. The cost of goods sold comes from this category.

Government Bonds Investment vehicles issued by local, state, or federal government agencies with a possible tax-free status.

Gross Earnings The amount earned by the employee in a pay period.

Gross Sales All the sales of a firm for a time period.

H

Home Mortgage An installment loan on a home. Loan may carry a fixed or variable rate, a fixed or variable payment, or a balloon payment.

Horizontal Analysis Comparing information on the same line. The difference is then changed into a percent.

I

I = P × R × T The basic interest formula.

Imprinting Checks The process of personalizing check forms with the name and account number of the depositor.

Improper Fraction The type of fraction that has a numerator larger than the denominator.

Income Statement A report that shows the financial activity of a firm over a period of time.

Income Tax Return A form that must be completed by every individual and firm that has income. On this form, the actual amount of tax owed is determined through calculations.

Inflation An increase in the price of most goods and services.

Insurance A way to share the risk of loss with another firm that specializes in estimating the probability of loss.

Interest A financial return to the lender for the risks undertaken during the term of a loan.

Interest Formula A method used to determine the amount of money that may be charged when money is borrowed; Interest = Principal × Rate × Time. (See I = P × R × T.)

Interest-bearing Note A promissory note that has an interest provision. The maturity value will be larger than the face value.

Internal Revenue Service An agency of the federal government that is responsible for collecting the income taxes due from persons and firms.

L

Leap Year A year that is evenly divisible by four. February has 29 days in a leap year.

Liabilities Amounts due others; financed obligations.

L.I.F.O. (Last In–First Out) A method of determining the value of ending inventory or cost of goods sold.

Limited Pay Life Policy A type of life insurance that has a savings factor and that involves a limited number of payments before the policy is paid up.

Line Graph A type of graph that uses a line to depict information.

Linear Measure Measurement in only one direction.

Liquidity Assets that are easily converted into cash are considered to have high liquidity.

List Price The price shown in a catalog. This price is often subject to trade discounts.

Lowest Common Denominator (LCD) The smallest value that will serve as a denominator for all of the fractions being considered.

M

Maker The person or organization that borrows money on a promissory note.

Markdown An amount or percent subtracted from the selling price.

Market Value The amount of money that an asset can be sold for at a point in time.

Markup An amount or percent added to the cost price of merchandise in order to cover overhead and profit. Can be based on cost or selling price.

Maturity Value The amount of money that the maker of a promissory note must pay at the end of the note term.

Mean The average of a data set.

Median The middle ranked value of a data set.

Metric Measurement System A system of measure that is used in nearly all of the world. The system uses a base of ten to convert measurements.

Mills A means of expressing one thousandth.

Minimum Payment Amount The smallest amount that can be paid on a credit card balance to maintain a good credit rating.

Minuend The larger of two numbers that are being subtracted.

Mixed Number A type of a fraction consisting of a whole number and a common fraction.

Mode The value of greatest frequency in a data set.

Modified Accelerated Cost Recovery System (M.A.C.R.S.) A depreciation method for tax use. For items placed into service after 1986.

Monthly Statement A record of the account balance from the viewpoint of the issuer of the credit card. This record is sent each month.

Multiplicand The top number of two numbers that are to be multiplied.

Multiplier The bottom number of two numbers that are to be multiplied.

N

Net Pay The amount due a person after deductions have been subtracted from gross pay.

Net Price The amount to be paid after all discounts have been subtracted.

Net Sales The dollar value of sales after sales returns and sales allowances have been subtracted from gross sales.

Noninterest-bearing Note A promissory note that has a maturity value that is the same as the face value.

Numerator The number on top in a fraction.

O

Ordinary Annuity A type of annuity that requires that the investments be made at the end of the investment period.

Ordinary Interest An interest rate that uses a 360-day year; used for commercial loans only.

Outstanding Checks Checks that have been written by the depositor, but that have not been received by the bank.

Owner's Equity The portion of the firm owned by the investors. Part of the basic accounting formula: Assets = Liabilities + Owner's Equity.

P

Participating A stipulation on a preferred stock that allows the stockholder to receive additional dividends after the holders of common stock have received a stated amount.

Pay Systems The methods that are available to determine gross earnings for an employee, including hourly, salary, piece-rate, and commission methods.

Payee, Check The person or organization to whom the check is made out.

Payee, Promissory Note The lender of the money on a promissory note.

Payment Amount A monthly payment on an installment loan. This payment will pay off the interest and principal over the term of the loan.

Payroll Record A form showing gross earnings, deductions, and net pay for all employees.

Percent A means of expressing some part of 100.

Place Values The names of the locations of digits in a number.

Points A point is one percent of the face value of the loan; points are usually paid by the buyer at closing.

Policy A contractual document explaining the financial commitment of an insurance company to the policy holder.

Preferred Stock A type of equity instrument that pays the owner a dividend before common stock is paid.

Premiums Payments for insurance protection.

Present Value The current value of an investment due some date in the future. The present value is found by discounting back the future value at prevailing interest rates.

Prime Number A number that is divisible only by itself and one.

Principal The amount borrowed.

Principal Balance The amount remaining to be paid on an installment loan.

Proceeds The amount due the payee when a note is discounted.

Product The result of multiplication; the answer.

Promissory Note A legally enforceable written contract; an "IOU."

Proprietorship The portion of the firm that is owned by the investors.

Purchases The merchandise that is bought during the period.

Q

Quotient   The number found as a result of division.

R

Range   The difference between the largest and smallest value of a data set.

Rate   A percent.

Rate, Interest   The annual charge for the use of money.

Rate of Return   The amount earned on an investment expressed as a percent of the amount invested.

Ratio   The comparison of one value to another value.

Ratio, Financial   A set of comparisons of one item to another, from either a balance sheet or an income statement.

Registered Bonds   A type of bond that requires that a new owner provide the firm with his or her name, address, etc., so that the firm may communicate with the new owner.

Regulation Z   The portion of the Truth-in-Lending Act that specifies that interest rates must be calculated using a set method and the rate be provided to consumer borrowers.

Remainder   The value that is left over after division of whole numbers.

Repayment Schedule   A record of payments to be made; includes the interest amounts and the principal balance.

Restrictive Endorsement   A type of endorsement of a check that determines the check's final use.

Retail Inventory Estimating Method   A method of providing an approximate value of the merchandise without a physical inventory.

Revenue   The result of selling goods or services.

Risk   The chance taken by a business person who makes a business decision.

R.O.G. (Receipt of Goods)   One of the terms of trade for a cash discount; the buyer computes the days of discount beginning with the date that the merchandise was actually received by the buyer.

Round Off   Directions that determine where an answer should be terminated.

Rounding Rule   Used to determine the final number for an answer. The number to the right must be 5 or larger to round up.

Rule of 78s   A formula used to determine the payoff amount on a short-term loan.

S

Sales Allowances   Adjustments made in the price of the merchandise in order to compensate the buyer for keeping damaged or incorrectly shipped merchandise.

Sales Returns   Merchandise returned to the seller because of damage, errors in ordering, etc.

Sales Tax   A tax that is determined by taking a percent of the selling price of items sold.

Salvage Value   The estimated market value of an asset at the end of its life.

Secured Bonds   Bonds that have certain properties pledged to the owner of the bond.

Selling Price   The price that the buyer (consumer) pays for the goods.

Service Charges   Expenses charged to the depositor by the bank for use of the account, imprinting of checks, etc.

Sinking Fund   The accumulation of payments into an interest-paying investment; a desired amount will then be available to the investor at a future date.

Special Endorsement   An endorsement of a check that limits what can be done with it in the next transaction.

Specific Duty   A customs duty that is a flat amount per unit.

Specific Identification Method   An inventory method in which each item is identified through a physical inventory.

Store Card   A credit card that is used in only one store or chain of stores.

Straight Commission   A percent of the dollar value of the sale paid to the sales person.

Straight Life Policy   A type of life insurance that has a small surplus factor built into the premium.

Straight-line Method   A type of depreciation that provides an equal amount of expense per year.

Subtrahend   The smaller of two numbers that are being subtracted.

Sum   The amount resulting from the addition of two or more numbers.

Sum-of-Years'-Digits (S.O.Y.D.) Method   A type of depreciation that is accelerated and that uses the years of expected life to determine the fraction of total depreciation for that year.

Surface Measure   Found by multiplying length by width.

T

Tax Rate   The percent that is used to find the Social Security and Medicare tax; set by Congress.

Taxable Base   The maximum amount of earnings that can be taxed for Social Security purposes for that year.

Taxable Income   The amount of income left after subtracting the amount for exemptions and deductions. This amount is used to find the tax from the tax table.

Term Life Policy   A type of life insurance that has no savings factor built into the premium. A best buy for the money if the face value needs to be large.

Term of Discount   The period of time that the promissory note is discounted by the third party.

Term of Note   The number of days between the date of the note and the due date.

Terms of Trade   A numerical statement of the percent discounts that are available and the time period for qualification for cash discounts.

Time   Used to compute interest or discounts. Usually a fraction of 360 days (ordinary interest) or 365 days (exact interest).

Time Card   An item of control that allows a payroll department to determine the number of hours worked by employees.

Time Line   A line that enables the asset life to be compared with the fiscal year to determine the proper amount of depreciation expense.

Total Cost   All of the costs associated with placing the item into service. Used in determining depreciation expense.

Total Depreciation   The amount that the asset will depreciate over its expected life.

Trade Discount   A reduction of the list price to the buyer based on large quantity purchases or on the assumption of marketing functions being performed by the buyer.

Travel or Entertainment Card   A credit card that is accepted primarily by firms in the travel or entertainment business.

Turnover   The number of times that the average inventory is sold. A measure of success for the merchandiser.

U

Units-of-Production Method   A method of depreciation that uses the amount of times an asset is used instead of the number of years it is used.

Universal Life Policy   A type of life insurance that allows for changes in premium and coverage amounts. Surplus is invested in fixed-rate vehicles.

V

Variable Life Policy   A type of life insurance that allows for changes in premium and coverage amounts. Surplus is invested in stocks and funds.

Variable Payment Amount   Payment on a home mortgage that begins as a small amount and then increases with time. May cause loan balance to increase in early years.

Variable Rate   Interest rate that changes; changes are based on an index of interest indicators—usually government securities.

Volume Measure   Found by multiplying length by width by depth.

W

W-2 Form   (*See* Form W-2.)

W-4 Form   (*See* Form W-4.)

Workers' Compensation   A type of insurance that provides coverage to workers who have been injured or who become ill because of work-related activities.

Answers to the Student Edition

Chapter One (Pages 1–20)
Exercise A
1. 8 **2.** 2 **3.** 3 **4.** 0 **5.** 1
Exercise B
1. thousands **2.** hundred thousands
3. tens **4.** hundreds **5.** ten thousands
Exercise C
Check number 1269. Five thousand, eight hundred forty-three and no/100
Check number 1270. One hundred sixty-five and no/100
Exercise D
1. 98,000 **2.** 2,640 **3.** 278,900
Exercise E
1. $1,900,000 **2.** $630,000
3. $320,000,000
Exercise F
A. 1. 124 **2.** 53 **3.** 112 **4.** 219
5. 90 **6.** 106 **7.** 99
B. 1. 772 **2.** 3,493 **3.** 15,215
4. 2,029 **5.** 10,655
C. 1. $8,116 **2.** $92,131 **3.** $16,417
4. $76,901 **5.** $89,140
D. 1. $114,635 **2.** $11,565
3. $84,744 **4.** $27,935
5. $23,980
Exercise G
Breakfast $13
Lunch $6
Dinner $49
Travel $353
Lodging $114
Tips $11
Misc. $9
Monday $245
Tuesday $91
Wednesday $70
Thursday $149
Week $555 (owe)
Exercise H
1269—$906, $286
1270—$286, $186
1271—$186, $117
1272—$117, $70
1273—$590, $526
Yes, pay Cindi.
Exercise I
A. 1. 19 **2.** 68 **3.** 24 **4.** 42 **5.** 28
6. 16 **7.** 62
B. 1. $17 **2.** $36 **3.** $160 **4.** $138
5. $86 **6.** $47
C. 1. $174 **2.** $98 **3.** $437 **4.** $155
5. $64 **6.** $544
Exercise J
A. 1. 67,300 **2.** 89,200 **3.** 5,460
4. 891,000 **5.** 8,940
B. 1. 8,190 **2.** 96,100 **3.** 5,700
4. 46,900 **5.** 162,000
C. 1. $134,937 **2.** $574,864
3. $228,690 **4.** $240,875

D. 1. $152,405 **2.** $251,805
3. $700,104 **4.** $138,678
Exercise K
$94, $279, $534, $46, $108, $27, $102, $120, $144, $192, $82, $104, $42, $100, $70, $60, $64, $126, $50, $20, $12, $48, $36, $68, $2,528
Exercise L
1. 157 **2.** 1,334 **3.** 118 **4.** 3,489 r32
5. 421 **6.** 2,033 r33 **7.** 224 r364
8. 19 r11
Exercise M
1. 7,930 **2.** 470 **3.** 906 **4.** 891 **5.** 60,050

Chapter One
Assignment
Skill Problems
1. 4 **3.** 7 **5.** billions **7.** hundred thousands **9.** One hundred forty-eight and 00/100 **11.** Two hundred fifty thousand and no/100 **13.** 5,450 **15.** D
17. G **19.** I **21.** 10 **23.** 599
25. 13 **27.** 352 **29.** 119 **31.** 906
33. 9,650 **35.** 1,028 **37.** $5,562
39. 13 r41 **41.** $27,636 **43.** 23 r18
45. $69,971 **47.** 80 **49.** $32,400
51. 3,824 **53.** 47,280 **55.** $142,968
57. $32,571 **59.** 31 r1
Business Application Problems
1. 116 **3.** 70 **5.** $11,898 **7.** 544
9. $16,740 **11.** $16,498 **13.** $1,200
15. 1,071 **17.** College **19.** $2,100
21. $325,300 **23.** 5 **25.** 48,028 **27.** 591
29. The cost of 23 chairs will be $2,921.

Chapter One
Mastery Test
1. 54,400 (Objective 3)
($64 \times 850 =$)
2. $10,873 (Objective 3)
($1,457 + 5,800 + 3,575 + 41 =$)
3. $830,079 (Objective 3)
($42,803 + 787,276 =$)
4. 4,214 r7 (Objective 3)
($75,859 \div 18 =$)
5. $10,773 (Objectives 3,4)
($64,638 \div 6 =$)
6. 102,663 (Objective 3)
($549 \times 187 =$)
7. 19 r8 (Objective 3)
($768 \div 40 =$)
8. $6,249,488 (Objectives 3, 4)
($6,185,005 + 54,980 + 7,681 + 1,822 =$)
9. 680 (Objectives 1, 2, 3, 4)
($891 - 147 - 63 =$)
10. $326,723 (Objectives 1, 3, 4)
($85,605 + 200,038 + 41,080 =$)

11. $197,217 (Objective 3)
($387 \times 651 = 456 \times 120 = 251,937 - 54,720 =$)
12. Forty-three thousand, six hundred seventy-nine dollars (Objective 1)
13. $367,000 (Objectives 2, 3)
($459,073 - 91,752 =$)
14. $11,760 (Objective 3)
($40 \div 2 = 20 \div 2 = 10 \times 98 = 980 \times 12 =$)
15. 685,200 (Objective 2)

Chapter Two (Pages 21–44)
Exercise A
1. common **2.** improper **3.** improper
4. improper **5.** mixed **6.** common
7. improper **8.** common **9.** common
10. common **11.** mixed **12.** mixed
Exercise B
1. $1\frac{5}{8}$ **2.** $5\frac{7}{12}$ **3.** $2\frac{4}{5}$ **4.** $3\frac{5}{14}$ **5.** $3\frac{1}{6}$
6. $2\frac{1}{2}$ **7.** $2\frac{1}{9}$ **8.** $9\frac{3}{10}$ **9.** $4\frac{3}{10}$ **10.** $1\frac{1}{8}$
11. 4 **12.** $7\frac{1}{8}$
Exercise C
1. $\frac{27}{4}$ **2.** $\frac{59}{6}$ **3.** $\frac{39}{5}$ **4.** $\frac{15}{8}$ **5.** $\frac{551}{5}$
6. $\frac{7}{2}$ **7.** $\frac{147}{8}$ **8.** $\frac{20}{3}$ **9.** $\frac{19}{4}$ **10.** $\frac{3}{2}$
11. $\frac{87}{16}$ **12.** $\frac{21}{2}$
Exercise D
1. $\frac{1}{2}$ **2.** $\frac{3}{4}$ **3.** $\frac{7}{10}$ **4.** $\frac{2}{5}$ **5.** $\frac{1}{2}$ **6.** $\frac{7}{8}$
7. $\frac{11}{25}$ **8.** $\frac{1}{2}$ **9.** $\frac{3}{4}$ **10.** $\frac{3}{5}$ **11.** $\frac{8}{11}$
12. $\frac{3}{8}$
Exercise E
1. 15 **2.** 33 **3.** 90 **4.** 28 **5.** 15 **6.** 27 **7.** 16
8. 5 **9.** 410 **10.** 12 **11.** 15 **12.** 5
Exercise F
1. $1\frac{1}{12}$ **2.** $2\frac{1}{8}$ **3.** $1\frac{16}{105}$ **4.** $2\frac{3}{8}$ **5.** $1\frac{1}{35}$
6. $1\frac{11}{40}$ **7.** $1\frac{11}{60}$ **8.** $1\frac{37}{80}$ **9.** $1\frac{1}{16}$ **10.** $1\frac{13}{14}$
Exercise G
1. $\frac{5}{8}$ **2.** $\frac{3}{8}$ **3.** $\frac{7}{30}$ **4.** $\frac{1}{20}$ **5.** $\frac{113}{180}$
6. $\frac{11}{24}$ **7.** $\frac{7}{36}$ **8.** $\frac{5}{12}$ **9.** $\frac{51}{80}$ **10.** $\frac{5}{16}$
Exercise H
1. $15\frac{1}{4}$ **2.** $14\frac{7}{24}$ **3.** $5\frac{1}{16}$ **4.** $10\frac{23}{24}$
5. $11\frac{4}{5}$ **6.** $8\frac{67}{72}$ **7.** $24\frac{81}{100}$ **8.** $8\frac{17}{24}$
9. $15\frac{11}{12}$ **10.** $17\frac{19}{24}$

Exercise I

1. $7\frac{1}{3}$ **2.** $2\frac{1}{4}$ **3.** $5\frac{11}{40}$ **4.** $52\frac{23}{24}$

5. $2\frac{11}{30}$ **6.** $5\frac{1}{40}$ **7.** $6\frac{7}{12}$ **8.** $173\frac{1}{6}$

9. $3\frac{1}{12}$ **10.** $6\frac{11}{18}$

Exercise J

1. $\frac{1}{6}$ **2.** $\frac{77}{80}$ **3.** $\frac{5}{8}$ **4.** $\frac{5}{42}$ **5.** $\frac{1}{2}$ **6.** $\frac{7}{18}$

7. $3\frac{1}{4}$ **8.** $19\frac{19}{24}$ **9.** $\frac{25}{84}$ **10.** $\frac{13}{224}$

Exercise K

1. $1\frac{5}{7}$ **2.** $1\frac{1}{2}$ **3.** $1\frac{5}{7}$ **4.** $2\frac{1}{2}$ **5.** $1\frac{11}{16}$

6. $1\frac{2}{7}$ **7.** $3\frac{1}{2}$ **8.** 6 **9.** $4\frac{2}{7}$ **10.** $1\frac{1}{4}$

Exercise L

1. $106\frac{47}{256}$ **2.** $4\frac{5}{7}$ **3.** $4\frac{2}{13}$ **4.** 4

5. $1\frac{71}{84}$ **6.** $2\frac{53}{76}$ **7.** $10\frac{19}{35}$ **8.** $1\frac{83}{162}$

Exercise M

1. 3:2 **2.** 1:10 **3.** 4:1 **4.** 5:3 **5.** 5:2
6. 1:4

Chapter Two
Assignment
Skill Problems

1. improper **3.** mixed **5.** $8\frac{1}{2}$ **7.** $12\frac{1}{2}$

9. $\frac{16}{3}$ **11.** $\frac{48}{7}$ **13.** $\frac{3}{4}$ **15.** $\frac{2}{5}$ **17.** $1\frac{1}{4}$

19. $2\frac{2}{5}$ **21.** 12 **23.** 18 **25.** A **27.** H

29. $9\frac{1}{4}$ hours **31.** $10\frac{1}{4}$ hours **33.** 3:2

35. 3:2 **37.** $\frac{27}{143}$ **39.** $\frac{24}{35}$ **41.** $1\frac{5}{12}$

43. $1\frac{11}{24}$ **45.** $\frac{1}{30}$ **47.** $2\frac{9}{40}$ **49.** $3\frac{27}{40}$

51. $1\frac{3}{8}$ **53.** $\frac{21}{32}$ **55.** $1\frac{11}{30}$ **57.** $\frac{1}{2}$

59. $1\frac{1}{5}$ **61.** $\frac{8}{75}$ **63.** $2\frac{1}{12}$

Business Application Problems

1. 3,871 **3.** $6\frac{5}{12}$ **5.** 240 **7.** $603

9. 1:136 **11.** $21 **13.** 2 **15.** 7:1
17. $1,400,000 **19.** 8:1 **21.** $18

23. $500 **25.** 30,488 **27.** $227\frac{7}{10}$

29. There are 130 units of the condominium that are unsold.

Chapter Two
Mastery Test
1. Whole number (8), numerator (4), denominator (7) (Objective 6)

2. 1. B
2. A
3. C
4. A
5. B
6. C (Objective 1)

3. $1\frac{11}{20}$ (Objective 4) $\left(\frac{15}{20} + \frac{16}{20} = \right)$

4. $\frac{5}{6}$ (Objective 4) $\left(1\frac{3}{6} - \frac{4}{6} = \right)$

5. 50 (Objective 4) $\left(15 \times \frac{10}{3} = \right)$

6. $4\frac{3}{8}$ (Objective 4) $\left(\frac{7}{8} \times \frac{5}{1} = \right)$

7. $5\frac{1}{3}$ (Objective 3)

8. $147,720 (Objective 4) (98,480 × 3 =, 295,440 ÷ 2 =)

9. 9:1 (Objective 5)

10. $5\frac{2}{5}$ (Objective 4) $\left(8\frac{5}{10} - 3\frac{1}{10} = \right)$

11. $11,592 (Objective 4) (57,960 ÷ 5 =)

12. $104 (Objective 4) (8 × 13 =, 104 ÷ 2 =, 52 × 2 =)

13. 270 (Objective 4)
$\left(800 \times \frac{1}{2} =, 400 \times \frac{3}{4} =, 300 \times \frac{9}{10} = \right)$

Chapter Three   (Pages 45–62)
Exercise A
1. 7 **2.** 8 **3.** 5 **4.** 3 **5.** 2 **6.** 9
Exercise B
1. hundred thousandths **2.** hundredths
3. millionths **4.** ten thousandths
5. tenths **6.** thousandths
Exercise C
Reading
Exercise D
Reading
Exercise E
1. Eight hundred twenty-three dollars and sixty-seven cents
2. Two thousand, one hundred five dollars and forty-six cents
3. Eight hundred sixty-seven dollars and twenty-eight cents
4. Two hundred thirty-four dollars and eighty-seven cents
5. Three thousand, eight hundred sixty-nine dollars and sixty-one cents
Exercise F
1. $935.49 **2.** $63.26 **3.** $572.21
4. $3,018.07 **5.** $51,307.36
Exercise G
1. .6 **2.** .47 **3.** .0 **4.** .546 **5.** 1
Exercise H
1. $605.84 **2.** $57.65 **3.** $486.94
4. $68,640.27 **5.** $1,754.11
Exercise I
1. 607.80 **2.** 800 **3.** 759.9
4. 550,000 **5.** 14.80

Exercise J
1. 15,576.1975 **2.** 889.1357
3. 745.22646 **4.** 1,165.7941
5. 840.48988
Exercise K
1. 612,026.0913 **2.** 346.2974
3. 79.8762 **4.** 502.2 **5.** 158,463.0286
Exercise L
1. 7.530 **2.** 16.433 **3.** .4383
4. 56.043 **5.** 41.863
Exercise M
1. $835.49 **2.** $530.05 **3.** $48.17
4. $596.99 **5.** $278.92
Exercise N
1. 8,445.2 **2.** 2.278 **3.** 49.53
4. 2.28015 **5.** 2.754
Exercise O
1. 35.7 **2.** .4 **3.** 308.7 **4.** 1.7 **5.** 0
Exercise P
1. .375 **2.** .5625 **3.** 1.6 **4.** 2.4 **5.** 2.1875

Chapter Three
Assignment
Skill Problems
1. 7 **3.** 5 **5.** 0 **7.** Eight hundred fifty-six dollars and forty-five cents **9.** .7
11. 548,658.459 **13.** .461 **15.** 52.0027
17. $10,319.65 **19.** $14,725.60
21. .75 **23.** 2.6 **25.** B **27.** $840
29. 92.964 **31.** 3.334 **33.** 6,175.8
35. 22.31 **37.** .02 **39.** .7 **41.** .1512
43. 1,458.6557 **45.** 4.35 **47.** 48.8
49. 72.1589 **51.** 1.17 **53.** .04644
55. 1,462.6193
Business Application Problems
1. $37.15 **3.** $32,145.12
5. $112,070.50 **7.** $3,234 **9.** $85
11. $219 **13.** $112.18 **15.** $74.12
17. 175 **19.** $33,513.99
21. $125,819.25 **23.** $8.20 **25.** $129.70
27. $12,286.30 **29.** The total cost of the small flower arrangements and cards will be $2,227.50 for the month.

Chapter Three
Mastery Test
1. a. 10
   b. .1
   c. 600
   d. 64.097
   e. 43.9 (Objective 2)
2. Three hundred eighty-five dollars and seventy cents (Objective 1)
3. 755.167 (Objective 3)
   (733.042 + 8.367 + 13.758 =)
4. 47.674 (Objective 3)
   (57.079 − 9.405 =)
5. 1 (Objective 3) (1.2 × .56 =)
6. $.47 (Objectives 2, 3) (3.72 ÷ 8 =)
7. a. .625 **b.** .3 **c.** 4.2 (Objective 5)
   (5 ÷ 8 =, 3 ÷ 10 =, 1 ÷ 5 =)
8. 19¢ (Objective 3) (1,560 ÷ 6,000 =)

9. $844,875.20 (Objective 3)
(324,952 × 2.6 =)
10. 10,835 (Objectives 2, 3)
(15,479 × .7 =)

Chapter Four   (Pages 63–82)
Exercise A
1. restrictive **2.** special **3.** blank
4. blank **5.** restrictive **6.** restrictive
7. blank **8.** special **9.** blank **10.** blank
Exercise B
1. $775.44 **2.** $159.80 **3.** $3,895.07
4. $1,563
Exercise C
1. $789 **2.** $1,178 **3.** $1,576
4. $728.88 **5.** $3,540

Chapter Four
Assignment
Skill Problems
1. blank **3.** restrictive **5.** blank **7.** E
9. C **11.** depositor **13.** deposit slip
15. deposits in transit **17.** service
charges **19.** $1,166.03 **21.** $1,699
23. $1,663.33 **25.** 214.00
27. 2,936.49 **29.** 2,962.04
31. 545.01 **33.** 2,607.12
35. 2,592.77 **37.** 2,580.33
39. $1,663.33
Business Application Problems
1. $3,559.50 **3.** $1,527.50 **5.** $25.00
7. $4,724.95 **9.** $7,258
11. $1,919.80 **13.** $124.30 **15.** $11.50
17. $331.28 **19.** $425 **21.** $429.40
23. The bank information indicates a
$5,256 balance. **25.** The real balance
is an overdraw of $338.

Chapter Four
Mastery Test
1. a. For deposit only, G. S. Student
b. Gerry S. Student
c. Payable to B. College G. S.
Student (Objective 1)
2. $1,197.55 (Objective 2)
(1,088 + 555.97 − 447.42 =,
1,063.08 − 100 + 234.47 =)
3. $10.98 (Objective 2)
(47 + 25 − 37.80 =,
92.61 − 47.43 =, 45.18 − 34.20 =)
4. $7.40 (Objective 2)
(89.70 − 97.10 =)
5. $754.54 (Objective 2)
(2,726.34 − 1,962.80 − 9 =)

Chapter Five   (Pages 83–106)
Exercise A
1. $400, $45, $445 **2.** $380, $28.50,
$408.50 **3.** $320, $12, $332
4. $880, $66, $946 **5.** $320, $0, $320
6. $684, $0, $684 **7.** $360, $13.50,
$373.50
Exercise B
1. $312.00, $252.00, $564.00
2. $295.20, $0, $295.20 **3.** $307.80,

$328.05, $635.85 **4.** $487.50, $37.50,
$525.00 **5.** $576.00, $48.00, $624.00
6. $361.00, $299.25, $660.25
7. $331.50, $229.50, $561.00
8. $390.00, $165.00, $555.00
Exercise C
1. $57,048 **2.** $1,052.04 **3.** $19,994.96
4. $396,541.50 **5.** $14,875
Exercise D
1. $2,504.70 **2.** $2,061 **3.** $2,641.80
4. $3,818 **5.** $2,943.50
Exercise E
1. $1,380 **2.** $1,840 **3.** $3,219.50
4. $1,255 **5.** $274.50
Exercise F
1. $8.70, $37.20, $467.10
2. $18.32, $78.33, $869.02
3. $13.02, $55.68, $564.42
4. $10.97, $46.92, $580.78
5. $10.07, $43.04, $509.37
Exercise G
1. $30, $6.31, $26.97, $371.72
2. $44, $7.55, $32.30, $437.15
3. $46, $7.06, $30.19, $403.75
4. $17, $7.35, $31.43, $451.22
5. $56, $7.19, $30.75, $402.06
6. $43, $6.70, $28.64, $383.66
7. $54, $8.63, $0, $532.37
8. $43, $6.67, $28.52, $381.80
9. $29, $6.19, $26.47, $365.34
10. $20, $6.95, $0, $452.05
Exercise H
1. $606.36 **2.** 1,593.40 **3.** 431.27
4. 2,926.40 **5.** 1,106.08 **6.** $141.81
7. 372.65 **8.** 100.86 **9.** 684.40
10. 258.68 **11.** $39.12 **12.** 102.80
13. 27.82 **14.** 188.80 **15.** 71.36
16. $264.06 **17.** 693.90 **18.** 187.81
19. 1,274.40 **20.** 481.68

Chapter Five
Assignment
Skill Problems
1. C **3.** I **5.** D **7.** J **9.** P **11.** H
13. $563.50 **15.** $287. **17.** $467.65
19. $12,361.92 **21.** $56,010.19
23. $2,527.20 **25.** $990
27. $19,304.20 **29.** $3,013.40
31. $960.25 **33.** $444.90
35. $332.01 **37.** $3,153.09
39. $2,561.31 **41.** $2,897.10
43. $103 **45.** $66.87
47. $176.90 **49.** $19.26
Business Application Problems
1. $50 **3.** $4,823.50 **5.** $992
7. $2,273.58 **9.** $4,095 **11.** $281.55
13. $7,347.15 **15.** $4,724.40
17. $3,648 **19.** $403.84 **21.** $9,360
23. $268.96 **25.** The difference will
be $17. **27.** The additional take home
pay will be $3,273.50.

Chapter Five
Mastery Test
1. F (Objective 4)
2. T (Objective 4)
3. F (Objective 4)
4. F (Objective 4)
5. F (Objective 4)
6. T (Objective 4)
7. $453.10 (Objective 1)
(394 × 1.15 =)
8. $45 (Objective 2)
9. $1,823.14 (Objective 3)
10. $407.05 (Objective 2)
(40 × 9.50 =, 7 × 1$\frac{1}{2}$ × 9.50 =,
479.75 × .062 =,
479.75 × .0145 =)
11. $396 (Objective 1) (40 × 9 =,
2 × 18 =)
12. $324.15 (Objective 1) (39.8 × 7 =,
584 × .078 =)

Chapter Six   (Pages 107–126)
Exercise A
1. 55 yd, 1 ft, 4 in.
2. 1 lb, 6$\frac{1}{3}$ oz.
3. 33 min, 5 sec
4. 424 lb, 13 oz
5. 1 hr, 26 min, 47 sec
Exercise B
1. 27 sq. yd, 1 sq ft, 7 sq. in.
2. 23 sq. yd, 6 sq. ft
3. 6 sq. yd, 7$\frac{3}{4}$ sq. ft
4. 21.76 sq. mi
5. 1 sq. yd, 7 sq. ft
Exercise C
1. 5 cu. yd
2. 8 cu. yd, 24 cu. ft
3. 10 cu. yd, 24 cu. ft
4. 20 cu. yd, 4 cu. ft
5. 4,000 cu. ft
Exercise D
1. 9.632 **2.** 8.56 **3.** .806329 **4.** 76
5. 473,570
Exercise E
1. 3.8354 **2.** 4.4196 **3.** 39.47896
4. 15.1408 **5.** 8.4539 **6.** 18.123
7. 3.1096 **8.** .1135749 **9.** 1.7577
10. 8.37748
Exercise F
1. 7,050.88 **2.** 126,044.07
3. 3,456,950 **4.** 339,020 **5.** 4,369.5
Exercise G
1. $96 **2.** $163.10 **3.** $87.65
4. $7,757.44 **5.** $1,315

Chapter Six
Assignment
Skill Problems
1. 3 yd, 1 ft, 11 in. **3.** 54 min, 5 sec
5. 1 hr, 42 min, 36 sec **7.** 15 sq. yd,
3 sq. ft **9.** 39.84 sq. mi **11.** 6 cu. yd
13. 19 cu. yd, 6 cu. ft **15.** 7 yd, 2 ft, 4 in.
17. 10 hr, 8 min, 40 sec **19.** 55 cu. yd,

11 cu. ft **21.** 43,620 **23.** 7.865
25. .807604 **27.** 374,800 **29.** 7.5438
31. 5.1816 **33.** 51.3992 **35.** 44.4761
37. 8.8255 **39.** 11,206.372
41. 316.86 **43.** 9,909.68 **45.** 726.31
47. 4,534.90 **49.** 607,600
Business Application Problems
1. $14,285.71 **3.** $208 **5.** $50,025
7. $36,727.68 **9.** 1,173 **11.** 4 hr, 14
min **13.** 10 **15.** 200 sq. yd **17.** $133\frac{1}{3}$
hr **19.** The cost of the material will be
$1,028.40.

Chapter Six
Mastery Test
1. 1 mile, 885 yd (Objective 2)
  (4 ÷ 3 =, 2,655 ÷ 3 =)
2. 486 (Objective 1) (60 × 5 =,
  300 × 60 =, 18,000 ÷ 37 =)
3. 3 (Objective 2) (30 × 8 =,
  42 × 8 =, 1,152 ÷ 500 =)
4. 56 (Objective 2) (400 × 2.5 × 1.5 =,
  1,500 ÷ 27 =)
5. 33.558355 (Objective 4)
  (53 × 1.196 =, 63.388 × 9 =,
  570.492 ÷ 17 =)
6. 5.81096 (Objective 4)
  (7.6 × .7646 =)
7. 583.2 (Objective 4) (90 × .4536 =,
  40,824 ÷ 70 =)
8. 3 yd, 10 in. (Objective 2) (8 − 5 =)
9. 14,300 (Objective 3)
10. $5,109.84 (Objective 5)
  (42,300 × .1208 =)
11. no (Objective 2) (60 × 120 × 12 =)
12. 36,270,000 (Objective 5)
  (3,900 × 9,300 =)

Chapter Seven   (Page 127–142)
Exercise A
1. .17 **2.** .56 **3.** .11 **4.** .09 **5.** 1.56
Exercise B
1. 41,.41 **2.** 62,62 **3.** 127,127 **4.** 10,
10 **5.** 21, .21
Exercise C
1. 8.2, .082 **2.** 1.375, .01375 **3.** .6,
.006 **4.** 52.9, .529 **5.** 29.5, 29.5
Exercise D
1. 36 **2.** 11 **3.** 59 **4.** 6 **5.** 320 **6.** 2
7. 6 **8.** 10 **9.** 25 **10.** 64
Exercise E
1. 100 **2.** 86% **3.** 26 **4.** 69 **5.** 11%
6. 500
Exercise F
1. $7.14 **2.** $180 **3.** 512% **4.** $23.20
5. $7.08 **6.** 43%
Exercise G
1. 25 **2.** 10 **3.** 14 **4.** 3 **5.** 11 **6.** 935
7. $400,000 **8.** $115,384.61 **9.** 20 **10.** 9

Chapter Seven
Assignment
Skill Problems
1. B **3.** 18 **5.** 41 **7.** .7 **9.** 175.9
11. 4.58 **13.** 30 **15.** .47 **17.** .17
19. 26 **21.** .102 **23.** 3.6 **25.** 254
27. 91 **29.** $64.80 **31.** $4.80
33. 150 **35.** $11.03 **37.** 84 **39.** 11
41. 14% **43.** 35 **45.** 6 **47.** 2 **49.** 2
51. 108 **53.** 22 **55.** $8,184
Business Application Problems
1. $4,248.10 **3.** 17% **5.** 10%
7. $51,239.90 **9.** 66% **11.** 68 **13.** 17%
15. $200,000 **17.** $24,000 **19.** 13%
21. $786.36 **23.** 120 **25.** $500
27. 30% **29.** The sales in excess of
quota were $325,000 for the month.

Chapter Seven
Mastery Test
1. .09 (Objective 2)
2. 87.5 (Objective 2)
3. 21 (Objective 2)
4. 9.5 (Objective 3)
5. T (Objective 5)
6. F (Objective 5)
7. F (Objective 6)
8. $150 (Objective 7) (600 × .25 =)
9. 2% (Objective 7)
  (12,856 ÷ 642,800 =)
10. $1.88 (Objective 7) (24 ÷ 1.085 =)
11. $240 (Objective 7) (1,500 × .16 =)
12. 595 (Objective 7) (85,000 × .007 =)
13. 8.9% (Objective 7)
  (7,000 ÷ 79,000 =)
14. $27,200,000 (Objective 7)
  (68,000,000 × .4 =)
15. 1.2% (Objective 7)
  (816,000 ÷ 68,000,000 =)

Chapter Eight   (Pages 143–160)
Exercise A
1. $111.60 **2.** $95.04 **3.** $194.04
4. $169.32 **5.** $66.83 **6.** $250.60
7. $132.30 **8.** $159 **9.** $138.13
10. $73.27
Exercise B
1. $11,739.13 **2.** $20,000 **3.** $5,783.13
4. $33,333.33 **5.** $27,692.31
6. $10,000 **7.** $50,000 **8.** $12,000
9. $22,500 **10.** $23,750
Exercise C
1. $299.60 **2.** $249.15 **3.** $2,202 **4.** $96
5. $1,219.80 **6.** $178.10 **7.** $538.40
8. $774.75 **9.** $93.40 **10.** $4,414.80
Exercise D
1. $6,208 **2.** $123,750 **3.** $4,500
4. 0 **5.** $134,600
Exercise E
1. $156,950 **2.** $80,400 **3.** $2,968,880
4. $4,895,100 **5.** $3,487,128
Exercise F
1. $1.83 **2.** $.60 **3.** $1.18 **4.** $9.45
5. $4.69

Chapter Eight
Assignment
Skill Problems
1. E **3.** C **5.** A **7.** $410 **9.** $1,050
11. $132.84 **13.** $199.20 **15.** $325
17. $193.20 **19.** $1,302 **21.** $226.50
23. $9,375 **25.** $1,010.40 **27.** $300
29. $232,050 **31.** $1,444,800
33. $24,500 **35.** $7,513,560
37. $1,222,650 **39.** $13,461.54
41. $35,000 **43.** $8,163.27
45. $1,011.49 **47.** $49,188.38
49. $20,294.12 **51.** $10,158.73
53. $4,088.12 **55.** $13,427.49
57. $31,250 **59.** $80,000
Business Application Problems
1. $5,778 **3.** $790 **5.** $160,000
7. $1,576.40 **9.** $1,951.60
11. $1,923.75 **13.** $78,688.52
15. $75,000 **17.** $38,000 **19.** $360,000
21. $39,375 **23.** $440,000 **25.** The
premium cost is $6 per hour.

Chapter Eight
Mastery Test
1. T (Objective 5)
2. T (Objective 5)
3. F (Objective 5)
4. T (Objective 5)
5. T (Objective 5)
6. T (Objective 5)
7. T (Objective 5)
8. F (Objective 5)
9. $136.80 (Objective 2) (18 × 7.6 =)
10. $169,000 (Objective 2)
  (70 × 12 =, 840 ÷ 4.98 =)
11. $27,200 (Objective 3)
  (34,000 × .75 =, 34,000 × .6 =,
  20,400/25,500 × 34,000 =)

Chapter Nine   (Pages 161–178)
Exercise A
1. 35 **2.** 44 **3.** 83 **4.** 78 **5.** 42
Exercise B
1. March 8 **2.** November 26 **3.** August
2 **4.** December 28 **5.** March 23
Exercise C
1. $5.60 **2.** $9.72 **3.** $245 **4.** $13.57
5. $9.71 **6.** $93.33
Exercise D
1. $5.92 **2.** $181.23 **3.** $7.80
 4. $11.32 **5.** $9.86 **6.** $12.95
Exercise E
1. $720 **2.** $30.56 **3.** 14% **4.** 27 **5.** $750
6. 12% **7.** $212.50 **8.** $1,080 **9.** 7
10. 13% **11.** $4,132.09 **12.** $6,000.02
13. 24 **14.** 12% **15.** 251 **16.** $1,155
17. 10% **18.** $4,615.38 **19.** 59
20. $1,757.15

Chapter Nine
Assignment
Skill Problems
1. E **3.** C **5.** G **7.** 211 **9.** 451
11. 135 **13.** January 23, 2003

15. August 6, 2004 **17.** 53 **19.** 64
21. 96 **23.** October 1 **25.** April 15
27. $40. **29.** $2,396.24 **31.** $10.79
33. $7.89 **35.** $9.64 **37.** $15.90
39. $20.83 **41.** $826.67 **43.** $1,496.25
45. $110. **47.** 20 **49.** 8% **51.** $780
53. 7% **55.** $60,000 **57.** 10%
59. 10% **61.** $3,500
Business Application Problems
1. $4,000 **3.** May 11th **5.** 10%
7. $164,133.33 **9.** 15% **11.** $350
13. $414.25 **15.** Mr. Severance
17. $2,024.56 **19.** $12,954.
21. $184.38 **23.** $3.77 **25.** The amount
of payoff on the loan is $1,542.50.

Chapter Nine
Mastery Test
1. c (Objective 1)
2. 148 (Objective 2)
$(25 + 30 + 31 + 30 + 31 + 1 =)$
3. April 21, 2002 (Objective 2)
$(8 + 31 + 20 =)$
4. $666.67 (Objective 2)
$(50,000 \times .16 \times 30 \div 360 =)$
5. $657.53 (Objective 3)
$(50,000 \times .16 \times 30 \div 365 =)$
6. June 26 (Objective 2)
$(31 - 12 + 26 =)$
7. $13.44 (Objective 3)
$(1,000 \times .1075 \times 45 \div 360 =)$
8. $1,013.44 (Objective 3)
$(1,000 + 13.44 =)$
9. $46.50 (Objective 3)
$(1,800 \times .155 \times 60 \div 360 =)$
10. $5,290.27 (Objective 3)
$(5,000 \times .13 \times 163 \div 365 =,$
$5,000 + 290.27 =)$
11. $5,037.26 (Objective 3)
$(5,000 \times .17 \times 16 \div 365 =,$
$5,000 + 37.26 =)$
12. $21.30 (Objective 3)
$(900 \times .12 \times 71 \div 360 =)$

Chapter Ten (Pages 179–194)
Exercise A
1. $6,662.50 **2.** $4,080 **3.** $13,985.33
4. $25,746.53 **5.** $9,000 **6.** $7,303.33
7. $8,623.96 **8.** $5,000 **9.** $61,703.33
10. $3,540
Exercise B
1. 23 **2.** 23 **3.** 69 **4.** 21 **5.** 26 **6.** 75
7. 22 **8.** 30
Exercise C
1. $453.60 **2.** $5,048.63 **3.** $5,120.81
4. $6,086.85 **5.** $1,849.20 **6.** $170.90
7. $807.81 **8.** $1,017.06 **9.** $1,020.90
10. $3,533.95

Chapter Ten
Assignment
Skill Problems
1. F **3.** T **5.** F **7.** E **9.** $20,466.67
11. $460,800 **13.** $8,806.94
15. $3,050 **17.** $964.68 **19.** $10.61

21. $259 **23.** 131 **25.** $7,408.17
27. $10,100 **29.** $84.17 **31.** $263.01
33. 10 **35.** $5,237.06 **37.** $1,140
39. $1.25 **41.** $65,878.94
43. $4,518.66 **45.** $40,361.88
47. $6,996.09 **49.** $413.92
Business Application Problems
1. $16,520 **3.** $16,038.17 **5.** Balentine
and Company is the maker. **7.** August
10 due **9.** $61,407.75 **11.** $3,953.33
13. $4,064.92 **15.** $12,980.24
17. $19,753.89 **19.** $1,221.10
21. $268.49 **23.** The proceeds were
$40,425.84.

Chapter Ten
Mastery Test
1. B (Objective 4)
2. E (Objective 4)
3. D (Objective 4)
4. K (Objective 4)
5. C (Objective 4)
6. A (Objective 4)
7. 62 (Objective 3) $(21 + 7 =,$
$90 - 28 =)$
8. $95.72 (Objective 3)
$(3,580 \times .14 \times 90 \div 360 =,$
$3,705.30 \times .15 \times 62 \div 360 =)$
9. $3,609.58 (Objective 3)
$(3,705.30 - 95.72 =)$
10. $3,705.30 (Objective 2)
$(3,580 \times .14 \times 90 \div 360 =)$
11. $893 (Objective 3) $(900 \times$
$.14 \times 20 \div 360 =, 900 - 7 =)$

Chapter Eleven (Pages 195–216)
Exercise A
1. $209.99 **2.** $430.44 **3.** $444.50
4. $517.50 **5.** $373.15 **6.** $39.56
7. $29.51 **8.** $279.47
Exercise B
1. $381.98 **2.** $892.58 **3.** $344.82
4. $680.01 **5.** $1,140.22 **6.** $108.75
7. $1,297.26
Exercise C
1. $29,346.13 **2.** $13,815.86
3. $105,866.19 **4.** $9,549.58
5. $14,220.04
Exercise D
1. $123.82 **2.** $931.43 **3.** $259.37
4. $263.85 **5.** $216.14

Chapter Eleven
Assignment
Skill Problems
1. $672.66 **3.** $235.04 **5.** $279.86
7. $156.75 **9.** $417.13 **11.** $244.44
13. $1,850.48 **15.** $33.19
17. $1,445.68 **19.** $1,294.17 **21.** $4.17
23. $418.39 **25.** $82.29 **27.** $2.80
29. $253.12 **31.** $83.67 **33.** $1.41
35. $85.08 **37.** $85.08 **39.** $37.05
41. $11.03 **43.** $50.77 **45.** $7,799
47. $25,602.36 **49.** $25,703.99
51. $95.67 **53.** $9,199.23

55. $48,730.90 **57.** $284.57
59. $381.43 **61.** $911.93
Business Application Problems
1. $852.23 **3.** $82,307.16 **5.** $145.40
7. $5,490 **9.** $1,500 **11.** $1,207.50
13. $318.67 **15.** $444.30
17. $74,651.40 **19.** The difference
over the term of the mortgage is $33,408

Chapter Eleven
Mastery Test
1. True (Objective 5)
2. False (Objective 5)
3. $22.48 (Objective 1) $(8.99 \times 2.5 =)$
4. $810.48 (Objective 1) $(86 \times 6.43 =,$
$1,200 \div 12 =, 1,890 \div 12 =)$
5. $27,819.33 (Objective 2) $(28,000 \times$
$.09 \times \frac{1}{12} =, 27,910 \times .09 \times \frac{1}{12} =)$
6. $156.34 (Objective 4) $(158.87 \times 6 =,$
$953.22 - 900 =, 158.87 - 2.53 =)$
7. $4,997.31 (Objective 4) $(4,907.35 \times$
$.10 \times \frac{1}{12} =, 4,907.35 \times .01 =)$

Chapter Twelve (Pages 217–234)
Exercise A
1. $38.68 **2.** $171.87 **3.** $16.87
$64.07 $153.59 $53.14
$77.31 $175.13 $55.47
$46.02
4. $53 **5.** $36.66
$109.03 $208.53
$100.12 $212.12
$184.24
Exercise B
1. 15.5 **2.** 23.0 **3.** $35.87 **4.** 14.25
5. 21.0
Exercise C
1. $24, $456 **2.** $85.50, $1,814.50
3. $12.01, $588.64 **4.** $36, $1,164
5. $3.35, $85.85

Chapter Twelve
Assignment
Skill Problems
1. C **3.** B **5.** F **7.** C **9.** $.37
11. $.48 **13.** $.39 **15.** $.24 **17.** $.09
19. $.55 **21.** $14.67 **23.** $14.14
25. $14.13 **27.** $0 **29.** $1.44
31. $2.10 **33.** $1.60 **35.** $2.19
37. $10.06 **39.** $4.85 **41.** $4.31
43. $.79 **45.** 17% **47.** 17% **49.** 21%
51. 14.25% **53.** 17% **55.** 23%
57. $1,241.60 **59.** $82.87
61. $81,499.40 **63.** $1,797.41
65. $1,532.70
Business Application Problems
1. $211.28 **3.** $20.47 **5.** $15.42
7. $1,684.30 **9.** $133.33 **11.** $287.63
13. $126.74 **15.** $9.20 **17.** $67.44
19. The total amount of the deposit is
$839.56.

Chapter Twelve
Mastery Test
1. T (Objective 5)
2. T (Objective 5)

3. F (Objective 5)
4. F (Objective 5)
5. $1,122.91 (Objective 4) (824.8 × .04 =, 327 + 4.10 + 791.81 =)
6. $44.13 (Objective 2) (160 − 20 =, 140 × .015 =, 92.10 × .015 =, 43.48 × .015 =)
7. 17.5% (Objective 3) (100 × 8 =, 50 ÷ 7.5 =)

Chapter Thirteen   (Pages 235–262)
Exercise A
1. $1,792.12 **2.** $7,234.90 **3.** $7,550.32
4. $10,494.58 **5.** $43,822.46
6. $8,704.06 **7.** $27,057.50
8. $6,348.67 **9.** $12,033.10
10. $12,854.22
Exercise B
1. $1,365.58 **2.** $1,672.77 **3.** $3,905.99
4. $1,531.70 **5.** $624.60 **6.** $9,970.67
7. $5,435.70 **8.** $6,015.68 **9.** $6,736.25
10. $10,053.29
Exercise C
1. $42,576.09 **2.** $6,603.39
3. $15,449.34 **4.** $22,337.43
5. $7,423.10 **6.** $9,702.61
7. $11,687.27 **8.** $73,088.95
9. $9,581.01 **10.** $24,984.23
Exercise D
1. $2,046.15 **2.** $1,439.30 **3.** $7,664.70
4. $3,168.78 **5.** $2,419.02 **6.** $8,088.26
7. $8,795.56 **8.** $4,695.14
9. $14,384.73 **10.** $1,670.23

Chapter Thirteen
Assignment
Skill Problems
1. F **3.** F **5.** T **7.** F **9.** T **11.** 3%
13. $6,047.62 **15.** 30 **17.** 5%
19. $78,663.61 **21.** 12 **23.** ³/₄%
25. $7,178.48 **27.** 14 **29.** $50
31. $4,050 **33.** $4,100.63 **35.** $51.26
37. $4,151.88 **39.** $4,203.78
41. $52.55 **43.** $4,256.33 **45.** $4,309.53
47. 3, 10, $40,553.12 **49.** 4, 30, $6,166.37 **51.** 1.5, 24, $11,192.70
53. $52,059.12 **55.** $16,159.49
57. $6,565.97 **59.** $3,683.48
61. $71,527.29 **63.** 1.5, 30, $24,024.76
65. 2, 17, $12,433.89 **67.** 5, 13, $158,773.32 **69.** $98,553.91
71. $879.56 **73.** $7,806.16
75. $13,783.21 **77.** $4,138.10
79. $3,409.65 **81.** $17,486.88
83. $478.41
Business Application Problems
1. $120,482.96 **3.** $5,232.73
5. $4,381.55 **7.** $5,837.11 **9.** $171,979
11. $6,937.82 **13.** $14,215.79
15. $2,867.25 **17.** $44,705.34
19. $16,082.17 **21.** $431,448.57
23. The profit was $2,019,600 for one year.

Chapter Thirteen
Mastery Test
1. $2,745.57 (Objective 1) (2,000 × 1.37278571 =)
2. $2,981.34 (Objective 3) (5,000 × .59626733 =)
3. $37,618.70 (Objective 1) (19,000 × 1.9799316 =)
4. $7,490.68 (Objective 3) (9,500 × .78849318 =)
5. $4,011.39 (Objective 4) (400 × (11.0284738 − 1) =)
6. $2,498.73 (Objective 5) (30,000 × .08329094 =)
7. $5,968.92 (Objective 4) (375 × 15.91712652 =)
8. $24,271.84 (Objective 5) (50,000 × .48543689 =)
9. $374.66 (Objective 2) (300 × 1.24886297 =)
10. $8,233.49 (Objective 4) (750 × (11.97798875 − 1) =)

Chapter Fourteen   (Pages 263–284)
Exercise A
1. $5,120 **2.** $40,800 **3.** $812.43
4. $4,320 **5.** $2,041.20 **6.** $432
7. $15,552 **8.** $684 **9.** $449.55
10. $4,968
Exercise B
1. $7,866 **2.** $4,590 **3.** $522.50
4. $54,432 **5.** $59,280 **6.** $6,361.20
7. $13,608 **8.** $1,988.35 **9.** $1,923.75
10. $266.12
Exercise C
1. $654.75 **2.** $441 **3.** $200 **4.** $64.35
5. $175 **6.** $1,400 **7.** $69.30 **8.** $490
9. $118.80 **10.** $88.11
Exercise D
1. $190 **2.** $150 **3.** $59.40 **4.** $735
5. $600 **6.** $252.20 **7.** $550
8. $1,514.70 **9.** $31.20 **10.** $693
Exercise E
1. $601.40 **2.** $83.30 **3.** $120 **4.** $450
5. $323.73 **6.** $693 **7.** $171.50
8. $76.44 **9.** $321.75 **10.** $802
Exercise F
1. $492.63 **2.** $1,339.20 **3.** $233.77
4. $704.95 **5.** $2,245.32
Exercise G
1. $2,067.41 **2.** $8.50 **3.** $2,619
4. $180 **5.** 30
Exercise H
1. $2,257.70 **2.** $3,559 **3.** $23,659.51
4. 150 **5.** $76

Chapter Fourteen
Assignment
Skill Problems
1. $10,965 **3.** $6,925.86 **5.** $26,732.50
7. $6,739.20 **9.** $1,574.63
11. $253,634.80 **13.** $10,771.14
15. $30,723 **17.** 21.5%, $1,306,774.74
19. 41.6%, $592.18 **21.** July 10
23. February 10 **25.** November 16

27. April 14 **29.** $929.61 **31.** $425.65
33. $1,930 **35.** $252.23 **37.** $259.58
39. $881.10 **41.** $117.60 **43.** $2,787.30
45. $2,018.55 **47.** $2,088.06
49. $208.17 **51.** $3,376.48
53. $6,111.99 **55.** $453.99
57. $1,345.96 **59.** $56,517.45
61. $444.72 **63.** $1,681.02
Business Application Problems
1. $41,055 **3.** $212.50 **5.** $1,016.75
7. $54,850.50 **9.** $182.51 **11.** $4,048
13. $850 **15.** $1,453.50 **17.** $86
19. $1,764 **21.** $4,888.80 **23.** The list price was $700. **25.** The company will save $747.77.

Chapter Fourteen
Mastery Test
1. L (Objective 5)
2. J (Objective 5)
3. E (Objective 5)
4. H (Objective 5)
5. O (Objective 5)
6. C (Objective 5)
7. D (Objective 5)
8. M (Objective 5)
9. N (Objective 5)
10. G (Objective 5)
11. I (Objective 5)
12. A (Objective 5)
13. K (Objective 5)
14. F (Objective 5)
15. $492 (Objective 1) (600 × .82 =)
16. $640.80 (Objective 1) (890 × .9 × .8 =)
17. $891 (Objective 4) (900 × .99 =)
18. $733.04 (Objective 4) (748 × .98 =)
19. $564.48 (Objective 3) (800 × .9 × .8 × .98 =)
20. $812.25 (Objective 3) (1,000 × .95 × .95 × .9 =)
21. $501.06 (Objective 3) (650 × .95 × .9 × .92 × .98 =)
22. $739.31 (Objective 3) (92 × 10 × .82 × .98 =)
23. $470.45 (Objective 3) (600 × .9 × .88 × .99 =)

Chapter Fifteen   (Pages 285–302)
Exercise A
1. $20, $60 **2.** $12, $72 **3.** $35, $135
4. $50, $175 **5.** $15.64, $83.64
Exercise B
1. $37.50, $187.50 **2.** $107.14, $357.14 **3.** $26.67, $106.67 **4.** $112, $160 **5.** $10.64, $23.64
Exercise C
1. $99 **2.** $42 **3.** $213.33 **4.** $365
5. $24 **6.** $20.30 **7.** $10.40 **8.** $1,131
9. $25.33 **10.** $153.33
Exercise D
1. $128, $32 **2.** $35, $15 **3.** $68, $28
4. $111.60, $18.60 **5.** $29.90, $16.10
6. $33.25, $14.25 **7.** 82, $17.95 **8.** 31, $20.50 **9.** $171, $228 **10.** $228.57,

$308.57 **11.** 140, $28 **12.** 25, $12
13. 37, $74 **14.** $44.80, $11.20
15. $189, $147
Exercise E
1. $59.50 **2.** $10.20 **3.** $26 **4.** $14.36
5. 25% **6.** 43% **7.** 15% **8.** 31%
Exercise F
1. 29% **2.** 33% **3.** 67% **4.** 100% **5.** 41%
6. 23% **7.** 67% **8.** 17% **9.** 31% **10.** 25%
Exercise G
1. $117.65 **2.** $42.74 **3.** $355.12
4. $95.19 **5.** $149.40 **6.** $102.78
7. $146.04 **8.** $10,000 **9.** $119.47
10. $180.72
Exercise H
1. $166.01, 6% **2.** $720.56, 16%
3. $501.96, 16% **4.** $166.52, 20%
5. $942.31, 6%

Chapter Fifteen
Assignment
Skill Problems
1. $52.25 **3.** $245 **5.** $27.70
7. $401.70 **9.** $34.30 **11.** $97.78
13. $64.62 **15.** $55.40 **17.** $521.68
19. $113.17 **21.** $660 **23.** $169.20
25. $578.03 **27.** $55.35 **29.** $17.50
31. $208.13 **33.** $1,512.28 **35.** $15
37. $49 **39.** $14.50 **41.** $25
43. $31.33 **45.** $210 **47.** $240
49. $108 **51.** $11 **53.** $74.40
55. $6.98 **57.** 22.57% **59.** 21.45%
61. 19.50% **63.** 51.00% **65.** 21.25%
67. 30% **69.** 150% **71.** $50
73. $71.65 **75.** $59.80 **77.** $36.43
79. $38.40 **81.** $140.63 **83.** $758.51
85. $135.04 **87.** $128.02
89. $1,120.40 **91.** $57.27
93. $44.58 **95.** $253.39
Business Application Problems
1. $362.60 **3.** $431.67 **5.** 31%
7. $206.75 **9.** 56% **11.** 56%
13. 57% **15.** $6,251.20 **17.** 89
19. $428.57 **21.** $90.13 **23.** The
store must sell $1,866,666.67 per month.
25. They will receive $1,094,846.10 in
revenue.

Chapter Fifteen
Mastery Test
 1. $65 (Objective 1) (50 × 1.3 =)
 2. $95 (Objective 1) (38 ÷ .4 =)
 3. $142.86 (Objective 2) (200 ÷ 1.4 =)
 4. $23.50 (Objective 2) (47 × .5 =)
 5. 140% (Objective 3) (60 − 25 =,
   35 ÷ 25 =)
 6. 43% (Objective 3) (35 − 20 =,
   15 ÷ 35 =)
 7. 67% (Objective 6) (40 ÷ 60 =)
 8. 33% (Objective 5) (50 ÷ 150 =)
 9. $746.67 (Objective 1) (80 ÷ .12 =,
   666.67 + 80 =)
10. $345 (Objective 2) (30 ÷ .08 =,
   375 − 30 =)

11. $258.06 (Objective 7) (120 ÷ .62
   =, 193.55 ÷ .75 =)
12. $188.96 (Objective 2) (380 × .85 =,
   323 × .78 =, 251.94 × .75 =)
13. $392 (Objective 1) (280 × 1.4 =)
14. $1.91 (Objective 2) (2.29 ÷ 1.2 =)
15. 15% (Objective 4) (16 ÷ 105 =)

Chapter Sixteen   (Pages 303–322)
Exercise A
1. 1,728   **2.** 1,150
  1,428   800
  $383.62 $1,743.75
  $1,855.14 $3,882.25
3. 370   **4.** 2,005
  190   1,273
  $6,992 $7,456.60
  $6,732.50 $14,245.65
Exercise B
1. $6,720.45   **2.** $264.25
  $11,415.55 $22,196.75
3. $1,597.12   **4.** $7,528.51
  $8,784.13 $36,701.49
Exercise C
1. $8,608, $17,669
  $1,820.41, $3,738
2. $46,018, $54,351
  $15,910.05, $18,784
3. $80,683, $117,595
  $9,394.77, $13,695
4. $136,062, $221,256
  $122,891.76, $199,824
Exercise D
1. $346,650.41   **2.** $90,330.77
  $52,117.25 $49,925
  6.7   1.8
3. $103,488   **4.** $17,777.78
  $45,650 $3,616.67
  2.3   4.9
5. $15,840
  $5,042
  3.1

Chapter Sixteen
Assignment
Skill Problems
1. A **3.** K **5.** H **7.** D **9.** $40,928.57
11. $10,044.44 **13.** $43,737.93
15. 2,075 **17.** $723.75 **19.** 1,727
21. $7,947 **23.** $357.70 **25.** $348.70
27. $343.70 **29.** $26,588
31. $26,530.40 **33.** $26,541.99
35. $1,492.60 **37.** $1,509.52
39. $1,498.10 **41.** $233,310.28
43. $233,627.06 **45.** $233,065.50
47. $8,017.20 **49.** $417 **51.** $1,311.70
53. $200.20 **55.** $163,896 **57.** $86,402
59. $95,080 **61.** $54,840
63. $11,474.90 **65.** $16,064.86
67. $2,093.25 **69.** $2,616.56
71. $12,906.29 **73.** 3.0 **75.** $70,334
77. 2.8 **79.** 2.6
Business Application Problems
1. 6.3 **3.** $71,768 **5.** .8 **7.** $302,187
9. $5,599.72 **11.** $35,760

13. $98,678.45 **15.** $27,600.22
17. $916 **19.** The turnover is 1.8 for
the month.

Chapter Sixteen
Mastery Test
 1. F (Objective 4)
 2. F (Objective 4)
 3. T (Objective 4)
 4. T (Objective 4)
 5. $185,000 (Objective 2)
   (42,000 + 175,000 − 32,000 =)
 6. $158,455 (Objective 2)
   (345,600 + 56,835 − 243,980 =)
 7. $17,800 (Objective 2)
   (39,800 + 156,000 − 178,000 =)
 8. $39,786 (Objective 1)
   (890 × 41 =, 80 × 41.20 =)
 9. $40,189 (Objective 1)
   (650 × 41.50 =, 300 × 41.30 =,
   20 × 41.20 =)
10. $39,979.77 (Objective 1)
11. 2.1 (Objective 3) (36,490 +
   39,786 ÷ 2 =, 79,329 ÷ 38,138 =)
12. $192,250, $385,830
   $62,215.14, $124,930 (Objective
   1) (38,500 + 153,750 =,
   84,600 + 301,230 =)
13. 1.0 (Objective 3) (96,800 +
   42,690 ÷ 2 =, 120,970 × .6 =,
   72,582 ÷ 69,745 =)
14. $230,885 (Objective 1) (34,980 ×
   .6 =, 251,873 − 20,988 =)

Chapter Seventeen   (Pages 323–346)
Exercise A
1. $2,318.33   **2.** $650
  $9,273.32 $4,550
  $21,756.68 $16,207
3. $100   **4.** $3,000
  $300 $6,000
  $600 $46,980
5. $100
  $300
  $825
Exercise B
1. $55,000 **2.** 7 **3.** 10 **4.** $5,000
5. $20,000
Exercise C
1. $586.93, $32,351.40
2. $227.50, $393.25
3. $2,625, $8,980
4. $198.91, $3,001.19
5. $562.07, $7,296.49
Exercise D

1. 105, $\frac{12}{105}$ **2.** 210, $\frac{19}{210}$ **3.** 91, $\frac{10}{91}$

4. 120, $\frac{10}{120}$ **5.** 55, $\frac{1}{55}$
Exercise E
1. $2,330.95   **2.** $2,137.14
  $32,845.24 $9,714.29
  $15,654.76 $11,285.71

3. $10,458.33   **4.** $287.27
$78,437.50 $7,756.36
$57,062.50 $1,143.64
5. $211.43
$6,871.43
$728.57
Exercise F
1. $950   **2.** $10,800
$2,850 $28,800
$950 $16,200
3. $2,376.56   **4.** $988.15
$9,770.31 $8,020.74
$7,129.69 $1,979.26
5. $6,995.97
$57,416.13
$27,983.87
Exercise G
1. $8,352   **2.** $1,748
$24,592 $1,748
$33,408 $6,992
3. $825.37   **4.** $18,857.50
$825.37 $28,782.50
$32,018.63 $169,717.50
5. $1,733.55
$3,033.42
$866.58
Exercise H
1. $96,606.44 **2.** $1,760 **3.** $90,644
4. $36,800 **5.** $183,450
Exercise I
1. $215 **2.** $208.20 **3.** $80 **4.** $70
5. $196.88 **6.** $125 **7.** $118.78
8. $652 **9.** $1,312.50 **10.** $81.25

Chapter Seventeen
Assignment
Skill Problems
1. $66,000 **3.** $14,990 **5.** $15,476
7. $1,446 **9.** $4,338 **11.** $375.52
13. $1,712.50 **15.** $15,425

17. $790.50 **19.** $81.43 **21.** $230.77
23. $933.33 **25.** $31,976.13
27. $119.92 **29.** $3,152 **31.** 13/136
33. 8/120 **35.** 22/903 **37.** 6/78
39. $48,297.44 **41.** $1,190.34
43. $12,808.22 **45.** $51,024.76
47. $900 **49.** $900 **51.** $3,072
53. $1,387.97 **55.** $4,163.90
57. $4,129.97 **59.** $9,748.48
61. $38,993.92 **63.** $2,600
65. $1,111.11 **67.** $2,822.22
69. $12,433.80 **71.** $101.25
73. $911.25 **75.** $2,247.50 **77.** $80
79. $710 **81.** $3,306.24 **83.** $50
85. $345 **87.** $600 **89.** $86.67
91. $576 **93.** $8,332.50 **95.** $16,667.50
97. $11,112.50 **99.** $5,555
101. $3,702.50 **103.** $1,852.50
105. $8,841.01 **107.** $1,654.90
109. $4,137.73 **111.** $3,926.75
113. $67.40 **115.** $1,280.60
117. $7,000.20 **119.** $3,052.80
121. $4,579.20 **123.** $8,500
125. $89,684 **127.** $64,417.50
129. $665 **131.** $1,041.67
133. $1,216.95 **135.** $10.50
Business Application Problems
1. $1,200 **3.** $1,248 **5.** $800 **7.** $1,275
9. $4,992 **11.** $5,950 **13.** $5,385.15
15. $2,866.67 **17.** $27,000 **19.** $2,136
21. $3,023.18 **23.** The accumulated
depreciation would be $1,512.

Chapter Seventeen
Mastery Test
1. F (Objective 1)
2. J (Objective 1)
3. A (Objective 1)
4. I (Objective 1)
5. K (Objective 1)

6. E (Objective 1)
7. C (Objective 1)
8. D (Objective 1)
9. $2,545.45
$12,363.64
$8,636.36 (Objective 2) (21,000 −
1,000 =, 20,000 × 7 ÷ 55 =)
10. $6,731.25
$15,706.25
$20,193.75 (Objective 2)
(35,900 × (1 ÷ 8 × 2 =) =)
11. $627.50
$10,667.50
$1,932.50 (Objective 2)
(12,600 − 50 =, 12,550 ÷ 20 =)
12. $8,000 (Objective 1)
(42,000 − 2,000 =,
40,000 × 20,000 ÷ 100,000 =)
13. $4,930 (Objective 1)
(5,800 × .15 =, 5,800 − 870 =)
14. $6,000 (Objective 1)
(150,000 × 2,000 ÷ 50,000 =)
15. $183,333.33 (Objective 2)
(260,000 − 30,000 =,
230,000 × 10 ÷ 30 =)
16. $6,382.50 (Objective 4) (74,000 ×
2 ÷ 20 =, 7,400 × 5 ÷ 12 =,
70,916.67 × .9 =, 63,825 × .1 =)

Index

Record of Performance on Mastery Tests

Chapter	Mastery Test Score
1. Whole Numbers	_____
2. Fractions	_____
3 Decimals	_____
4. Banking Records	_____
5. Payroll	_____
6. Business Measurements	_____
7. Percent	_____
8. Insurance	_____
9. Simple Interest	_____
10. Promissory Notes	_____
11. Installment Loans	_____
12. Consumer Credit	_____
13. Compound Interest and Present Value	_____
14. Discounts	_____
15. Markup	_____
16. Inventory and Turnover	_____
17. Depreciation	_____